Computational Models of Scientific Discovery and Theory Formation

The Morgan Kaufmann Series in Machine Learning

Edited by Pat Langley

Machine learning studies the mechanisms through which intelligent systems improve their performance over time. Research on this topic explores learning in many different domains, employs a variety of methods, and aims for quite different goals, but the field is held together by its concern with computational mechanisms for learning. The Morgan Kaufmann series in machine learning includes monographs and edited volumes that report progress in this area from a variety of perspectives. The series is produced in cooperation with the Institute for the Study of Learning and Expertise, a nonprofit corporation devoted to research on machine learning.

Readings in Machine Learning

By Jude W. Shavlik (University of Wisconsin-Madison) and Thomas G. Dietterich (Oregon State University).

Computational Models of Scientific Discovery and Theory Formation

Edited by Jeff Shrager (Xerox Palo Alto Research Center) and Pat Langley (NASA Ames Research Center).

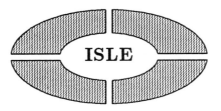

COMPUTATIONAL MODELS OF SCIENTIFIC DISCOVERY AND THEORY FORMATION

Edited by
Jeff Shrager
and Pat Langley

Morgan Kaufmann Publishers, Inc.
San Mateo, California

Sponsoring Editor   *Michael B. Morgan*
Production Editor   *Sharon E. Montooth*
Cover Designer   *Jo Jackson*
Copyeditor   *Larry Olsen*
Composition   *Technically Speaking Publications*
Proofreader   *Martha Ghent*
Cover Mechanical   *Victoria Ann Philip*

Library of Congress number: 90-4907

CIP data is available for this book.

MORGAN KAUFMANN PUBLISHERS, INC.
Editorial Office:
    2929 Campus Drive
    San Mateo, California
Order from:
    P.O. Box 50490
    Palo Alto, CA 94303-9953
©1990 by Morgan Kaufmann Publishers, Inc.

93 92 91 90 5 4 3 2 1

Symposium on Computational Models of Scientific Discovery and Theory Formation

List of Participants

Peter Cheeseman
AI Research Branch
Mail Stop 244-17
NASA Ames Research Center
Moffett Field, CA 94035 USA

Lindley Darden
University of Maryland
Committee on the History and
Philosophy of Science
1131 Skinner Hall
College Park, Maryland 20742 USA

Kevin Dunbar
McGill University
Department of Psychology
Stewart Biology Building
1205 Dr. Penfield Avenue
Montreal, Quebec, Canada H3A 1B1

Brian Falkenhainer
Xerox Palo Alto Research Center
3333 Coyote Hill Road
Palo Alto, CA 94304

Anne L. Fay
Baker Hall 446A
Department of Psychology
Carnegie Mellon University
Pittsburgh, PA 15213 USA

Margot Flowers
Department of Computer Science
University of California
Los Angeles, CA 90024 USA

John R. Josephson
Laboratory of Artificial Intelligence
Department of Computer Science
Ohio State University
Columbus, Ohio 43210 USA

Peter D. Karp
Department of Computer Science
Stanford University
Palo Alto, CA 94304 USA

David Klahr
Baker Hall 345F
Department of Psychology
Carnegie Mellon University
Pittsburgh, PA 15213 USA

Deepak Kulkarni
AI Research Branch
Mail Stop 244-17
NASA Ames Research Center
Moffett Field, CA 94035 USA

Pat Langley
AI Research Branch
Mail Stop 244-17
NASA Ames Research Center
Moffett Field, CA 94035 USA

Dale Moberg
Laboratory for Artificial Intelligence
Research
Department of Computer and
Information Science
Ohio State University
228 Bolz Hall
2036 Neil Avenue Mall
Columbus, Ohio 43210 USA

Steven Morris
Department of Information
& Computer Science
University of California
Irvine, CA 92717 USA

Bernd Nordhausen
Department of Information
& Computer Science
University of California
Irvine, CA 92717 USA

Greg Nowak
Cognitive Science Laboratory
Princeton University
221 Nassau Street
Princeton, New Jersey 08542 USA

Paul O'Rorke
Department of Information
& Computer Science
University of California
Irvine, CA 92717 USA

Michael J. Pazzani
Department of Information
& Computer Science
University of California
Irvine, CA 92717 USA

Shankar A. Rajamoney
Department of Computer Science
University of Southern California
Los Angeles, CA USA

David Schulenberg
Department of Information
& Computer Science
University of California
Irvine, CA 92717 USA

Herbert A. Simon
Baker Hall 339
Department of Psychology
Carnegie Mellon University
Pittsburgh, PA 15213 USA

Jeff Shrager
Xerox Palo Alto Research Center
3333 Coyote Hill Road
Palo Alto, CA 94304 USA

Paul Thagard
Cognitive Science Laboratory
Princeton University
221 Nassau Street
Princeton, New Jersey 08542 USA

Ryan D. Tweney
Department of Psychology
Bowling Green State University
Bowling Green, Ohio 43403-0228 USA

Jan M. Zytkow
Computer Science Department
Wichita State University
Wichita, Kansas 67208 USA

Contents

Preface

JEFF SHRAGER

PAT LANGLEY

Scientific reasoning is among the most revered activities of the western intellectual tradition, and scientific discovery is often considered the pinnacle of science. Until recently, the formal study of science has been the domain of philosophers and historians, and the processes of discovery have been surrounded in mystery. However, three decades of research in cognitive science have provided new methods for studying this intriguing area of cognition, and a variety of theoretical frameworks have arisen within which formal theories of scientific reasoning and discovery may be cast.

One of the principal frameworks of cognitive science is the computational paradigm. A growing number of researchers are using the tools and metaphors of computation to develop accounts of the scientific process. Only five years ago, work in this area was sequestered in one or two laboratories; in contrast, today it occupies the attention of numerous researchers at many different sites. This growing community consists of computer scientists, psychologists, philosophers, and historians, including many who cross two or more of these disciplines. We are witnessing the birth of a new field: The computational study of science.

Our own interest in discovery arose during our graduate studies in psychology and artificial intelligence at Carnegie Mellon University. Scientific reasoning and discovery can be viewed as intricate examples of human learning, and protocol analysis and cognitive simulation seemed to provide the tools for the direct study of such behavior. In this regard, the guidance of Herbert Simon played a central role, although we were also influenced by David Klahr, John Anderson, and many others. Our post-dissertation careers have diverged, with one of us (Langley) focusing on machine learning and the other (Shrager) on cognitive and developmental psychology. However, we remain active and interested in each other's field, and we believe that the crossroads of cognitive psychology and machine learning are an ideal site for the growth of a rich understanding of discovery and theory formation.

In order to explore the intersection of these disciplines, we organized a symposium on computational models of discovery and theory formation, and invited fellow cognitive scientists to share their work in this area. The chapters in this volume are based on the talks presented at that gathering, which was held at Stanford University in January of 1989. These contributions are richly connected, suggesting a convergence of both goals and methods. Such interconnections also suggest many possible ways in which to organize the chapters, from which we had to select only one.

The collection begins with a general introduction to computational models of scientific discovery that summarizes previous research and exposes difficulties, attempting both to provide a sense of the field's unity and to focus attention on important issues. The next two contributions focus on the evaluation of theories. Thagard and Nowak describe a computational framework for evaluating competing theories in terms of support and internal consistency, whereas Cheeseman argues for the importance of statistical evaluation based on probability theory. After this come two chapters on the discovery of empirical laws that summarize data. Nordhausen and Langley describe an AI system that integrates taxonomy formation, the discovery of qualitative laws, and the induction of numeric laws. Zytkow focuses on the role of structural information in parsing empirical laws into useful components.

The following six chapters deal with many aspects of theory formation and revision, which invoke unobserved processes to explain observed phenomena. Falkenhainer describes a framework that integrates induction, deduction, and abduction, drawing on similarity as the central mechanism. O'Rorke, Morris, and Schulenburg focus on the revision of existing theories, emphasizing the role of abduction in this process. Rajamoney follows a similar path, but combines theory revision with experimentation, using each process to constrain the other. Kulkarni and Simon also integrate theory revision and experimentation, but focus on the role of surprises in directing attention and resources. Karp deals with analogous issues, casting theory revision as a design problem and using experiments to constrain the hypothesizing of initial conditions. Darden takes a similar view but examines the effect of different types of anomalies.

The final four chapters share a direct concern with psychological issues. Klahr, Dunbar, and Fay examine strategies for experimentation exhibited by human subjects. Pazzani and Flowers demonstrate the close relationship between everyday learning and scientific discovery. Shrager describes a theory of conceptual combination, framed within a theory of sensori-motor representation, to explain aspects of human reasoning and theory formation. In the final chapter, Tweney takes a historical perspective, presenting direct challenges to computational theory that suggest future research in this challenging area and provide a fitting conclusion.

The 1989 meeting was the first in a series of symposia organized by the Institute for the Study of Learning and Expertise, a non-profit research institute that focuses on computational approaches to learning. The meeting was supported by the National Science Foundation under Grant Number IRI-8822762, the Office of Naval Research under Grant Number N00014-89-J-1133, and the American Association for Artificial Intelligence. The Xerox Palo Alto Research Center provided supporting services and equipment, and Stanford Conference Services provided excellent assistance in organizing the meeting. We owe thanks to many people, but especially to Julia Hough for encouraging us to undertake the project, Nils Nilsson and Peter Karp for making arrangements with Stanford University, and Sharon Montooth and Shirley Jowell for their careful production of the collection.

Our greatest debt is to the contributors to the current volume and other researchers concerned with the computational study of scientific discovery. As this collection reveals, the confluence of these researchers' diverse backgrounds provides rich soil for the growth of a science of discovery and theory formation. Still, much work remains to be done, and we invite others to join us in nurturing this fertile field.

CHAPTER 1

Computational Approaches To Scientific Discovery

JEFF SHRAGER
PAT LANGLEY

1. Introduction

Science is perhaps the most complex of intellectual activities, and its study has traditionally been the realm of historians and philosophers. However, recent advances in cognitive science—particularly in artificial intelligence and cognitive psychology—have provided new approaches and fresh insights into the nature of science. Whereas early work in the philosophical tradition emphasized the *evaluation* of laws and theories (e.g., Popper, 1965), recent research in the paradigm of cognitive science has emphasized scientific *discovery*, including the activities of theory formation, law induction, and experimentation. Moreover, the early philosophical approaches focused on the *structure* of scientific knowledge, whereas recent work has focused on the *process* of scientific thought and on describing these activities in *computational* terms. The aim of this chapter is to provide an overview of this computational research on scientific discovery.

Three basic developments have led to progress in this area during the past decade. First, cognitive psychology has made significant advances in its understanding of complex human behavior, which have encouraged psychologists to study domains such as scientific reasoning (e.g., Gholson, Shadish, Neimeyer, & Houts, 1989; Mynatt, Doherty, & Tweney, 1978; Shrager & Klahr, 1986). Second, the field of artificial intelligence has evolved into a mature discipline and has explored a variety of computational approaches to representation, performance, and learning (e.g., Weld & de Kleer, 1990). Finally, many philosophers of science have adopted a historical and psychological perspective on science, focusing less on normative and structural theories and more on how discoveries actually take place (e.g., Darden, this volume; Kuhn, 1962; Lakatos, 1970; Thagard & Nowak, this volume).

These advances have supplied the data and techniques needed to construct detailed computational models of the acquisition of knowledge in scientific domains. Research goals and methods differ, with some researchers giving detailed accounts of historical discoveries, others studying subjects' behavior in simulated scientific settings, and still others—caring less for historical or psychological adequacy—proposing algorithms with desirable computational properties. Taken together, these different emphases provide a multifaceted view of scientific discovery, giving a broader and deeper understanding than was possible even a few years ago.

We begin our survey of computational models of discovery by identifying some components of scientific behavior and proposing an associated vocabulary. We then review recent progress in computational approaches to discovery, using our framework to describe developments during the past five years. Finally, we consider some open problems in scientific discovery that do not fall within the framework and that have not been modeled in existing systems. We argue that these issues should receive significant attention in future research.

2.   Components of Scientific Behavior

In order to discuss computational theories of scientific behavior, we need a vocabulary with which to describe their components. In English, such terms as *discovery* and *theory formation* describe the diverse and complex behavior of a scientist at work, but in a vague and ill-defined manner.[1] One advantage of computational approaches is that they force the researcher to provide precise specifications of data structures and algorithms. Unfortunately, the goal of implementation often leads one to adopt narrow definitions of concepts that potentially have a much wider scope.

---

1. We will assume that the scientist is working *alone* in a given domain and that he or she has instruments available to manipulate and observe the domain. Later, we will reconsider these assumptions.

Following the tradition in artificial intelligence, we divide scientific behavior into *knowledge structures* and the *processes* or activities that transform them. Although narrow definitions are necessary to produce computational models, they are not required in a survey. Therefore, we will restrict ourselves to definitions of knowledge structures that are independent of particular representations, and to definitions of activities that focus on input/output relations rather than on specific methods. Even at this level, clear definitions are difficult to provide, and the reader should treat the statements that follow as tentative formulations. In addition, the list of components is clearly incomplete, being limited to aspects that have been addressed in existing models.

In the following discussion, we assume that the scientist is working in some particular *field* and more specifically on some problem in a particular *domain* within that field. For instance, the domain of neutrino interactions lies within the field of nuclear physics. We further assume that the scientist is operating in a laboratory or in some other relatively controlled *setting* (as opposed to field work), and we refer to particular arrangements of the setting, such as a specific experimental arrangement, as a *situation*. All of this together will be called the scientific *environment*.

2.1 Scientific Knowledge Structures

Before we can talk about activities, we must identify the knowledge structures that are inspected and manipulated. Together with the physical setting, these components constitute the raw materials and the products of science. In a given computational model, a number of these structures are cast in some specific representational framework, but in our quest for generality we will avoid commitment to particular representational assumptions. We describe the basic knowledge structures below.

Observations (or *data*) represent recordings of the environment made by sensors or measuring instruments. For instance, in his studies of heat, Joseph Black (1728–1799) recorded the temperatures of objects before and after he heated them. Each of these recordings was an observation.

Taxonomies define or describe concepts for a domain, along with specialization relations among them. One example is the taxonomy for biological organisms, which are grouped into species, genera, families, and so forth. Another is the grouping of chemical substances into acids, bases, and salts, and the subdivision of bases into alkalis and metals. Taxonomies specify the concepts used in stating laws and theories, and in giving units to observations.

Laws are statements that summarize relations among observed variables, objects, or events. For example, Black's heat law states that if one mixes two substances, the temperature of one substance increases and the temperature of the other decreases until they reach equilibrium. It also describes a precise numeric relation among the initial and final temperatures. The first statement is qualitative in form, whereas the latter is quantitative. Some laws may be quite general, whereas others may be very specific, potentially composed entirely of constants or ground terms.

Theories represent hypotheses about the structures or processes in the environment. They differ from laws in making reference to unobservable objects or mechanisms. For instance, the caloric theory stated that all material objects contained a substance called *caloric* and that heating involved a transfer of caloric to the heated object. A theory is stated in terms of concepts from the taxonomy.

Background knowledge is a set of beliefs or knowledge about the environment aside from those that are specifically under study. Such knowledge differs from theories or laws, in that the scientist holds background knowledge with relative certainty rather than as the subject of active evaluation. Statements that begin as theories or laws may eventually come to act as background knowledge. For instance, Black probably assumed that placing a flame under an object would increase its temperature.

Models are descriptions of the environmental conditions, both overt and hidden, for an experimental or observational setting. Thus, a model is required to indicate the manner in which a law or theory applies to a particular situation. For instance, one might attempt to understand a particular physical situation in terms of frictionless pulleys connected by massless strings, thus enabling the application of simple Newtonian mechanical theories.

Explanations are narratives that connect a theory to a law by a chain of inferences appropriate to the field. In such cases, we say that the theory *explains* the law. For instance, the caloric theory explains Black's observation that objects of different temperature move toward equilibrium when placed in contact. In some disciplines, inference chains must be deductive or mathematical, but many fields sanction other forms of explanation.

Predictions represent expectations about the behavior of the environment under specific conditions. One prediction that follows from the caloric theory is that a heat source will eventually stop transferring heat since ultimately the source will run out of caloric. For instance, if rubbing two objects together adds heat to the surrounding air, eventually this heating effect will halt. *Postdictions* are analogous to predictions, except that the scientist generates them after making the observations he or she intends the postdictions to explain. Successful predictions and postdictions lend support to the theory or law that produced them.

Anomalies describe laws that cannot be explained by a theory, or observations that cannot be predicted by a law. For instance, suppose one finds that the heating effect continues no matter how long one rubs two objects together. This finding is an anomaly with respect to the caloric theory since that theory leads to no laws that accord with the observation.

Although each of the above concepts plays an important role in scientific thought and action, many developers of specific discovery systems have collapsed some of them and left others implicit. For instance, rather than being stored as a separate structure, a theory might be implemented as an active subset of the background knowledge. Similarly, predictions need not be explicitly represented for one to obtain observations that violate the theory. To our knowledge, no existing discovery system explicitly incorporates all of these concepts.

Before proceeding to the activities of the scientist, we should note some concepts that we have explicitly left out of the foregoing analysis. These include *hypotheses*, *explorations*, *instruments*, and *representations*, as well as many others. Although these are important aspects of science, we believe that the concepts described above provide a satisfactory basis for a concrete discussion of scientific behavior.

2.2   Scientific Activities

Knowledge structures alone cannot provide a complete account of science. Processes and activities that use the structures in the setting under study are essential to the production of scientific knowledge. In this subsection, we propose a set of activities that describe the classical view of science, though we will broaden this set later in the chapter. Many philosophers have either explicitly or implicitly proposed categories of scientific activities (e.g., Feyerabend, 1975; Hacking, 1983; Lakatos, 1976; Popper, 1965; Suppe, 1977), but only a few computationalists have explicitly addressed this issue (e.g., Falkenhainer & Rajamoney, 1988).

We have attempted to describe activities that lie at approximately the same level. We have also aimed for functional definitions that are specified in terms of the knowledge structures each activity inspects and affects. Any given computational system will use a specific method to implement such activities, but we have intentionally avoided giving particular methods in our definitions. We describe the basic scientific activities below.

The *observation* process inspects the environmental setting by training an instrument, sometimes simply the agent's senses, on that setting. The result is a concrete description of the setting, expressed in terms from the agent's taxonomy and guided by the model of the setting. Since one can observe many things in any given situation, the observer must select some aspects to record and some to ignore. For example, Joseph Black observed a setting in which two fluids were brought into contact. Using a thermometer and a clock, he measured the temperature of each fluid at successive points in time. From this activity, he obtained data providing a set of concrete descriptions of the setting.

Taxonomy formation (and revision) involves the organization of observations into classes and subclasses, along with the definition of those classes. This process may operate on, or take into account, an existing taxonomy or background knowledge. For instance, early chemists organized certain chemicals into the classes of acids, alkalis, and salts to summarize regularities in their taste and behavior. As time went on, they refined this taxonomy and modified the definitions of each class. Another example of changing taxonomies involves the distinction between heat and temperature, which scientists had initially confounded (Carey & Wiser, 1983).

Inductive law formation (and revision) involves the generation of empirical laws that cover observed data. The laws are stated using terms from the agent's taxonomy and are constrained by a model of the setting and possibly by the scientist's background knowledge. In some cases, the scientist may generate an entirely new law; in others, an existing law may be modified or extended. For instance, Black arrived at his law of specific heat to summarize the temperature changes he observed in his heat experiments. Similarly, based on systematic experiments with the pressure and volume of gases in containers, Robert Boyle (1627–1691) induced a law that related these two variables.

Theory formation (and revision) stands in the same relation to empirical laws as does law formation to data. Given one or more laws, this activity generates a theory from which one can derive the laws for a given model by explanation. The theory is stated using terms from the domain's taxonomy and may be influenced by its background knowledge. Thus, a theory interconnects a set of laws into a unified theoretical account. For example, Boyle's law describes the inverse relation between the pressure and volume of a gas, whereas Charles' law states the direct relation of its temperature and pressure. The kinetic theory of gases provides an elegant explanation for both laws in terms of Newtonian interactions among molecules. Theory revision takes into account an anomalous phenomenon or law that cannot be explained by an existing theory. The revised theory should explain the anomalous phenomenon while maintaining the ability to cover existing laws, although this is often not possible.

Deductive law formation produces laws by a second route, starting with a theory and using an explanatory framework to deduce both a law and an explanation of how that law derives from the theory. Recall that laws can be composed entirely of ground terms, so this process can create very specific laws that lend themselves to prediction and thus aid in theory evaluation. For instance, Einstein's theory of general relativity led to an inferred law about the orbit of Mercury. However, not all such derived laws will be testable.

The *explanation* process connects a theory to a law by a narrative whose general form is given by the field's explanatory framework. In the context of evaluation (described below), if such a narrative can be produced, support may be lent to the theory or law from which the prediction arose. If no such narrative can be produced—that is, if explanation fails—then an anomaly results. The explanation process can also aid theory revision by verifying that the revised theory can be connected to known laws in the domain.[2] Explanation differs from deductive law formation, in that explanation attempts to account for a law that is already known.

The *prediction* process takes a law and a model of the setting, and produces a prediction about what will be observed. This often involves the results of intentional experimental manipulation, but it can also occur in observational domains. For example, one can use the ideal gas law to predict that, upon compressing a cylinder of gas, its temperature will rise. One can also use Kepler's laws of planetary motion to predict that an eclipse will occur at a certain time. The analogous process of *postdiction* takes place in cases where the scientist must account for an existing observation. Prediction and postdiction stand in the same relation to each other as deductive law formation and explanation.

Experimental design generates models of settings in which observations are to be made. Typically, selected aspects of the model (the independent variables) are systematically varied to determine their effect on other aspects (the dependent variables). This design process may take existing laws or theories into account, or it may be more exploratory in nature. Thus, Black decided to systematically vary the substances used in his experiments to determine their effects on rates of temperature change. If competing theories are considered in experimental design, they generally make different predictions.

The *manipulation* process constructs a physical setting that corresponds (to whatever extent possible) to a desired model. Thus, the scientist manipulates the environment in order to implement a given experimental design. For instance, Black instantiated his experimental design for studying temperature phenomena by physically heating various substances.

Evaluation, comparing a prediction with observations, generally follows experimental design and observation. Since predictions can vary in their level of detail, evaluation may vary in what is accepted. This produces either a successful *postdiction* or an *anomaly*, which may serve to stimulate further theory or law formation or revision. For instance, the

---

2. A subtlety of the present definition arises from the fact that we have defined the explanatory process to operate on laws, whereas one may sometimes want to explain precise observations as well. However, recall that laws can vary in their level of generality, so that one can easily transform observations into very specific laws, and vice versa.

anomalous behavior of rubbed objects (as described above) shed doubt on the caloric theory.

For the sake of simplicity, we have omitted a number of important activities from the above framework. These include: the process of *accepting* a tentatively held theory, thus adding it to one's background knowledge; the process of *scientific revolution*, in which one revises an entire theoretical framework; *model formation and revision*, in which one generates or revises a model that connects a theory and its laws to an experimental setting; and activities attending the important *social* and *embodied* aspects of scientific activity, such as communication, note taking, perception, and the construction of measurement instruments. In Section 4, we will return to the last of these topics in an effort to expand the traditional view of scientific behavior.

In any particular research endeavor, many of the activities described, as well as those that we have omitted, will be composed into greater units at various levels, ranging from daily actions to weekly plans to research programmes that cover months or years. Specific computational models implement certain combinations of these activities. In surveying the past decade of research on computational models of discovery, we will discuss the particular knowledge structures and activities that researchers have implemented.

3. Recent Research on Machine Discovery

We have chosen to divide research on scientific discovery into two broad periods. The first interval, during which cognitive scientists developed the first computational models of the discovery process, extends from the late 1970s through 1984. Below we provide a brief review of work from this period. During the second period, from 1984 through the present, researchers expanded on this early work along a variety of dimensions. We review this work in more detail, drawing from the concepts specified in the previous section.

3.1 Early Computational Research on Discovery

Early work on computational approaches to discovery focused on finding empirical regularities such as taxonomies and laws. This was a natural starting point, since empirical discovery tends to occur in the early stages of a scientific discipline. Thus, it should require less domain knowledge and permit the use of general heuristics.

Lenat's (1979) AM was one of the earliest discovery systems, operating in the domain of elementary number theory. This domain is unusual when viewed in the light of more recent work, in that one can generate data internally rather than observing them in a real or simulated environment. The user provided AM with an initial taxonomy of mathematical concepts, which it proceeded to extend and revise by mutation. Upon defining a new concept, the system used the definition to generate examples, which it then used to direct the search for other concepts. AM could also posit that two concepts were equivalent even though they had different definitions, as well as notice relations among different concepts. Thus, the system could discover certain classes of qualitative laws, revise its taxonomy, create new terms, and observe examples of these terms. However, it lacked components for experimentation, explanation, prediction, theory formation, and evaluation.

Another early discovery system was Langley, Zytkow, Bradshaw, and Simon's (1983) BACON, which focused on the induction of numeric laws from experimental data.[3] This program was provided with a set of independent and dependent variables, which it used to carry out simple experiments drawing on simulated data, and which it used to organize results into a taxonomic hierarchy. Once BACON had gathered data for a given node in its hierarchy, it searched for constant values of dependent terms or relations between independent and dependent terms. In the former case, it augmented the node's description with that constancy; in the latter case, it defined new terms as products or ratios of existing terms and continued the search. The system propagated constant values to higher levels in its hierarchy, where it treated them as dependent values in its search for higher-level numeric laws. BACON's main contribution was in the area of quantitative discovery and term definition, though it also included user-specified methods for experimentation, taxonomy formation, and observation. Like AM, it contained no explicit components for explanation, prediction, theory formation, or evaluation.

---

3. Langley et al.'s approach was influenced by earlier work on function discovery by Huesmann and Cheng (1973) and by Gerwin (1975). Langley, Simon, Bradshaw, and Zytkow (1987) provide a more detailed description of the BACON system.

Langley et al. (1983) described two additional systems that address different aspects of the discovery process. GLAUBER carried out a form of taxonomy formation that also produced simple qualitative laws relating the categories it defined.[4] STAHL formulated simple structural theories of chemical substances based on observed reactions, carrying out a revision process upon encountering anomalous observations that could not be explained by existing theories. Neither system contained explicit methods for experimentation, prediction, or evaluation.

At the level of our framework, AM and BACON cover similar aspects of the scientific process. Although both systems tackled important aspects of scientific discovery, they also ignored many components of the overall process and thus constituted initial forays rather than integrated models. During the past five years, research on computational approaches to scientific discovery has produced a number of advances over this early work. One can divide these developments into progress in knowledge representation, progress on methods for discovery, and progress on the integration of these methods. In the remainder of this section, we discuss each of these in turn, providing examples from the recent literature.

3.2   Progress on Scientific Knowledge Structures

The most basic advances in machine discovery have involved the representation of observations, laws, models, and theories. Early work assumed simple descriptions of objects and events in terms of numeric attributes or, at best, relations among objects. Qualitative and quantitative representations were entirely separate, and there existed no explicit representation for temporal information. However, a number of recent discovery systems have drawn heavily on Forbus' (1985) work on qualitative process representations. This approach represents events as a sequence of qualitative states, with each state describing an interval of time during which the signs of derivatives remain constant. Forbus' framework also lets one represent theories about processes in qualitative terms and provides mechanisms for making qualitative predictions.

At least four researchers have incorporated this qualitative process representation directly into their discovery systems. For instance, Falkenhainer's PHINEAS (this volume) uses qualitative data to retrieve and

---

4. More recently, Jones (1986) has described an incremental version of GLAUBER that contains explicit components for experimentation, prediction, and evaluation.

match against promising background knowledge, then forms a new process theory by analogy with this knowledge. O'Rorke, Morris, and Schulenberg (this volume) represent data and theories in a similar form but use anomalies to drive the process of theory revision. Rajamoney's COAST (this volume) uses a qualitative representation for models but uses qualitative anomalies to constrain the experimentation process.[5] Finally, Nordhausen and Langley's IDS (this volume) uses Forbus' formalism to represent both observations and qualitative laws, including ones that involve relations among successive states.

Another representational advance involves the storage of justifications on theories that aid in the processes of theory evaluation and revision. For instance, Thagard and Nowak (this volume) explicitly represent the arguments for and against competing theories, using this information in their evaluation mechanism. In a similar manner, Pazzani and Flower (this volume) make an analogy between theory evaluation and argumentation, proposing the use of explicit arguments and counterarguments in evaluating theories. Rose and Langley (1986) take a related approach in their STAHLp system, indexing observations by the theories they support and retrieving them when anomalies call the theory into question. Rajamoney's COAST (this volume) employs a similar strategy but stores only some of the evidence for a given theory to use during later revisions.

A final representational innovation concerns the role of imagery. Miller (1986) and Tweney (this volume) argue for the central role of imagery in scientific thinking and call for research on computational approaches to this topic. A number of researchers in qualitative reasoning are explicitly working on the problem of spatial reasoning (e.g., Nielsen, 1988), and Shrager's work (this volume) constitutes a novel approach, introducing a representation of scientific knowledge that is grounded in sensory-motor operations. The use of qualitative process formalisms also bears on this topic, in that one can "run" qualitative simulations to "envision" what may follow from given starting conditions. These are only beginnings, but they considerably extend the simplistic mathematics-based schemes that predominated in the early work on discovery.

---

5. Kulkarni and Simon (this volume) and Karp (this volume) also employ qualitative representations in the design of experiments, but they do not explicitly work in Forbus' framework.

3.3   Progress on Discovery-Related Activities

In terms of scientific activities, the most impressive advances have occurred with respect to the formation and revision of theories. Falkenhainer's work on analogy describes one approach to theory formation, in which knowledge of other domains is transferred to the one under study. Kulkarni and Simon, O'Rorke et al., and Rajamoney all focus on theory revision, showing how anomalies can lead to modification of an initial theory and its gradual improvement over time. Karp's HYPGENE uses similar methods to deal with the related problem of model revision, and Darden (this volume) discusses similar issues in her historical analysis. Rose (1989) describes a unified approach to incrementally revising both theories and observations. Contrasting approaches to theory revision have been proposed that rely on conceptual combination (Holland, Holyoak, Nisbett, & Thagard, 1986; Shrager, 1987), and Shrager's work (this volume) follows this approach in novel directions.

Another area of progress has involved experimentation. Klahr, Dunbar, and Fay (this volume), following upon the theory formation studies of Shrager and Klahr (1986), have carried out detailed studies of the experimentation strategies of humans in understanding complex devices, extending previous work (e.g., Mynatt, Doherty, & Tweney, 1978) in important ways. The computational models of Kulkarni and Simon, Rajamoney, and Karp have all focused on experimentation, and their approaches share some important similarities. Each of their systems makes predictions, notes anomalies, uses the latter to generate alternative hypotheses, and then designs experiments to discriminate among the competitors.

Although a smaller fraction of researchers have focused on empirical discovery than in earlier days, advances have also occurred along this front. One development is the work on "conceptual clustering" by Stepp (1984), Lebowitz (1987), and Fisher (1987), which organizes observations into taxonomies of concepts described at varying levels of abstraction. Another area concerns improved methods for discovering numeric laws, such as those described by Falkenhainer and Michalski (1986), Kokar (1986), and Zytkow (1987). More recently, Nordhausen and Langley (this volume) have reported novel methods in both areas, along with techniques for discovering qualitative laws. Zytkow (this volume) outlines a method for quantitative discovery that takes advantage of domain models to parse numeric laws into useful components. Both approaches rely on more powerful representations of observations and laws than were used in earlier work. Another line of research by Epstein (1987), Shen (1990), and Sims and Bresina (1989) has continued in the AM tradition, refining Lenat's approach and applying it to new mathematical domains.

Finally, research has also progressed in the area of evaluation. Thagard and Nowak (this volume) describe a method for evaluating the relative quality of two theories in terms of each theory's ability to explain a variety of phenomena. Taking a different approach, Cheeseman (this volume) proposes Bayesian probabilistic criteria for evaluating taxonomies and laws. Both approaches seem likely to find their way into future discovery systems, where they could be used to direct the search for improved laws and theories.

3.4   Progress on Integrated Approaches to Discovery

Another important trend has been the evolution toward *integrated* discovery systems. A number of researchers have combined nontrivial components of the discovery process, producing synergistic effects from their interactions. One can view these efforts as steps along the path toward a complete theory of scientific discovery that describes not only basic activities but also the relations among them.

One relatively complete integration of activities is embodied in Shrager's (1987) IE system, which carried out experiments on simulations of a complex device and formed "mental models" of the device by conceptual combination. The system performed explorations and experiments (both involving prediction) on the simulated device and carried out exercises in order to test the completeness of its theory. Although Shrager was concerned mainly with IE's "view application" method for theory reformulation, the model also included simple versions of analogical theory extension and postdiction.

Another example is Nordhausen and Langley's work, which integrates taxonomy formation, qualitative law discovery, and numeric law discovery. Their IDS system incrementally organizes observed qualitative states into a taxonomic hierarchy and then formulates qualitative laws in terms of temporal relations between classes of states. It also uses these qualitative laws to provide context for numeric relations and to constrain the search for the latter.

A third case is Kulkarni and Simon's KEKADA, which integrates theory revision, prediction, experimentation, and evaluation. Their system begins with a partial theory and an anomaly, which KEKADA attempts to explain by elaborating the theory. This leads to a number of alternative hypotheses, which the system evaluates by designing and running experiments. If KEKADA encounters some new anomaly along the way, it shifts attention and follows this path instead.[6]

These three systems are not the only ones that attempt to integrate aspects of the discovery process, but they provide prototypical examples of this trend. If one compares the above descriptions of IE, IDS, and KEKADA with the earlier characterizations of AM and BACON, the recent progress toward integrated models of scientific discovery becomes apparent. However, it is also clear that much work remains before we arrive at a model that fully integrates even the incomplete set of processes included in the framework from Section 2.

4. Open Issues in Scientific Discovery

In closing, we consider two important aspects of intellectual activity—*embedding* and *embodiment*—that have significant bearing on science but that have not been addressed by existing computational models. Briefly, science takes place in a world that is occupied by the scientist, by the physical system under study, and by other agents, and this world has indefinite richness of physical structure and constraint. Thus the scientist is an embodied agent embedded in a physical and social world.

Embodiment brings to the fore components of scientific behavior that are easily ignored when the model exists entirely within a computer, where all aspects of the environment are controllable, where observation can take place by direct reference to data structures, and where the environment has finite and known complexity. Embedding highlights issues that have been traditionally ignored by models that focus on the intellectual activity of individual scientists rather than on communities. In this section, we consider some results of embedding and embodiment that have generally been ignored in computational models of scientific behavior (see also the critique of Tweney, this volume). We end by

---

6. Rajamoney's COAST and Karp's HYPGENE also combine prediction, experimentation, and revision, but they focus on individual steps in this process rather than on the continuing cycle.

discussing some promising approaches toward creating computational accounts for these components.

4.1   External Representations and Research Programs

As scientific domains become increasingly data intensive, external representations come to play a central role in the research process. For instance, notebooks and graphics are widely used in many disciplines as memory aids and, more importantly, as aids to discovery through data organization. In addition, such records help in research planning, in which scientists sequence their activities within the larger scientific context. Several researchers (Darden, this volume; Gorman & Carlson, in press; Tweney, this volume) have studied the use of laboratory notes and records, along with their influence on scientific reasoning. Their analyses suggest that these external records have a major influence on the discovery process. Kulkarni and Simon (this volume) are concerned with programs of research, but they do not model the role of external records in the planning process.

Given the importance of notebooks, graphics, and similar records, it may seem astonishing that none of the existing computational models of discovery incorporate such devices. Part of the reason comes from unrealistic assumptions about the memory and speed of computational systems. For instance, Langley et al.'s BACON has no need to plot its data since it can retain as many observations as necessary in working memory and scan the data rapidly. Although no computational models of discovery have taken seriously the function of external representations, some work has been done in other areas of cognitive science (e.g., Larkin & Simon, 1987; Shrager, 1989). Also, Shrager's theory of grounded representation (this volume) partially addresses this issue, in that it is designed to operate with external stimuli as well as with internal sensory content.

4.2   Perception and Measurement Instruments

The measurement process alone occupies a major fraction of scientists' time and energy, leaving precious little remaining time for the intellectual activities that we considered in Section 2. However, existing models of scientific discovery are disembodied; they assume immediate and unproblematic access to observations. Even in models that explicitly separate the external setting from the agent's internal knowledge, the environment is sufficiently constrained that issues of attention and perception are avoided. In addition to measuring simple quantities, scientists must also connect observables to theoretical terms if the latter are to be operational. Recent research on attention in concept learning (Billman & Heit, 1988; Gennari, 1989) has started to address some of these issues, but much more remains to be done.

Moreover, even the earliest histories of discovery involve some forms of instrumentation. Many authors (e.g., Feyerabend, 1975; Giere, 1988; Hacking, 1983) have noted the importance of measurement instruments in the scientific process, but computational models have ignored this aspect of research. As with perception, this oversight is understandable, in part because instrument construction is largely a physical phenomenon that is difficult to model without solving difficult problems in robotics or building rich simulations. One approach that shows some promise is Nordhausen and Langley's (this volume) method for postulating *intrinsic properties*, which provides a method for computing features of new objects based on their behavior in familiar qualitative histories. Effectively, these abstract histories describe "instruments" that let one measure properties like boiling point or specific heat.

4.3 Laboratories, Collaboration, and Communication

Most modern science is too large and too expensive an undertaking for an independent researcher to succeed, making it essential that scientists collaborate. Although there are many alternative organizations for joint research, the most common is the laboratory, in which a small number of researchers collaborate on a small set of problems. Laboratories generally exist at a single location and include scientists at different levels of expertise, from students to senior researchers. In addition, different laboratories often work on the same or closely related problems. In some cases, this work is competitive, but in other cases there is significant cooperation, with division of labor and open interactions.

Collaboration of any sort requires some form of communication among scientists, and it takes no statistical sophistication to conclude that scientists spend much of their time talking, reading, and writing. These sorts of communication provide another example of external representations in which graphics, mathematical expression, and language play central roles. Within a laboratory, communication often occurs verbally before anything reaches the formal scientific literature. Tweney (this volume) has argued that Faraday enriched his understanding of one domain from his ongoing research in other domains, but such enrichment is surely is not restricted to the mental activities of individual scientists. Formal communication is essential for the broader dissemination of ideas, making reading and writing central scientific activities.

Existing computational models of discovery have avoided the collaborative and communicative aspects of scientific research, focusing on individual scientists' behavior and ignoring group interactions. This was a natural development, given the traditional focus of cognitive science on the cognitive processes of individuals. However, the social organization of science in the laboratory and in broader contexts has a major influence on the nature of science, and future modeling efforts should move toward incorporating aspects of this structure.[7]

4.4 Toward a Fuller Computational Account of Discovery

In summary, actual science occurs in the context of a physical world and in the context of other agents, but existing computational accounts of discovery have avoided these major issues. The reason for this bias is straightforward. The methods and theories of cognitive science were originally designed to model individual cognition, and the computational work on discovery has relied heavily on these tools. A deeper understanding of embodiment will require considerable research in AI and cognitive psychology, and the embedded nature of science awaits additional work in sociology, anthropology, and psychology. Nevertheless, some preliminary results hold out hope for advances in these areas.

For instance, the active research area of "distributed artificial intelligence" focuses on understanding the ways that multiple agents can interact in communities. Several collections are available on this topic (see Gasser & Huhns, 1989; Huberman & Hogg, 1988), and researchers

---

7. Excellent collections have recently appeared on the sociology of scientific practice and knowledge (see Fuller, De May, Shinn, & Woolgar, 1989). Readers of this chapter will be particularly interested in a special issue of *Social Studies of Science* (volume 19, number 4), in which several authors respond to Slezak (1989), who argues that the success of BACON and similar discovery programs "[provide] dramatic confirmation [of the view that] there are principles of rationality and a 'scientific method' which are independent of social factors."

in this field explicitly draw upon results in the social sciences, especially from economics and scientific reasoning. There is also hope that psycho-anthropological approaches (e.g., Latour & Woolgar, 1979; Lynch, 1985; Pickering, 1984) will explain certain social aspects of science, especially the role of communication. However, to date these accounts have been descriptive rather than computational. The literature on distributed artificial intelligence also deals with issues of communication but focuses on the nature of the information passed rather than on the processes of individual agents acting in the community. Overall, there has been little computational work on the communicative interactions of agents with one another. Thagard and Nowak's work (this volume) on the acceptance of revolutions most closely speaks to the issues of interactions among researchers, but their paradigm does not model the richness and detailed functions of scientific communication.

Research on embodied agents has also made progress, not only in traditional approaches to robotics but also in the interface between AI, machine learning, and robotics. For instance, Laird, Yager, Tuck, and Hucka (1989) describe a system that improves its ability to use a robot arm with experience. The work of Iba and Langley (1987) on motor learning provides an additional example of this encouraging trend. More relevant to scientific discovery are recent attempts (Zytkow, Zhu, & Hussam, in press) to employ AI methods to control robotic equipment for chemical experimentation. In addition, some researchers (see Shrager, this volume) have taken perception as a central problem and have attempted to explain complex intellectual activity in terms of sensation and action. Other researchers have even attempted to deal with the issues of physical and social environments simultaneously, as Cohen, Greenberg, Hart, and Howe (1989) have done in their work on cooperative fire fighting in a simulated (burning) forest.

We believe that an important source for models of embedding and embodiment in science will come from an unexpected direction: the developmental psychology of *socialization*, which studies the ways in which a child learns to become a part of his or her culture (e.g., Bruner, 1985; Kuhn, Amsel, & O'Loughlin, 1988; Vygotsky, 1962). Insights into this process may provide hypotheses about the paths through which graduate students and junior scientists become members of their scientific community—mastering the ways of thinking, operating, and communicating that constitute the institution of science.[8]

5.   Conclusion

In effect, this chapter has attempted to define a new field of study—the computational modeling of scientific behavior. Despite its relatively recent development, this research area has already made significant progress on issues that philosophers of science have traditionally ignored. In particular, the field has emphasized the nature of *discovery* rather than evaluation, and it has dealt with the *processes* that underlie science as well as the representation of knowledge. The result has been a rapidly growing set of computational models that deal with many facets of the scientific enterprise.

Although the existing models are best viewed as embodying tentative hypotheses about the nature of science, it is also clear that the past decade has seen real progress. Current systems still ignore many important aspects of discovery and theory formation, but idealizations are a central part of science; we should no more expect our computer simulations to account for *every* aspect of discovery than we expect our physical or chemical theories to explain every aspect of the physical world. What we can expect is incremental progress toward fuller models and deeper understanding, and that is precisely what has occurred in the developing computational "science of science."

The past few years have seen notable developments, not only in the representations and processes used to model scientific discovery and theory formation but also in their integration into a coherent framework. We will not make specific predictions about the outlook for extending the computational paradigm into the more difficult areas of embedded and embodied science. However, the paths toward these goals seem lined with fertile research questions waiting to be addressed. Progress along these paths will certainly tax our existing theories and methodology, but it should also bear rich rewards.

---

8. Luhrmann's (1989) insightful psychoethnography of British witchcraft provides a carefully researched example of a sort of socialization that she calls "interpretive drift." The analogy between becoming a scientist and becoming a witch runs more deeply than one might think. Both deal with belief and action, and both have significant rites of passage. Most of the structures and activities that we have identified as typical of science apply equally well to witchcraft, and even to more commonplace activities (Pazzani & Flowers, this volume).

Acknowledgements

We owe thanks to many colleagues for discussions that led to the ideas in this chapter. In particular, Mike Morgan suggested that we develop an encompassing vocabulary; Rik Belew and Michael Gorman encouraged us to think more about the social aspects of science; and Deepak Kulkarni provided helpful comments on an earlier draft.

References

Billman, D., & Heit, E. (1988). Observational learning from internal feedback: A simulation of an adaptive learning method. *Cognitive Science*, *12*, 587–626.

Bruner, J. (1985). *Child's talk*. New York: W. W. Norton.

Carey, S., & Wiser, M. (1983). When heat and temprature were one. In D. Gentner & A. L. Stevens (Eds.), *Mental models*. Hillsdale, NJ: Lawrence Erlbaum.

Cohen, P. R., Greenberg, M. L., Hart, D. M., & Howe, A. E. (1989). Trial by fire: Understanding the design requirements for agents in complex environments. *AI Magazine*, *10*, 32–48.

Epstein, S. L. (1987). On the discovery of mathematical theorems. *Proceedings of the Tenth International Joint Conference on Artificial Intelligence* (pp. 194–197). Milan, Italy: Morgan Kaufmann.

Falkenhainer, B. C., & Michalski, R. S. (1986). Integrating quantitative and qualitative discovery: The ABACUS system. *Machine Learning*, *1*, 167–402.

Falkenhainer, B. C., & Rajamoney, S. (1988). The interdependencies of theory formation, revision, and experimentation. *Proceedings of the Fifth International Conference on Machine Learning* (pp. 353–366). Ann Arbor, MI: Morgan Kaufmann.

Feyerabend, P. (1975). *Against method*. London: Verso.

Fisher, D. H. (1987). Knowledge acquisition via incremental conceptual clustering. *Machine Learning*, *2*, 139–172.

Forbus, K. D. (1985). Qualitative process theory. In D. G. Bobrow (Ed.), *Qualitative reasoning about physical systems*. Cambridge, MA: MIT Press.

Fuller, S., De May, M., Shinn, T., & Woolgar, S. (1989). *The cognitive turn: Sociological and psychological perspectives on science.* Boston: Kluwer Academic Publishers.

Gasser, L., & Huhns, M. N. (Eds.). (1989). *Distributed artificial intelligence* (Vol. 2). San Mateo, CA: Morgan Kaufmann.

Gennari, J. H. (1989). Focused concept formation. *Proceedings of the Sixth International Workshop on Machine Learning* (pp. 379–382). Ithaca, NY: Morgan Kaufmann.

Gerwin, D. (1975). Information processing, data inferences, and scientific generalization. *Behavioral Science, 19*, 314–325.

Gholson, B., Shadish, W. R., Neimeyer, R. A., & Houts, A. C. (1989). *Psychology of science: Contributions to metascience.* Cambridge: Cambridge University Press.

Giere, R. N. (1988). *Explaining science.* Chicago: University of Chicago Press.

Gorman, M. E., & Carlson, B. W. (in press). Interpreting invention as a cognitive process: Alexander Graham Bell, Thomas Edison, and the telephone, 1875–1878. *Science, Technology, and Human Values.*

Hacking, I. (1983). *Representing and intervening.* Cambridge: Cambridge University Press.

Holland, J. H., Holyoak, K. J., Nisbett, R. E., & Thagard, P. R. (1986). *Induction: Processes of inference, learning, and discovery.* Cambridge, MA: MIT Press.

Huberman, B. A., & Hogg, T. (1988). The behavior of computational ecologies. In B. A. Huberman (Ed.), *The ecology of computation* (pp. 77–115). New York: North-Holland.

Huesmann, L. R., & Cheng, C. M. (1973). A theory for the induction of mathematical functions. *Psychological Review, 80*, 126–138.

Iba, W., & Langley, P. (1987). A computational theory of motor learning. *Computational Intelligence, 3*, 338–350.

Jones, R. (1986). Generating predictions to aid in the scientific discovery process. *Proceedings of the Fifth National Conference on Artificial Intelligence* (pp. 513–517). Philadelphia, PA: Morgan Kaufmann.

Kokar, M. M. (1986). Determining arguments of invariant functional descriptions. *Machine Learning*, *1*, 403–422.

Kuhn, D., Amsel, E., & O'Loughlin, M. (1988). *The development of scientific thinking skills.* New York: Academic Press.

Kuhn, T. S. (1962). *The structure of scientific revolutions.* Chicago, IL: University of Chicago Press.

Laird, J., Yager, E. S., Tuck, C. M., & Hucka, M. (1989). Learning in teleautonomous systems using SOAR. *Proceedings of the 1989 NASA Conference on Space Telerobotics.* Pasadena, CA.

Lakatos, I. (1970). Falsification and the methodology of scientific research programs. In I. Lakatos & A. Musgrave (Eds.), *Criticism and the growth of knowledge.* Cambridge: Cambridge University Press.

Lakatos, I. (1976). *Proofs and refutations: The logic of mathematical discovery.* J. Worrall & E. Zahar (Eds.). Cambridge: Cambridge University Press.

Langley, P., Simon, H. A., Bradshaw, G. L., & Zytkow, J. M. (1987). *Scientific discovery: Computational explorations of the creative processes* (pp. 251–283). Cambridge, MA: MIT Press.

Langley, P., Zytkow, J., Bradshaw, G., & Simon, H. A. (1983). Three facets of scientific discovery. *Proceedings of the Eighth International Joint Conference on Artificial Intelligence* (pp. 465–468). Karlsruhe, West Germany: Morgan Kaufmann.

Larkin, J. H., & Simon, H. A. (1987). Why a diagram is (sometimes) worth ten thousand words. *Cognitive Science*, *11*, 65–99.

Latour, B., & Woolgar, S. (1979). *Laboratory life: The social construction of scientific facts.* London: Sage.

Lebowitz, M. (1987). Experiments with incremental concept formation: UNIMEM. *Machine Learning*, *2*, 103–138.

Lenat, D. (1979). On automated scientific theory formation: A case study using the AM program. In J. Hayes, D. Michie, & L. I. Mikulich (Eds.), *Machine intelligence* (Vol. 9). New York: Halstead Press.

Luhrmann, T. M. (1989). *Persuasions of the witch's craft.* Cambridge, MA: Harvard University Press.

Lynch, M. (1985). *Art and artifact in laboratory science: A study of shop work and shop talk in a research laboratory.* London: Routledge & Kegan Paul.

Miller, A. I. (1986). *Imagery in scientific thought.* Cambridge, MA: MIT Press.

Mynatt, C. R., Doherty, M. E., & Tweney, R. D. (1978). Consequences of confirmation and disconfirmation in a simulated research environment. *Quarterly Journal of Experimental Psychology, 30,* 395–406.

Nielsen, P. (1988). A qualitative approach to mechanical constraint. *Proceedings of the Seventh National Conference on Artificial Intelligence* (pp. 270–274). Saint Paul, MN: Morgan Kaufmann.

Pickering, A. (1984). *Constructing quarks.* Chicago: University of Chicago Press.

Popper, K. (1965). *Conjectures and refutations: The growth of scientific knowledge* (2nd ed.). New York: Basic Books.

Rose, D. (1989). Using domain knowledge to aid in scientific theory formation. *Proceedings of the Sixth International Conference on Machine Learning* (pp. 272–277). Ithaca, NY: Morgan Kaufmann.

Rose, D., & Langley, P. (1986). Chemical discovery as belief revision. *Machine Learning, 1,* 423–451.

Shen, W. M. (1990). Functional transformation in AI discovery systems. *Artificial Intelligence, 41,* 257–272.

Shrager, J. (1987). Theory change via view application in instructionless learning. *Machine Learning, 2,* 247–276.

Shrager, J. (1989). Reinterpretation and the perceptual microstructure of conceptual knowledge: Cognition considered as a perceptual skill. *Proceedings of the Eleventh Annual Conference of the Cognitive Science Society* (pp. 876–883). Ann Arbor, MI: Lawrence Erlbaum.

Shrager, J., & Klahr, D. K. (1986). Instructionless learning about a complex device: The paradigm and observations. *International Journal of Man-Machine Studies, 25,* 153–189.

Sims, M. H., & Bresina, J. L. (1989). Discovering mathematical operator definitions. *Proceedings of the Sixth International Conference on Machine Learning* (pp. 308–313). Ithaca, NY: Morgan Kaufmann.

Slezak, P. (1989). Scientific discovery by computer as empirical refutation of the strong programme. *Social Studies of Science, 19,* 563–600.

Stepp, R. (1984). *Conjunctive conceptual clustering: A methodology and experimentation.* Doctoral dissertation, Department of Computer Science, University of Illinois, Urbana-Champaign.

Suppe, F. (1977). *The structure of scientific theories* (2nd ed.). Urbana, IL: University of Illinois Press.

Vygotsky, L. S. (1962). *Thought and language.* (E. Hanfmann & G. Vakar, trans.). Cambridge, MA: MIT Press.

Weld, D. S. , & de Kleer, J. (Eds.). (1990). *Qualitative reasoning about physical systems.* San Mateo, CA: Morgan Kaufmann.

Zytkow, J. M. (1987). Combining many searches in the FAHRENHEIT discovery system. *Proceedings of the Fourth International Workshop on Machine Learning* (pp. 281–287). Irvine, CA: Morgan Kaufmann.

Zytkow, J. M., Zhu, J., & Hussam, A. (in press). Automated discovery in a chemistry laboratory. *Proceedings of the Eighth National Conference on Artificial Intelligence.* Cambridge, MA: AAAI Press.

CHAPTER 2

The Conceptual Structure of the Geological Revolution

PAUL THAGARD
GREG NOWAK

1. Introduction

This chapter is part of a general research program to develop a computational understanding of what scientific revolutions are and how they occur. Historians of science often use the term *revolution* to refer to the major theoretical developments associated with the names of Copernicus, Newton, Einstein, and Darwin, as well as to the chemical revolution of the eighteenth century and the quantum and geological revolutions of the twentieth century (Cohen, 1985; Kuhn, 1970). Using the example of the development of ideas about continental drift and plate tectonics that constitute the geological revolution, we shall propose a general computational framework for understanding revolutionary conceptual change.

We begin by comparing our approach with a selection of other computational research programs on discovery. Then, to provide a quick overview of our general approach, we advance six theses about the understanding of conceptual revolutions. These set the stage for a discussion of the development of ideas in the geological revolution and a consideration of what led to their acceptance.

2. Computational Investigations of Discovery

During the past decade, machine learning has grown from a very small subfield of artificial intelligence to an enterprise with its own conferences and journal. As part of this explosion, research on computational models of discovery has expanded rapidly. We shall not attempt a general taxonomy or survey of machine discovery research here but will make some distinctions that serve to compare our work with other investigations. These distinctions are not invidiously intended to suggest the "right" and the "wrong" way to proceed; scientific thinking has many features, and there are many different approaches to the understanding of discovery that should proceed in parallel.

2.1   Psychological Versus Nonpsychological Approaches

The first distinction is between psychological and nonpsychological approaches to machine discovery. Since its inception, AI has included researchers who view the modeling of human cognition as the best route to the understanding of intelligence as well as researchers who prefer to think that such understanding is best achieved by a purer theoretical and engineering approach. Similarly, some computational models of human discovery are intended to apply to human thought, whereas other models are intended to develop discovery methods that exploit computational techniques that differ from those available directly to humans. In the first category, we can place, for example, the BACON program (Langley, Simon, Bradshaw, & Zytkow, 1987), and research reported in this volume by Klahr, Shrager, and Kulkarni. Our project to understand scientific revolutions also falls in this category since we are trying to model the actual thought processes with which great scientists initiated scientific revolutions. For some discovery purposes, however, it is desirable to use mathematical techniques that are not part of human cognition. Nonpsychological approaches to machine discovery include the work of Cheeseman (this volume), Glymour, Scheines, Spirtes, and Kelly (1987), and Buchanan (1983).

2.2   Theory Generation Versus Empirical Discovery

Most research in machine learning is concerned with deriving laws or concepts from data, not with developing theories that provide *explanations* of events and empirical laws. For example, the BACON program (Langley et al., 1987) takes numerical values as inputs and generates equations that fit those values. In this volume, research described by Nordhausen and Langley, Zytkow, and Cheeseman clearly falls into the category of empirical discovery from data. Much scientific reasoning, however, is concerned with forming hypotheses that *explain* empirical laws of unusual observations. Following the general practice in the philosophy of science, we distinguish between *empirical laws* based directly on observations, such as Kepler's laws, and *theories* that postulate unobserved entities or processes. In the latter category, for example, we have Newton's theory of gravitation, which uses the notion of gravitational force (which is not simply derived from observations) and Darwin's theory of evolution, which postulated natural selection. Explanation-driven theory formation is exemplified in this volume by the discussions of Darden, Rajamoney, Falkenhainer, O'Rorke, and our own analysis of the geological revolution to be presented below.

2.3   What Gets Discovered?

Most work on machine discovery involves the formation of propositions that constitute laws or hypotheses. In contrast, the research of Cheeseman (this volume) and other AI work on conceptual clustering constructs classes or concepts from data. Lenat's (1977) AM program for mathematical discovery constructed both concepts and propositions. We shall argue that understanding of scientific revolutions requires mechanisms for *both* concept and hypothesis formation.

2.4   Evolutionary Versus Revolutionary Change

Virtually all research on machine discovery has been concerned with gradual, evolutionary changes in the propositions and concepts taken as known. Since most developments of scientific knowledge have the character of accretional additions to what is already known, this focus is perfectly appropriate. However, there are cases in the history of science when transformations in theoretical frameworks are major, involving rejections of large parts of what was previously accepted. Kuhn (1970) has suggested that such changes are akin to Gestalt switches. Our concern is with such revolutionary changes and we shall next outline our approach.

3. Six Theses on Conceptual Revolutions

We now present six claims about what is involved in scientific revolutions from a computational point of view. We are not yet in a position to argue that these claims should be accepted. They are best understood as working hypotheses to be evaluated with respect to their historical adequacy and the richness of the computational models they inspire.

The concept of revolution was originally applied to scientific developments by analogy with political and social developments. Political revolutions involve major transformations in political structures; in the American Revolution, for example, power was transferred from the British monarchy and their representatives to American citizens. Social revolutions involve major transformations in social structure, with some social classes wresting wealth and power from other social classes—for example, during the Chinese revolution of 1949 (Skocpol, 1979). Similarly, to understand scientific revolutions, we need to have an understanding of the kinds of structures undergoing transformation. Thesis 1 accordingly states:

1. *Scientific revolutions involve major transformations in conceptual and propositional networks.*

How are concepts organized? Since the pioneering work of Quillian (1968), it has been common in AI systems to have much of the organization provided by KIND or IS-A hierarchies. For example, Tweety is a canary, which is a kind of bird, which is a kind of animal, which is a kind of thing. Similarly, psycholinguists have noticed the importance of kind hierarchies for organizing the mental lexicon (Miller & Johnson-Laird, 1976), and in addition have emphasized the role that *part-whole* hierarchies play. Part hierarchies have different inferential properties from kind hierarchies: Because canaries are a kind of bird, and because birds have feathers, one can generally infer that canaries have feathers, but one cannot infer that beaks have feathers because beaks are parts of birds. Part hierarchies have not received nearly as much attention in AI, although Darden and Rada (1988) show their importance in the development of the notion of a gene. Nevertheless, part hierarchies are important for organizing concepts because they provide such orderings as a toe is part of a foot, which is part of a leg, which is part of a body. We therefore conjecture that:

2. *Conceptual networks are primarily structured via kind hierarchies and part hierarchies.*

From theses 1 and 2 follows the conjecture that all scientific revolutions involve transformations in kind and/or part-whole relations. Preliminary historical investigations suggest that this conjecture is true. A thorough analysis of the chemical revolution brought to light major changes in both kind and part-whole relations (Thagard, in press a). One major ingredient in the revolution wrought by Copernicus is the reclassification of Earth as a kind of planet instead of as a unique body. Newton's theory, using gravitation to explain both Kepler's and Galileo's laws, showed that heavenly and terrestrial bodies are from a mechanical perspective the same kind of thing. Darwin not only proposed a major reorganization of kind hierarchies by reclassifying humans from being a special kind of creature to being a kind of animal; he also transformed the meaning of "kind" by substituting a historical conception based on common descent for a notion of kind based on superficial similarity. Einstein's relativity theory brought with it a conceptual organization very different from that found in Newtonian mechanics since for the first time mass and energy were viewed as interconvertible and therefore as the same kind of thing, and the meaning of part-whole relations was changed dramatically by the substitution of the notion of an integrated space-time for commonsense notions of space and time. Finally, with quantum theory we collapse the distinction between wave and particles since light waves are quantized and particles have wavelengths, in contrast with the previous two hundred years of physics that had seen combat between particle and wave theories of light. Scientific revolutions involve rejection of previous assumptions, not just additions. Later we shall describe the very interesting changes in kind relations and part relations that took place during the geological revolution.

Empirical laws are usually framed in the same vocabulary as the observational descriptions on which they are based, or, as in the *intrinsic properties* of the BACON system (Langley et al., 1987), in terms directly derived from observational ones. In contrast, theories often invoke entities and processes that are unobservable or at least unobserved. Examples of theoretical entities postulated because of the explanatory power of the hypotheses that state their existence include electrons, quarks, and mental processes. The distinction between theoretical and observational concepts is not absolute since better instruments can render a theoretical entity observable, as the electron microscope did for the gene reconceived as a sequence of DNA. Concepts referring to theoretical entities or processes cannot be derived from observation, so how can they arise? Thagard (1988) contends that theoretical concepts can be formed by *conceptual combination*, in which new concepts, as represented by frames in the sense of Minsky (1975), can be formed from old ones by using heuristics to select slots from existing concepts. For example, the concept of sound waves, which are not observable, is the result of conjoining the concept of sound with the concept of a wave, both derived from observation. We therefore conjecture that:

  3. New theoretical concepts generally arise by mechanisms of conceptual combination.

This conjecture seems to fit well with additional cases from the history of science, such as natural selection and continental drift (discussed below), but we have not conducted a sufficiently complete canvas of scientific theories to feel confident that it is true. Other mechanisms are probably necessary.

  Theses 2 and 3 deal with the structure and origin of conceptual networks; theses 4 and 5 address the same questions for propositional networks. Thagard (1989) has developed a computational theory of explanatory coherence designed to account for revolutionary and nonrevolutionary cases of theory evaluation. On this account, the most important relations between propositions concern explanation and contradiction. On the basis of that theory, we conjecture that:

  4. Propositional networks are primarily structured via relations of explanatory coherence.

More will be said about the nature of these relations below in the context of an analysis of the acceptance of plate tectonics.

  Where do new hypotheses come from? In the nineteenth century, the philosopher C. S. Peirce coined the term *abduction* to refer to the formation of explanatory hypotheses, and computational research on abduction is growing rapidly (Falkenhainer, this volume; Josephson, Chandrasekaran, Smith, & Tanner, 1987; O'Rorke, this volume; Pople, 1977; Reggia, Nau, & Wang, 1983; Thagard, 1988). The simplest case of abduction is one in which one wants to explain that an object *a* has some property G, and the knowledge that F's are G's enables one to conjecture that *a* is F to explain why it is G. We will save further discussion of abduction for geological examples but here merely conjecture that:

  5. New theoretical hypotheses generally arise by abduction.

Because theoretical hypotheses contain theoretical concepts, they cannot be formed by generalization from observations.

Finally, we must address the question of how scientific revolutions take place. Revolutions involve such major transformations in conceptual and propositional organization that there is no way in which simple techniques for operating on initial representations could have brought about the change gradually. Rather, a new theoretical and conceptual framework must be developed somewhat independently of the existing one. But now we face the difficult question, by what mechanism does the new framework supplant the old? We conjecture that:

> 6. *The transition to new conceptual and propositional networks occurs because of the greater explanatory coherence of the new propositions that use the new concepts.*

Thagard (1989) shows how Lavoisier's argument for the superiority of his oxygen theory over the accepted phlogiston theory, and Darwin's argument for evolution by natural selection, can be understood in explanatory coherence terms, and we shall apply a similar analysis below to plate tectonics. With six claims about scientific revolutions on the table, it is time to look at the geological revolution.

4.   The Conceptual Development of Plate Tectonics

4.1   Historical Sketch

The history of ideas about continental drift and plate tectonics has been told elsewhere (Glen, 1975; Hallam, 1973; Menard, 1986), so we shall give only a brief outline. In 1915 Alfred Wegener, a German meteorologist, published a book in which he brought forward geological and paleontological evidence supporting the hypothesis that the continents had once been joined together and drifted apart. Although Wegener's book went through four German editions and was translated into English and French, his theory of continental drift was not generally accepted. The dominant global theory for explaining geological phenomena at the time was based on the hypothesis that the Earth was cooling and therefore contracting, with contraction causing such phenomena as mountain building.

Although continental drift had some supporters during the following decades, it remained on the fringe of geology until the early 1960s. During the 1950s, the U.S. Navy sponsored expeditions to map the sea floor in both the Atlantic and Pacific oceans. These expeditions brought to light a great deal of new phenomena, in particular information about mid-ocean ridges. In 1960, Harry Hess (1962) and, independently, Robert Dietz (1961) proposed the hypothesis of *seafloor spreading*, according to which the ridges were caused by an upward flow of mantle material that then spreads out, creating new sea floor and carrying the continents away from each other. This is a very different mechanism from the one Wegener proposed, as Hess (1962, p. 609) takes pains to make clear: "The continents do not plow through the oceanic crust impelled by unknown forces; rather they ride passively on mantle material as it comes to the surface at the crest of the ridge and then moves laterally away from it."

In 1963 F. Vine and D. Matthews used the hypothesis of seafloor spreading to predict magnetic anomalies that could be expected to appear parallel to oceanic ridges if seafloor spreading occurs. The initial evidence collected by ocean surveys did not clearly confirm the Vine-Matthews prediction, but in 1965 the survey ship *Eltanin* found striking magnetic patterns that fit perfectly with what Vine and Matthews had predicted. The new view of the continents and sea floor was elaborated by J. Tuzo Wilson's (1965) idea of a transform fault and his proposal that the crust consisted of a set of large, rigid plates, along with Jason Morgan's (1968) development of a mathematical framework for what has come to be called plate tectonics. By the end of the 1960s, plate tectonics provided the generally (but not universally) accepted framework for understanding diverse geological phenomena. More historical details are provided below.

What would be required for a full computational understanding of these important developments? At the very least, we need a set of mechanisms sufficient for generating the key concepts and hypotheses. In addition, we need algorithms for evaluating continental drift and seafloor spreading; to be historically adequate, a model incorporating these algorithms will have to account for the rejection of Wegener's theory and the acceptance of the 1960s view. In addition, we need an account of the conceptual structures and transformations that took place as part of the geological revolution.

4.2 The Discovery of Continental Drift

The third and fifth theses on conceptual revolutions conjecture that theoretical concepts and hypotheses arise primarily by the mechanism of conceptual combination and abduction. We shall now sketch how these mechanisms, as implemented in the system PI (Thagard, 1988), can be applied to the early development of continental drift. Although we shall describe an actual run of PI, we stop well short of claiming to having simulated the discovery of continental drift since we gave the system only the minimal amount of information needed to illustrate the operation of the relevant mechanisms.

Fortunately, we have a direct quote from Wegener that reports how he first came to think of continental drift: "The first concept of continental drift came to me as far back as 1910, when considering the map of the world, under the direct impression produced by the congruence of the coastlines on either side of the Atlantic" (Wegener, 1929/1966, p. 1). Wegener noticed the large extent to which the coastline of Africa can be fitted to the coastline of South America. The hypothesis that the continents were once joined together and subsequently drifted apart provides an explanation for the degree of fit. Wegener later collected a vast amount of paleontological and geological evidence in favor of continental drift, but it is clear that the congruence of the coastlines was the stimulus for the initial discovery.

A crude version of this discovery has been implemented in the processing system PI (which is short for "processes of induction" and is pronounced "pie"). In PI, various kinds of induction occur in the context of problem solving. Explanation is understood as a kind of problem solving where the facts to be explained are treated as goals to be reached, and hypotheses can be generated to help provide the desired explanations. Problem solving proceeds by the spreading activation of concepts and the firing of rules. To get it started, PI is given the goals of explaining two propositions in predicate calculus whose English representations are:

E1. South America and Africa fit together.

E2. South America and Africa are apart.

The concepts FIT-TOGETHER and ARE-APART have associated rules that can potentially provide explanations of these jointly puzzling facts.

R1. If z split into x and y, then x and y fit together.

R2. If x and y were together and drift, then x and y are apart.

Abduction in PI, in the simplest version, has the form:

> To explain G(a), given the rule that if x is an F then it is a G, infer F(a), i.e., that a is an F.

The abduction using E1 and R1 is more complicated since it requires positing that there was a z that split into South America and Africa. PI does abduction with *n*-place relations as well as monadic predicates. In a kind of inference called *existential abduction*, PI makes this inference, concluding that some z (what Wegener called Pangaea) split into South America and Africa; this hypothesis explains why they now fit together. Similarly, R2 can be used to explain E2 by forming the hypotheses that South America and Africa drift. PI does not need to abduce that South America and Africa were together since it deduces this fact using an additional rule:

R3. If z split into x and y, then x and y were together.

Table 1 shows the representations, based on predicate calculus, used for the relevant propositions.

After having abduced that South American and Africa drift, PI can generalize that continents drift because it was also informed that they are continents. In PI, generalization that All A are B occurs when there are sufficient numbers of A's that are B's to justify the generalization, given background information about variability (see Holland, Holyoak, Nisbett, & Thagard, 1986, chap. 8). Our continental drift simulation would require background information about how things like continents vary with respect to properties like drifting, which is normally calculated on the basis of additional knowledge in the data base. Rather than build up the data base, we merely told the program that in this case variability is low, so two instances (and the absence of counterexamples) suffice for the generalization that continents drift. This inference is triggered by the simultaneous activation of the propositions that South America and Africa are continents and the propositions that each drifts.

Similarly triggered in PI is the attempt to combine old concepts into new ones. In this case, PI joins CONTINENT and DRIFT into CONTINENT-DRIFT, which can be understood as the concept of a continent that drifts. This is still short of the full concept of continental drift, which refers to a process rather than to instances of drifting continents, but it is a start. At the end of its short run, PI has thus formed

Table 1. Input to PI's simulation of the discovery of continental drift.

```
R1.     IF          (split ($z $x $y))
        THEN        (fit-together ($x $y))

R2.     IF          (were-together ($x $y))
                    (drift ($x))
                    (drift ($y))
        THEN        (are-apart ($x $y))

R3.     IF          (split ($z $x $y))
        THEN        (were-together ($x $y))

Problem Description:
    Starting conditions:
        (continent (south-america))
        (continent (africa)))
    Explain:
        (fit-together (south-america africa))
        (are-apart (south-america africa))
```

several hypotheses about South America and Africa and the general rule that continents drift, as well as the new combined concept.

This simulation has many limitations. PI was provided with only very limited information. It lacks understanding of the various geological processes involved. A representation scheme richer than predicate calculus may be needed, perhaps the qualitative process system of Forbus (1985). We have not provided detailed description of PI's mechanisms of problem solving, abduction, and conceptual combination, since these are available elsewhere (Thagard, 1988). Nevertheless, we hope this sketch indicates the relevance of those mechanisms to the problem of giving a computational account of the discovery of continental drift.

4.3 Changes in Conceptual Structure

There were two major kinds of conceptual change in the geological revolution. First, a host of new concepts arose, more in the 1960s with the development of plate tectonics than with Wegener's early theory. Wegener gave us continental drift, but the 1960s brought such important new concepts as seafloor spreading, transform fault, plate, subduction, lithosphere, and asthenosphere. The developments of the 1960s involved many more additions to the geological vocabulary than did Wegener's theory.

Conceptual change is not merely a matter of adding new concepts but also of *reorganizing* them. According to the first two theses on scientific revolutions stated above, revolutions involve transformations in the kind hierarchies and part hierarchies that are the main organizing backbone of conceptual systems. The geological revolution in the 1960s displays a particularly interesting transformation that involves both kind relations and part relations. During the 1950s, extensive surveys of the ocean floor began to cause problems for the old view that continents and sea floor are both parts of crust that differed primarily only in that the latter is submerged. This view was important for the contractionist alternative to Wegener's continental drift theory, which explained the similarity of flora and fauna in Africa and South America by postulating that there had been continental land bridges that subsided and became sea floor. Observations of the sea floor suggested, however, that sea floor and continents differed structurally and in composition, challenging the notion of land bridges: If the sea floor and the land masses were fundamentally different, one would not expect that parts of the continents would occasionally sink and become sea floor.

Hess' (1962) proposals about seafloor spreading explained why the sea floor, which in his view was continuously being produced by molten material rising up at mid-ocean ridges, should be younger and different in constitution than continents. Once continents and sea floor were seen as more different than alike, it became more appropriate to think of sea floor and the continents as *kinds* rather than *parts* of crust. Moreover, the notion of crust became much less important because what really mattered was that continental crust and oceanic crust were merely surface features of the huge plates, which, according to plate tectonics, were the primary structural components of the Earth's surface. In contrast to older views, which saw the continents and sea floor as parts of a uniform crust, plate tectonics depicted continents and ocean floors as different kinds of surfaces of plates; "crust" became a concept that referred to the surfaces of plates but played no dynamic role in geophysics. The geological revolution is the only case we have found of a transformation *from* part relations *to* kind relations. This transformation is summarized in Figure 1, in which (a) shows the part relations before plate tectonics, and (b) shows them after plate tectonics.

The conceptual transformation just described is only part of the general development that occurred in the twentieth century. To give a fuller picture of what happened, we shall now fill in some of the relevant history, presenting diagrammatic summaries of several important stages in the development of ideas from Wegener and his opponents to the 1960s.

4.3.1 Wegener's Concepts

Figure 2 is a rough attempt to portray the central concepts and explanatory relations of Wegener's system. Wegener wanted to account for the formation of mountains, the existence of fossils of land animals of the same species on land masses separated by oceans, and the apparent fit of the continents on opposite sides of the Atlantic. He believed that the crust originally contained a protocontinent that was fractured by the two mechanisms he posited to account for continental motion: "Westwanderung," or the gravitational influence of the moon and sun, which tended to drag continents westward; and the "Pohlflucht," or tangential force experienced by continents, which moved them away from the poles and toward the equator.

Figure 2 shows that Wegener's forces—Westwanderung and Pohlflucht —were intended to be responsible for the motion of the continents and protocontinent, causing the protocontinent to split into the continents as we know them. Soon after Wegener's work was first published, it was pointed out that both of these forces were too weak by about five orders of magnitude to accomplish their intended effect. Wegener was able to explain the apparent "fit" of the continents, the creation of mountains, the separation of fossil species, and glacial evidence, but there were some problems. For one, the relationship between the sea floor and the continents was unclear—the continents had to be denser than the sea floor so they could move through it but lighter than the sea floor so they could be compressed by it into mountains. Second, Wegener misjudged the relationship between certain glacial moraines found in both Europe and Greenland, assuming them to be related. Thus he conjectured an unreasonably recent date for the fission of Greenland and Europe and based his calculations of the rate of drift on this date. Faulty longi-

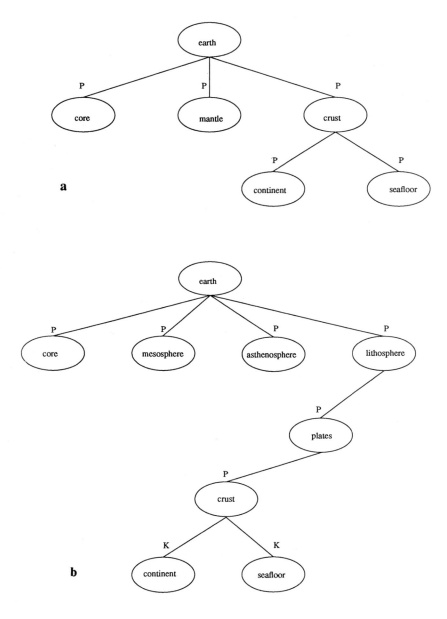

Figure 1. Part relations and kind relations before (a) and after (b) plate tec-
tonics. Part relations are labeled P, and kind relations are labeled
K, reading upward. For example, in (a), continent is a part of the
Earth's crust.

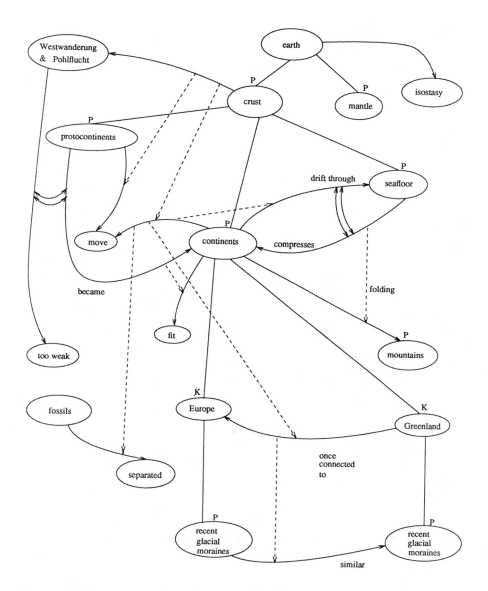

Figure 2. Wegener's conceptual system. Part relations are labeled P, and kind relations are labeled K. Curved solid lines with arrows represent rules involving other kinds of relations. The dotted lines indicate explanatory relations between propositions formed out of the concepts. For example, that continents move explains why fossils are separated. Doubled lines with arrows on both ends join contradictory propositions.

tudinal observations actually seemed to corroborate Wegener's claimed
figure for the rate of drift, but later, more accurate observations in the
1930s showed no sign of drift more significant than statistical error.

4.3.2   CONCEPTS OF WEGENER'S OPPONENTS

The opponents of Wegener were part of a long tradition of uniformitar-
ianism, or the belief in gradual change in the Earth's structure, which
dated from the 1830s with Charles Lyell. The basic motive force of
uniformitarian geophysics was the cooling of the Earth. Since it was
cooling, the crust was shrinking, and this caused lateral compression
in the crust. Figure 3 provides a rough sketch of the conceptual system of Wegener's contractionist opponents. After a sufficient buildup of
stress, the crust fractured into blocks. This description is represented
by the sequence of explanatory arrows at the upper left of Figure 3.
One of the puzzles that a comprehensive account of the Earth's geology
had to account for was the presence of sedimentary rocks (known to
have formed under water and containing aquatic fossils) in mountains.
The puzzle was solved by positing that sedimentary rocks formed over
subsided blocks that were later elevated by further compression of the
crust. This also explained the formation of the mountains themselves.
Several pieces of evidence that would later be claimed as support for
continental drift found explanations within the contractionist system—
for example, evidence for ancient glaciation in what are now temperate
zones was explained by assuming that the Earth's pole had wandered
over the course of time. In other words, the rotational axis of the Earth
had not always passed through the current North and South Poles.

   The presence of nearly identical fossils of early land-based flora and
fauna on opposite sides of oceans was explained by the prior existence
of land bridges between continents that spanned the oceans. These
land bridges are no longer in evidence since they were part of blocks
that had subsided. Note that the assumption that parts of the crust
considered land could become parts of the crust considered sea floor
required the assumption that the continental crust and the sea floor were
effectively identical in composition. Also, the ability of a large mass of
crust to significantly change its relative altitude was in conflict with an
accepted principle of geophysics. The principle of isostasy asserts that
the crust is usually in gravitational equilibrium; a large piece of the
crust that has not recently been disturbed has equal forces holding it

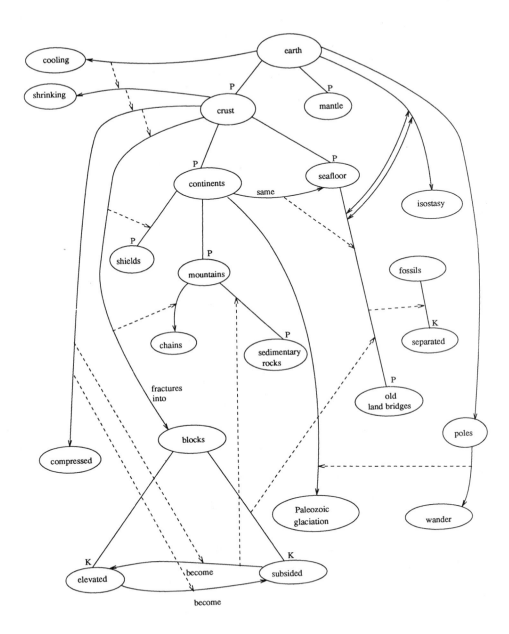

Figure 3. Conceptual system of Wegener's critics. See Figure 2 for description of the notation.

up (buoyancy with respect to denser strata below) and pulling it down (gravity). The conflict between the ideas of isostasy and land bridges was tolerated because they were the concerns of different disciplines; only geophysicists were concerned with the validity of isostasy, and only geologists were concerned with the existence of land bridges.

The contractionist program was not stagnant after the initial rout of Wegener. By 1959, it had been expanded to account for most of the major features of the Earth's surface. The authors of a major contractionist text (Jacobs, Russell, & Wilson, 1959) continued the geophysical tradition of ignoring the problem of the land bridges in their discussion of isostasy but were aware of one problem in their synthesis—the fact that the average heat flow from the ocean floor and the continents was the same, when one would expect higher continental heat flow due to the greater amount of radioactive material in the continents. This conflict was discussed but not explained by the text. Contemporaneous accounts that did discuss the problem assumed either that isostasy was a rough enough approximation that land bridges could exist for significant amounts of time, that the forces governing continental uplift and subsidence were stronger than isostasy, or that the land bridges were little more than isthmuses that were not large enough to be subject to isostasy.

4.3.3   CONCEPTS OF HESS AND PLATE TECTONICS

The 1950s saw a massive accumulation of hard data, especially about the nature of the sea floor. The sea floor was found not only to have central ridges (known since the laying of the transoceanic cables) but also a rift down the middle of the ridges. These rifts were almost invariably locations of high earthquake activity; in fact, maps of earthquake epicenters were used to detect midocean rifts. One other form of data that became available was heat flow profiles, measured in terms of units of energy per unit of area. The sea floor heat flow profiles peaked at the central rifts, but no theory had yet made an attempt at explaining the peaks.

Magnetic data also became significant in the 1950s. When some molten rocks cool, they are magnetized by the Earth's magnetic field. Like a bar magnet, they have a direction associated with them; this direction points toward the location of the north magnetic pole at the time the rocks were magnetized. It was apparent that the Earth's magnetic field had periodically reversed over time. Although magnetized rocks that appear to have been magnetized in the opposite direction from the current magnetic field of the Earth had been known since the 1920s, it was not until the 1950s that the fact of periodic reversals was accepted.

In 1959, Harry Hess first proposed his theory of seafloor spreading, which was eventually published in 1962. It was accompanied by the appearance of similar ideas by Robert Dietz, who coined the term *seafloor spreading* but gave Hess priority. Hess' theory took account of the new data gathered about the sea floor during the 1950s and attempted to explain it in a way that supported the idea of continental drift. Figure 4 gives a rough idea of the conceptual relations in Hess' theory.

Hess hypothesized that convection in the mantle caused rising areas of mantle material to surface at the crust at mid-ocean ridges; this rising material spread open the existing sea floor, cooled, and became new sea floor. In turn, old sea floor was consumed by the mantle at subduction zones located at oceanic trenches. One major difference between Hess' work and previous views was the fact that the ocean floor was no longer considered part of the crust; since it was composed of cooling mantle, the sea floor was seen as "crustless," and the only crustal material was continental. This is represented in Figure 4 by the absence of a part link joining "crust" and "sea floor." Note the central importance of the relation between the concepts "sea floor" and "spreads" in the diagram; this proposition was used to explain most of the major features of the sea floor and continents.

Shortly after Hess' paper appeared, new magnetic evidence seemed to confirm it. Strata on land revealed the possibility that the Earth's magnetic field had reversed polarity in the past; magnetic profiles of the sea floor revealed linear patterns parallel to the midocean ridges, and cores of sediment taken from the sea floor also gave evidence of magnetic reversals. Vine and Matthews (1963) hypothesized that, if new sea floor was being created at the midocean ridges, then there should be patterns of magnetic stripes parallel to the ridges that record the direction of the Earth's magnetic field at the time the sea floor was being created. After many ambiguous measurements, relatively clear data were obtained that confirmed the Vine-Matthews hypothesis that sea floor was constantly being created and recording the direction of the Earth's

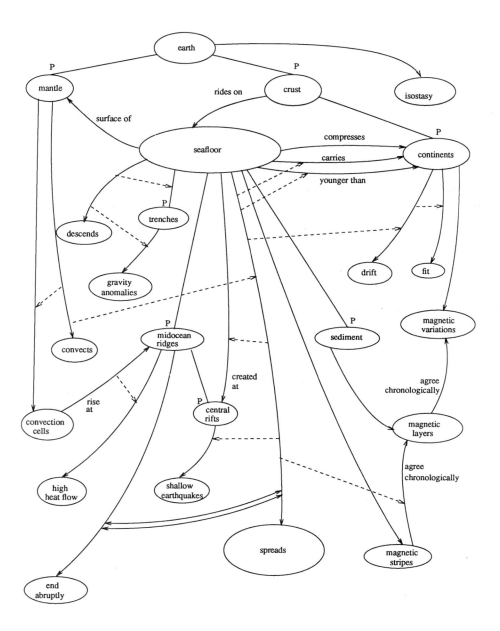

Figure 4. Conceptual system of Hess.

magnetic field as it cooled. A further corroboration of the magnetic evidence was obtained by the fact that the time scales of magnetic reversals obtained by three different methods—land-based sampling, horizontal striping around the midocean ridges, and reversals recorded in columns of sediment above the sea floor—were in general agreement. Other concerns such as mountain building were addressed, too. Mountains were said to be created when continents crashed into each other or resisted the motion of the sea floor; and all the paleontological and geological data that supported continental drift originally also supported seafloor spreading since seafloor spreading allowed for drift.

Note that many of the facts explained by seafloor spreading were relatively new pieces of evidence—gravitational and magnetic anomalies, the high heat flow of the midocean rift—that were only discovered in the 1950s. Hess' main goal was to account for features of the sea floor, not to prove continental drift, but his theory entailed continental drift as well. One minor internal conflict did appear in the theory—some midocean ridges ended abruptly in relatively undisturbed sea floor. If sea floor was being created by spreading at midocean ridges, how could these ridges come to an end?

Hess' seafloor spreading sufficed to show the benefits of a theory that embraced continental drift; the task remained to elaborate the idea beyond the level of the "geopoetry" proposed by Hess. In 1965, J. Tuzo Wilson proposed the existence of a new kind of fault, which he called the *transform fault*. Study of Hess' work had led him to realize that, if Hess were correct, a hitherto unobserved type of fault would be required in order to allow any motion to parts of the Earth's crust. Along with the existence of transform faults, Wilson described the crust as being divided into large "plates," which were composed of both seafloor and continental material; the mineralogical differences of the two types of crustal material were now considered less important than their structural unity. Plates could be bounded only by midocean ridges, trenches or subduction zones, and transform faults.

In addition, the geophysicists borrowed a new organization of the interior of the Earth from the seismologists. Seismic waves exhibit a change in behavior at a depth of about 100 kilometers. Immediately below this level is a relatively plastic zone over which plates could slip with less resistance than would otherwise be suspected. Thus the depth of 100 kilometers could be interpreted as the thickness of the plates.

The plates were the structural components of the lithosphere, which was the part of the Earth's surface that moved; the asthenosphere was the portion of the mantle in which convection occurred. Once again the crust included sea floor; it was the surface of the lithosphere and was thus broken into plates as well. Although many of the ideas of Hess were preserved in plate tectonics, their organization was different, and different points were emphasized. The development of plate tectonics has as one of its themes the gradual lessening of geophysical importance of the continents and gaining of importance of the sea floor, until both concepts were regarded as of small significance compared with the idea of plates. The mature theory of plate tectonics can be dated from the late 1960s, when Jason Morgan (1968) developed a mathematical framework for plate tectonics that yielded quantitative predictions. Figure 5 depicts the qualitative concepts of plate tectonics.

The richness of the history we have just sketched shows that a great deal of work remains to be done in modeling the conceptual development of the key ideas in plate tectonics. Instead of pursuing that project further here, however, we want to move on and address questions concerning the *evaluation* of geological theories.

5.   Evaluating Theories of Continental Drift and Plate Tectonics

So far, we have been addressing questions of discovery and conceptual development. But computational modeling of scientific thinking should also provide insights into how theories are evaluated and accepted. The key historical question that needs to be answered is, *why was Wegener's theory of continental drift largely rejected in the 1920s, and why, in contrast, were the new ideas about seafloor spreading and plate tectonics largely accepted in the 1960s?* We shall answer these questions using the computational theory of explanatory coherence that is implemented in a connectionist computer program called ECHO.

5.1   Explanatory Coherence

Since a complete statement of the theory of explanatory coherence and a full description of the algorithms of ECHO have been given elsewhere (Thagard, 1989), we shall be very brief here. According to the theory, a hypothesis is accepted if it coheres better with the evidence and

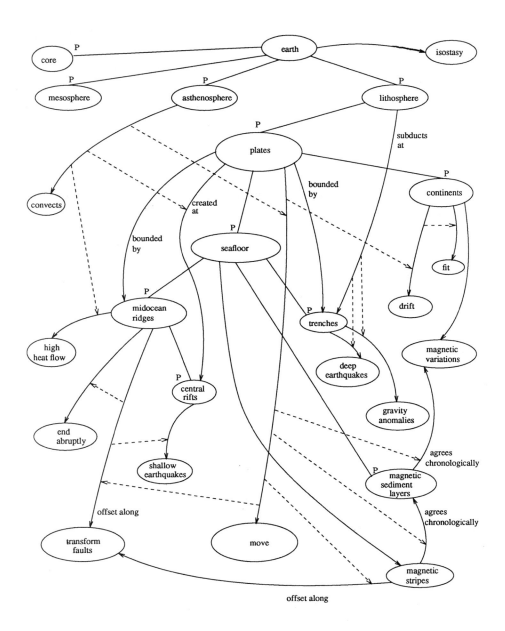

Figure 5. Conceptual system of plate tectonics.

with other hypotheses than do competing hypotheses. A hypothesis coheres with other propositions that explain it, that are explained by it,
that participate with it in explaining other propositions, or that participate in analogous explanations. Pieces of evidence have a degree of
acceptability on their own, although this can be overruled by coherence
relations. Propositions that contradict each other are incoherent with
each other.

Connectionist networks consist of units that are connected by excitatory and inhibitory links. ECHO is a COMMON LISP program running on
Sun workstations that constructs networks for evaluating the explanatory coherence of sets of propositions. Propositions that cohere are
represented by units connected by excitatory links; propositions that
incohere have units connected by inhibitory links. For input, ECHO
is given formulas describing the explanatory relations of propositions.
If two hypotheses H1 and H2 together explain a piece of evidence E1,
ECHO is given the LISP input:

(EXPLAIN (H1 H2) E1).

ECHO then sets up symmetric excitatory links between units representing H1 and E1, H2 and E1, and H1 and H2. If H1 and H3 are contradictory, ECHO gets the input:

(CONTRADICT H1 H3).

This sets up a symmetric inhibitory link between H1 and H3. That E1
and E2 are to be treated as pieces of evidence is represented by the
input:

(DATA (E1 E2))

To provide activation to E1 and E2, links to them are then set up from
a special evidence unit that is always active.

Connectionist networks make decisions by repeatedly updating the
activation of units in parallel until the whole network settles into a stable
state in which the activation of each unit has reached an asymptote.
ECHO adjusts the activation of a unit u by considering all the units to
which it is linked. An excitatory link with an active unit will increase
the activation of u, whereas an inhibitory link with a unit with positive
activation will decrease it. Activation of units starts at 0 and is allowed
to range between 1 and -1. Repeated adjustments of activations result

in a stable state where some units end up with high activation and others with activation below 0. Tables 2 and 3 summarize the algorithms used by ECHO to create and run networks. After input concerning explanations and contradictions has been used to set up the network, the network is run in cycles that synchronously update all the units. For each unit j, the activation is a continuous function of the activation of all the units linked to it, with each unit's contribution depending on the *weight* w_{ij} of the link from unit i to unit j. At the end of an ECHO run, each unit has an asymptotic activation, of which a value greater than 0 represents acceptance of a hypothesis, and a value less than 0 represents rejection.

ECHO has some points in common with recent computational models of scientific theory evaluation proposed by Rose and Langley (1988) and by Rajamoney (1988). Both of these models also attempt to take into account explanatory breadth and simplicity, although they do so in very different ways from ECHO. Rose and Langley's REVOLVER model and Rajamoney's COAST model use numerical weights to combine criteria for the best explanation, in contrast to ECHO's use of a relaxation algorithm. REVOLVER and COAST have the advantage over ECHO that they get their input directly from a discovery program, not from a programmer. Correlatively, however, they are dependent on particular kinds of domain representations, whereas ECHO applies to any domain where hypotheses can be represented by propositions.

5.2   The Rejection of Continental Drift

Elsewhere we have presented an ECHO analysis of the explanatory coherence of continental drift from Wegener's own perspective (Thagard & Nowak, 1988). From his perspective, the theory of continental drift was highly coherent and explained a very broad range of facts. But most geologists did not share Wegener's conclusion that continental drift should be accepted. In 1928, the American Association of Petroleum Geologists published a symposium on Wegener's views (van der Gracht et al., 1928). Most (but not all) symposiasts were highly critical of Wegener. We have accordingly compiled a composite picture of Wegener's opponents by amalgamating the various criticisms of his account, most of which asserted that his hypotheses contradicted well-established evidence. The input to ECHO based on our analysis of the criticisms Wegener received is presented in Appendix A, which shows the relevant propositions, ex-

Table 2. Algorithms for processing input to ECHO.

---

1. Input: (PROPOSITION NAME SENTENCE)
   Create a unit called NAME.
   Store SENTENCE with NAME.

2. Input: (EXPLAIN LIST-OF-PROPOSITIONS PROPOSITION)
   Make excitatory links between each member of LIST-OF-PROPOSITIONS and PROPOSITION.
   Make excitatory links between each pair of LIST-OF-PROPOSITIONS.
   Record what explains what.
   Note: Weights are additive, so that if more than one EXPLAIN statement creates a link between two proposition units, then the weight on the link is the sum of the weights suggested by both statements. The weights are inversely proportional to the length of LIST-OF-PROPOSITIONS, implementing a principle of simplicity.

3. Input: (CONTRADICT PROPOSITION-1 PROPOSITION-2).
   Make an inhibitory link between PROPOSITION-1 and PROPOSITION-2.

4. Input: (DATA LIST-OF-PROPOSITIONS)
   For each member of LIST-OF-PROPOSITIONS, create an excitatory link from the special evidence unit with the weight equal to the data excitation parameter, unless the member is itself a list of the form (PROPOSITION WEIGHT). In this case the weight of the excitatory link between the special unit and PROPOSITION is WEIGHT.

Table 3. Algorithms for network operation.

---

1. Running the network:

   Set all unit activations to an initial starting value (typically 0.01), except that the special evidence unit is clamped at 1.

   Update activations in accordance with 2 below.

   If no unit has changed activation more than a specified amount (usually 0.001), or if a specified number of cycles of updating have occurred, then stop.

   Print out the activation values of all units.

2. Synchronous activation updating at each cycle:

   For each unit u, calculate the new activation of u, a_j in accord with equations 1 and 2 below, considering the old activation a_i of each unit linked to u.

   Set the activation of u to the new activation.

Equations:

$$a_j(t+1) = a_j(t)(1-\theta) + \begin{cases} net_j(max - a_j(t)) & \text{if } net_j > 0 \\ net_j(a_j(t) - min) & \text{otherwise} \end{cases} \quad (1)$$

Here θ is a decay parameter that decrements each unit at every cycle, min is minimum activation (-1), max is maximum activation (1), and net_j is the net input to a unit. This is defined by:

$$net_j = \sum_i w_{ij} a_i(t) \quad (2)$$

---

planatory relations, contradiction relations, and evidence. The texts of the propositions are given for information only: ECHO only processes the proposition names. Given this input, ECHO produces the network shown in Figure 6. The evidence statements that are named E1–E18 in Appendix A are shown as 1–18 in the figure. Note the large number of NE statments—pieces of "negative evidence"—that are consequences of Wegener's view that contradict actual evidence according to his critics. Hypotheses W1–W6 are hypotheses of Wegener, and C1–C7 are hypotheses of the alternative contractionist view. CW1 is common to both Wegener and the contractionists. Excitatory links, indicating that two propositions cohere, are represented by solid lines. Inhibitory links are represented by broken lines. This figure is structured very differently from earlier figures, where the nodes represented concepts rather than propositions and the solid lines represented proposition-forming relations rather than explanatory links. There are 44 units and 110 symmetric links.

The result of running this network—updating the activation of each node based on the activation of the nodes with which it is connected—is that the contractionist hypotheses are accepted and the Wegener hypotheses that contradict them are rejected. Figure 7 shows the unit W5 and the other units to which it is connected. Figure 8 was produced like Figure 7, using a graphics program that runs with ECHO, and shows the activation histories of the units representing hypotheses. Each graph is on a scale of 1 to −1, with the starting point 0 indicated by the horizontal line. Activation values are shown over the 50 cycles of updating that it takes the network to settle. This run had the same parameter values for excitation and inhibition weights used in the examples presented in Thagard (1989).

5.3   The Acceptance of Plate Tectonics

Acceptance of the continental drift theory fared much better several decades later. Our ECHO analysis is based on the seminal paper of Harry Hess (1962) that first developed the concept of seafloor spreading. Hess explicitly presented his views as a set of propositions and indicated numerous phenomena that seafloor spreading could explain. His proposition that the sea floor spreads (S16) directly competes with the contractionist proposition that the crust is contracting (C2) since they function as opposing accounts of many crustal features. Hess him-

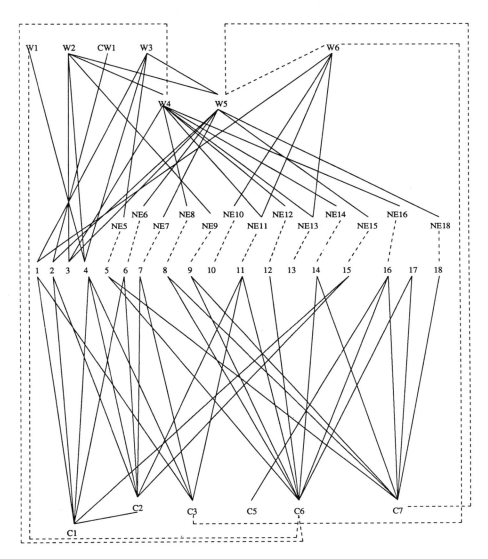

Figure 6. The explanatory coherence of contractionist hypotheses versus Wegener's hypotheses, from the point of view of Wegener's critics. W1–W6 represent Wegener's hypotheses, whereas C1–C7 represent contractionist hypotheses. CW1 is common to both. Numbers 1–18 represent the pieces of evidence labeled E1–E18 in Table 2. NE5–NE18 are pieces of "negative evidence" that contradict pieces of evidence. Solid lines indicate excitatory links representing coherence, and broken lines indicate inhibitory links representing incoherence.

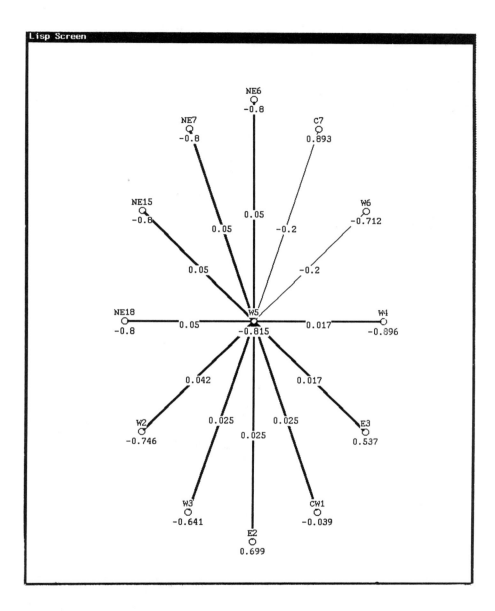

Figure 7. Connectivity of unit W5. Here thick lines indicate excitatory links and thin lines represent inhibitory ones; the number on the line indicates the weight of the links. Numbers under the unit names indicate the asymptotic activation of the unit after the network has settled.

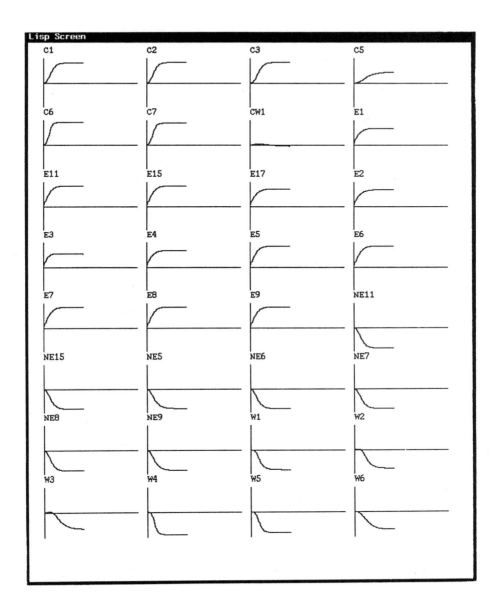

Figure 8. Activation histories of selected units in simulation of Wegener's opponents. Each graph shows the activation of a unit over 50 cycles of updating, on a scale of 1 to −1, with the horizontal line indicating the initial activation of 0. Note that contractionist hypotheses C1–C7 are all accepted, and drift hypotheses W1–W6 are all rejected.

self did no systematic comparison of his account against the contractionist theory, but, since that was undoubtedly in the background, we have included standard 1950s contractionist explanations in our analysis. These are taken mostly from Wilson (1954). The input given to ECHO is shown in Appendix B. Figure 9 shows the coherence relations of the most important hypotheses of Hess and the contractionists. C2, asserting that the Earth contracts, contradicts S4, the main hypothesis of the seafloor spreading model. Several hypotheses of the contractionists are explained by C1, the hypothesis that the Earth's mantle has been cooling, but this hypothesis would imply "negative evidence" such as NE14 and NE20 about the measured heat flow of the crust; these counterfactual statements contradict the actual observations represented by E14 and E20, which the spread model explains. Both the contractionists and spread models have about the same number of mutually coherent hypotheses; as this abbreviated diagram suggests, the spread model derives its advantage by explaining pieces of evidence that the contractionist model does not and by explaining actual observations that contradict the implications of the contractionist model.

Running the network produced by the input in Appendix B produces the activation histories shown in Figure 10. Hess' seafloor spreading hypotheses win out, and those contractionist hypotheses that contradict Hess are deactivated. (Proposition C8, that the Earth's magnetic pole has wandered over time, is not rejected, but it is consistent with Hess' views.) Clearly, then, at least from Hess' point of view, seafloor spreading and its consequences for continental drift should be accepted as possessing more explanatory coherence than the contractionist view. This network is comprised of 63 units and 267 symmetric links; it settles after 107 cycles.

The simulations of Wegener's opponents and Hess use the same parameter values used in almost all other ECHO simulations—excitation (default weight on links representing coherence) 0.05, inhibition (weight on links representing incoherence) -0.2, decay 0.05, data excitation (weight on links between the special evidence unit and data units) 0.1. Sensitivity analyses have determined that ECHO's performance does not depend on a narrow range of values. Holding everything else constant, excitation can range as high as 0.27 and 0.06 for the Wegener's opponents' simulation and the Hess simulation respectively, and inhibition can range as low as -0.021 and -0.17 for the two simulations. Cutting

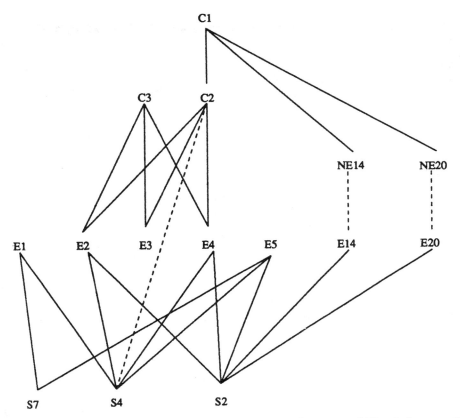

Figure 9. Partial depiction of the explanatory coherence of Hess' theory of seafloor spreading. The notation is the same as in Figure 6.

excitation and inhibition in half causes no significant changes in the simulation of Wegener's opponents.

The Hess simulation, however, is somewhat more sensitive to increases in excitation or decreases in inhibition. Both of these operations tend to decrease the likelihood that alternative hypotheses will be rejected. The ranges given above mark the points at which some contractionist hypotheses are no longer rejected; a broader range of parameters can be given if we wait until all hypotheses are accepted. Thus it is plausible to suggest that the narrower range of parameters necessary to keep the Hess simulation from accepting some contractionist hypotheses merely reflects a historical reality, that no geological theory was conclusively preferable in the early 1960s.

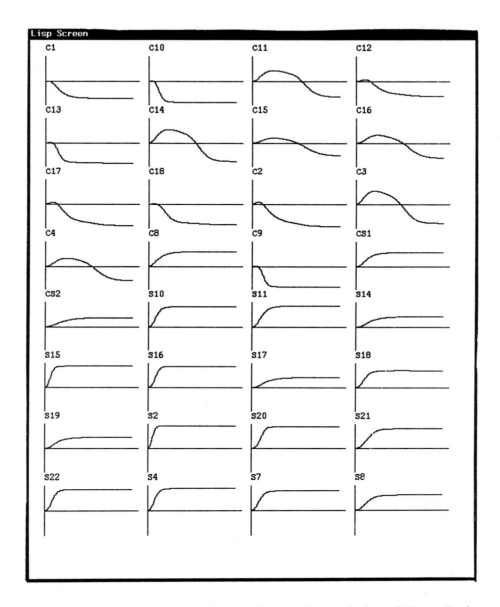

Figure 10. Activation histories of selected units in simulation of Hess. Each graph shows the activation of a unit over 107 cycles of updating, on a scale of 1 to −1, with the horizontal line indicating the initial activation of 0. Note that contractionist hypotheses C1–C8 are all rejected, and seafloor speading hypotheses S2-S21 are all accepted.

However, it is important to stress that we are modeling here only Hess, not the entire geological community, which did not immediately accept Hess' views. For some geologists not as familiar with seafloor phenomena as Hess, later developments were very important in making seafloor spreading acceptable. We have already mentioned developments such as the confirmation of the Vine-Matthews hypothesis by the *Eltanin* magnetic data and Wilson's ideas of transform faults and plates. In our view, the importance of the magnetic data has been overemphasized by philosophers of science who think that prediction and confirmation are more important than explanatory coherence (Frankel, 1979; Giere, 1988). For some geologists, such as Menard (1986), the magnetic data were important in tipping the balance in favor of the seafloor spreading hypothesis, but for other leaders in the field, such as Hess (1962), Dietz (1961), and Wilson (1963), seafloor spreading was already accepted before the magnetic data were in. Wilson is a particularly interesting example since he had been a leading exponent of the contractionist view but was an early convert to the new ideas. We have enhanced the above simulation by adding explanation by seafloor spreading of the magnetic data; the additional excitation enhances the activation of the node representing seafloor spreading in comparison with that possessed by its competitors, but this is an improvement only in degree. The *Eltanin 19* data by no means constituted a crucial experiment.

We can now answer the question with which this section began: Why was Wegener's theory of continental drift rejected and the 1960s theory accepted? The short answer is that the 1960s theory had a great deal more explanatory coherence than Wegener's theory. There are three principal reasons for the relative superiority of the later theory. First, by the 1960s a great deal more evidence had been collected that was explained by hypotheses of seafloor spreading and continental drift. Many facts had been brought to light by the explorations of the sea floor in the 1950s that were much more easily explained by seafloor spreading than by the contractionist theory, even though the latter had seemed superior to Wegener's theory. Second, Hess' theory of seafloor spreading, as well as the fully developed theory of plate tectonics, did not have the *incoherence* that Wegener's theory had with various geological facts and physical theory. The mechanisms postulated for seafloor spreading and plate movement do not contradict accepted principles in the way that Wegener's views of the continents plowing through the sea floor did. The third reason for the relative preferability of the 1960s theory over

Wegener's is that the alternative contractionist theory was on the wane. Accumulating geophysical evidence (for example, detailed information about the amount of heat-generating radioactivity in the Earth's crust) made the view that the crust was cooling less tenable. Thus explanatory coherence theory and ECHO appear adequate for accounting for the evaluation and acceptance of theories in the geological revolution.

6. Conclusion

This chapter has provided only fragments of a computational account of the geological revolution. A full account would have three components. First, it would simulate the formation of all relevant hypotheses and concepts. We provided an example of how this might be done if mechanisms of conceptual combination and abduction suffice. Second, it would simulate the use of those hypotheses in providing explanations of pieces of evidence. We have not attempted this in any serious way—the inputs to ECHO are provided by the programmer. It would be desirable to have all the hypotheses and evidence represented in a way that enabled a processing system to decide for itself what explains what. Finally, an integrated simulation would take the results of the first two components and pass them to ECHO for theory evaluation. There is clearly an enormous amount of work still to be done to develop a full computational understanding of the structure, development, and acceptance of the theories in the geological revolution.

Acknowledgements

We are grateful to Thomas James and Jason Morgan for helpful discussions of geology, and to Jeff Shrager and Pat Langley for valuable comments on an earlier draft.

References

Buchanan, B. (1983). Mechanizing the search for explanatory hypotheses. In P. Asquith & T. Nickles (Eds.), *PSA 1982*, Vol. 2. East Lansing, MI: Philosophy of Science Association.

Cohen, I. B. (1985). *Revolution in science.* Cambridge, MA: Harvard University Press.

Darden, L., & Rada, R. (1988). Hypothesis formation using part-whole interrelations. In D. Hellman (Ed.), *Analogical reasoning.* Dordrecht: Reidel.

Dietz, R. S. (1961). Continent and ocean basin evolution by spreading of the sea floor. *Nature, 190,* 854–857.

Forbus, K. (1985). Qualitative process theory. In D. Bobrow (Ed.), *Qualitative reasoning about physical systems.* Cambridge, MA: MIT Press.

Frankel, H. (1979). The career of continental drift theory: An application of Imre Lakatos' analysis of scientific growth to the rise of drift theory. *Studies in the History and Philosophy of Science, 10,* 21–66.

Giere, R. (1988). *Explaining science: A cognitive approach.* Chicago: University of Chicago Press.

Glen, W. (1975). *Continental drift and plate tectonics.* Columbus, OH: Charles E. Merrill.

Glymour, C., Scheines, R., Spirtes, P., & Kelly, K. (1987). *Discovering causal structure.* New York: Academic Press.

Hallam, A. (1973). *A revolution in the earth sciences.* Oxford: Clarendon Press.

Hess, H. H. (1962). History of ocean basins. In A. E. J. Engel, H. L. James, and B. F. Leonard (Eds.), *Petrologic studies: A volume to honor A. F. Buddington.* New York: Geological Society of America.

Holland, J., Holyoak, K., Nisbett, R., & Thagard, P. (1986) *Induction: Processes of inference, learning, and discovery.* Cambridge, MA: MIT Press.

Jacobs, J. A., Russell, R. D., & Wilson, T. Z. (1959). *Physics and geology.* New York: McGraw-Hill.

Josephson, J., Chandrasekaran, B., Smith, J., & Tanner, M. (1987). A mechanism for forming composite explanatory hypotheses. *IEEE Transactions on Systems, Man, and Cybernetics, 17,* 445–454.

Kuhn, T. S. (1970). *Structure of scientific revolutions* (2nd ed.) Chicago: University of Chicago Press. First published 1962.

Langley, P., Simon, H. A., Bradshaw, G. L., & Zytkow, J. M. (1987). *Scientific discovery: Computational explorations of the creative processes.* Cambridge, MA: MIT Press.

Lenat, D. B. (1977). The ubiquity of discovery. *Artificial Intelligence*, *9*, 257–285.

Menard, H. W. (1986). *The ocean of truth.* Princeton, NJ: Princeton University Press.

Minsky, M. (1975). A framework for representing knowledge. In P. H. Winston (Ed.), *The psychology of computer vision.* New York: McGraw-Hill.

Morgan, W. J. (1968). Rises, trenches, great faults, and crustal blocks. *Journal of Geophysics Research*, *73*, 1959–1982.

Pople, H. (1977). The formation of composite hypotheses in diagnostic problem solving. *Proceedings of the Fifth International Joint Conference on Artificial Intelligence* (pp. 1030–1037). San Mateo, CA: Morgan Kaufmann.

Quillian, M. (1968). Semantic memory. In M. Minsky (Ed.), *Semantic information processing.* Cambridge, MA: MIT Press.

Rajamoney, S. (1988). *Explanation-based theory revision: An approach to the problems of incomplete and incorrect theories.* Doctoral dissertation, University of Illinois, Urbana-Champaign.

Reggia, J., Nau, D., & Wang, P. (1983). Diagnostic expert systems based on a set covering model. *International Journal of Man-Machine Studies*, *19*, 437–460.

Rose, D., & Langley, P. (1988). A hill-climbing approach to machine discovery. *Proceedings of the Fifth International Conference on Machine Learning* (pp. 367–373). San Mateo, CA: Morgan Kaufmann.

Skocpol, T. (1979). *The state and social revolutions.* Cambridge: Cambridge University Press.

Thagard, P. (1988). *Computational philosophy of science.* Cambridge, MA: MIT Press.

Thagard, P. (in press). The conceptual structure of the chemical revolution. *Philosophy of Science.*

Thagard, P. (1989). Explanatory coherence. *Behavioral and Brain Sciences*, *12*, 435–502.

Thagard, P., & Nowak, G. (1988). The explanatory coherence of continental drift. In A. Fine & J. Leplin (Eds.), *PSA 1988*, Vol. 1 (pp. 118–126). East Lansing, MI: Philosophy of Science Association.

van der Gracht, W., et al. (1928). *Theory of continental drift.* Tulsa, OK: American Association of Petroleum Geologists.

Vine, F. J., & Matthews, D. H. (1963). Magnetic anomalies over ocean ridges. *Nature, 199,* 947–949.

Wegener, A. (1966). *The origin of continents and oceans* (J. Biram, trans.). New York: Dover. (Original work published 1929.)

Wilson, J. Tuzo. (1954). The development and structure of the crust. In G. P. Kuiper (Ed.), *The Earth as a planet.* Chicago: University of Chicago Press.

Wilson, J. Tuzo. (1963). Continental drift. In *Continents adrift.* San Francisco: W. H. Freeman and Company. (Reprinted from *Scientific American,* April 1963.)

Wilson, J. Tuzo. (1965). A new class of faults and their bearing on continental drift. *Nature, 207,* 343–347.

APPENDIX A

Input to ECHO for simulation of Wegener's opponents

Evidence:

E1	Mountains occur in chains, especially in the Tertiary.
E2	Large shields on the continents are flat and stable.
E3	The outlines of continents on the opposite sides of the Atlantic are similar.
E4	Many mountains were formed prior to the Cretaceous period.
E5	North America and Eurasia are still north of the equator.
NE5	Continents originally north of the equator should be equatorial by now.
E6	Many earthquakes occur under the oceans.
NE6	A plastic ocean floor would not exhibit signs of stress such as earthquakes.
E7	There are trenches and ridges on the ocean bottom.
NE7	A plastic ocean floor would gradually lose its surface features.
E8	The continents bordering the Atlantic cannot be made to fit together consistently.
NE8	The parts of an early supercontinent would fit together very well.
E9	There is no faulting due to tensile strain on the eastern edges of continents.
NE9	The drag on westward-moving continents would create stress faults in their eastern edges.
E11	Pre-Cretaceous mountain chains parallel current coastlines.
NE11	Features associated with the Atlantic coastlines would date from the Cretaceous or later periods.
E12	Epi-continental seas are ancient.
NE12	Features associated with continental boundaries would date from the Cretaceous or later periods.
E13	Australia experienced folding before but not after the Cretaceous period.
NE13	Australia would exhibit Cretaceous and later folding and mountain building.
E14	Rock formations on opposite sides of the Atlantic do not match.
NE14	Rock formations on corresponding coasts of the Atlantic would be nearly identical.

E15   Basalt at the ocean bottom is denser and more rigid than deeper basalt.

NE15   The ocean floors are less rigid than the continents.

E16   There are many differences between pre-Cretaceous flora and fauna on different continents.

NE16   Fossil remains of pre-Cretaceous flora and fauna should be identical on corresponding coastlines of the Atlantic.

E17   Islands in the Guinea Gulf are older than the Cretaceous period.

E18   Measurements made in 1926 show no movement of Greenland.

NE18   Measurement would indicate motion of Greenland.

E10   There are no mountains on the ocean floor.

NE10   Rigid ocean floors would also be compressed and form mountains.

Hypotheses of Wegener:

W1   The crust originally covered the earth, then folded together in the Paleozoic to create Pangaea.

W2   Tidal forces break up the continents and move them westward.

W3   Pohlfluchtkraft moves the continents towards the equator.

W4   Pangaea broke up in the Cretaceous period.

W5   Continents move through the less rigid ocean floors.

W6   Continents are folded and form mountains due to the resistance of the more rigid ocean floors.

CW1   Large sections of the Earth's crust are in gravitational equilibrium (isostasy).

Hypotheses of Contractionists:

C1   The Earth has been contracting since birth due to planetesimal settling.

C2   The crust is compressed and fractures into blocks.

C3   Blocks can override one another and tip.

C5   Isthmus connections between continents served as land bridges.

C6   Oceans and continents remain relatively the same.

C7   Continental crust can move vertically but not laterally.

Contradictions (LISP keyword "contradict" is not shown):

(W5 W6)   (NE6 E6)   (NE12 E12)
(C3 W6)   (NE7 E7)   (NE13 E13)
(W1 C6)   (NE8 E8)   (NE14 E14)
(W4 C6)   (NE9 E9)   (NE15 E15)
(W5 C7)   (NE10 E10)   (NE16 E16)
(E5 NE5)   (NE11 E11)   (NE18 E18)

Explanations (LISP keyword "explain" is not shown):

((W3) NE5)   ((W4) NE16)   ((C1 C2) E6)
((W5) NE6)   ((W5) NE18)   ((C2 C3) E7)
((W5) NE7)   ((W2 W3) W5)   ((C6 C7) E8)
((W4) NE8)   ((W2) W4)   ((C6 C7) E9)
((W2) NE9)   ((W3 W6) E1)   ((C1 C2 C3) E4)
((W6) NE10)   ((W5 CW1) E2)   ((C2 C3 C6) E11)
((W4) NE11)   ((W1 W2 W3) E4)   ((C6) E12)
((W6) NE11)   ((W2 W4 W5) E3)   ((C6 C7) E14)
((W4) NE12)   ((C1) C2)   ((C1 C2) E15)
((W6 W4) NE13)   ((C1 C3) E1)   ((C6 C7 C5) E16)
((W4) NE14)   ((C1 C2) E2)   ((C6 C7) E17)
((W5) NE15)   ((C6 C7) E5)   ((C7) E18)

APPENDIX B

Input to ECHO for simulation of Hess

Evidence:

E1	Many fossil species are divided by water.
E2	There are sedimentary strata at high elevations.
E3	Oceanic earthquakes produce steeply dipping faults.
E4	There are nearly horizontal faults in mountain ranges.
E5	The geology of opposite sides of the Atlantic is similar.
E6	Transoceanic continental margins are congruent.
E7	Scandinavia is rising 1 cm/yr.
E9	Existence of mountains.
E10	Structure of Alps requires major compression.
E11	Seismic refraction indicates that oceanic crust is very uniform.
E12	There are gravity deficiences at oceanic trenches.
E13	Deep earthquakes occur near oceanic trenches, at an angle to surface.
E14	Trenches are colder than the rest of the ocean floor.
NE14	Trenches over a cooling mantle should be as warm as the rest of the ocean floor.
E15	Midoceanic ridges have high heat flow.
E18	There are no pre-Cretaceous rocks on the ocean floor.
NE18	An ancient ocean floor should include rocks of all epochs.
E19	Apparent polar wander paths indicate that the north magnetic pole was not always where it is today.
E20	Average heat flow in oceans is about same as that of continents, despite radioactive rock.
NE20	A cooling mantle would leave the ocean floor cooler than the continents.
E21	There is a relatively thin veneer of sediment on the ocean floor.
NE21	An ancient ocean floor should be covered with a thick layer of sediment.
E22	Apparent polar wander paths differ between continents but are consistent within continents.
E23	There is a worldwide system of midocean ridges.
E24	Most oceanic earthquakes occur on central rift.

E25	Seismic velocities under the crests of ridges are lower than normal but become normal again on ridge flanks.
E16	There is a Mid-Pacific Mesozoic Ridge.
E17	There are relatively few volcanic seamounts on the ocean floor.
NE17	An ancient ocean floor should display many volcanic seamounts.

Contractionist hypotheses:

C1	The upper mantle of the earth has been gradually cooling.
C2	The Earth has been contracting since birth.
C3	The crust is compressed and fractures into blocks.
C4	Blocks subside and become elevated.
C8	Earth's magnetic pole has wandered over time.
C9	Ocean basins are older than continents.
C10	Ocean basins are parts of the original crust that have been least altered.
C11	Continents grow over former ocean floors, starting with accretion of deltas.
C12	Topography of ocean floors is complex, more rugged than that of continents.
C13	There are large numbers of islands and seamounts that originated as volcanoes on the ocean basin.
C14	After shelf accretion, a conical fracture zone forms farther offshore than the shelf, located by earthquake foci.
C15	Volcanism starts as a result of the fracturing and produces a volcanic arc.
C16	Continental material accumulates from volcanic residue and erosion of the continents, creating shallow inland seas.
C17	Further depression, compression, and uplift convert inland seas to land and volcanic arcs into mountain arcs.
C18	Primary arcs of mountain building meeting at an acute angle create a short mountain range of great height.
C19	Measures of the Earth's radioactivity are too high.
C20	Ocean trenches are due to overriding by inner blocks of arcs.

Hess' seafloor spreading hypotheses:

CS2   Ice sheets are heavy enough to depress continental margins.

CS1   Mantle is viscous.

S2   The mantle is convecting at the rate of 1 cm/yr.

S4   The continents are carried passively on the mantle at a uniform rate by convection and do not plow through oceanic crust.

S7   Continents were once all connected.

S8   The Earth's pole of rotation has moved with respect to the crust.

S10   Rising limbs coming up under continental areas fracture them and move the fragmented parts away from one another.

S11   Mountains form on the leading edges of continents underthrusting one another.

S14   Continental and oceanic crust are very different.

S15   The mantle's convection cells have rising limbs under midocean ridges.

S16   Mantle material comes to the surface at the crest of midocean ridges and becomes oceanic crust.

S17   The uniform thickness of the oceanic crust results from the maximum height that the 500 degree C isotherm can reach under the midoceanic ridge.

S18   Higher temperatures and intense fractures under ridge crests explain lower seismic velocities there.

S19   Midocean ridges are ephemeral features having a life of 200 to 300 million years, the life of a convecting cell.

S20   The whole ocean is virtually swept clean (replaced by new mantle material) every 300 to 400 million years.

S21   The leading edges of continents are strongly deformed when they impinge upon the downward moving limbs of convecting mantle.

S22   The oceanic crust, buckling down into the descending limb, is heated and loses its water to the oceans.

Contradictions (LISP keyword "contradict" is not shown):

(E18 NE18)	(E14 NE14)	(C17 S11)
(E21 NE21)	(C9 S20)	(C18 S21)
(E20 NE20)	(C10 S16)	(C20 S22)
(E17 NE17)	(C13 S20)	(C2 S4)

Explanations (LISP **keyword "explain" is not shown):**

((C1) C2)
((C1) NE20)
((C1) NE14)
((S4 S7 S10) E1)
((C2 C3 C11 C14 C17) E2)
((S2 S4 S11) E2)
((C2 C3 C14) E3)
((S2 S15 S18 S22) E3)
((C2 C3 C14 C15 C16 C17 C18) E4)
((S2 S4 S11 S21) E4)
((S2 S4 S7 S10) E5)
((S2 S4 S7 S10) E6)
((CS2 CS1) E7)
((C2 C3 C4 C17) E7)
((S2 S4 S11) E9)
((C2 C3 C4 C11 C14 C16 C17) E9)
((S2 S4 S11) E10)
((C11 C14 C16 C17 C18) E10)
((S2 S14 S15 S16 S17) E11)
((C9 C10) E11)
((S2 S22) E12)
((C2 C3 C4 C14 C15 C20) E12)

((S2 S21 S22) E13)
((C11 C13 C14) E13)
((S2 S22) E14)
((S2 S15) E15)
((C9 C10 C12 C13) E16)
((S2 S15 S19) E16)
((C12 C13) NE17)
((S2 S15 S16 S20) E17)
((S2 S15 S16 S20) E18)
((C9 C10) NE18)
((S4 S7 S8 S10) E19)
((C8) E19)
((S2 CS1 S15) E20)
((C19) E20)
((S2 S15 S16 S20) E21)
((C9 C10) NE21)
((S4 S7 S8 S10) E22)
((C8) E22)
((S2 S10 S15) E23)
((C2 C3 C12) E23)
((S2 S15 S16 S18) E24)
((S2 S15 S16 S18) E25)

CHAPTER 3

On Finding the Most Probable Model

PETER CHEESEMAN

1. Introduction

The question addressed in this chapter is, given data and weak prior domain knowledge, what is the most probable model[1] of the given domain? This question is at the opposite end of the knowledge spectrum from approaches such as explanation-based generalization. The latter knowledge-intensive approaches pose the question, given a strong theory of a domain, what are the interesting or important consequences of that theory? In this chapter, I consider only simple inductive models as a guide to the much more complex case of typical scientific discovery. For this reason, I focus on bottom-up model discovery, often called *inductive* or *empirical* learning, in which the basic problem is distinguishing the model from noise. At least since the time of William of Ockham (*ca.* 1285–1349), people have known that one can always find a sufficiently complex "theory" to "explain" any outcome. For example, a set of N points can always be fitted exactly by a polynomial of order N. However, if there is error in the measurement of the points, as is usually the case, then the polynomial is partially fitting the inherent noise. Ockham's razor tells us not to "overfit" the data.

The solution to this well-known "overfitting" problem is to find a suitable trade-off between the fit to data and the complexity of the model. Admonitions to keep the model "simple" are too vague for a

---

1. In this chapter, a model is any formal description (mathematical or logical) that assigns probabilities (including 1 or 0) to experimentally observable outcomes. The term *model*, as used in this chapter, can be interchanged with *assumptions*, *hypothesis*, *concept*, or *theory*; the important property these terms share is the ability to make (probabilistic) predictions about possible observations.

computer. A model as complex as the data itself can fit the data exactly, but this has very little predictive value for new data. Conversely, models with very little structure do not predict the given data or new data very well. The basic problem then is to find the *appropriate* trade-off. Some researchers in machine learning treat this as an empirical question (Iba, Langley, & Wogulis, 1988; Quinlan, 1986), whereas others avoid the problem entirely by investigating only deterministic domains with no noise in the data.

It is not widely known that the overfitting problem was solved in principle over fifty years ago. Jeffreys (1983), building on work going back to Bayes and Laplace, provided the theory and mathematical tools for finding the most probable model in noisy data, and he successfully applied this method to many complex inductive problems. Bayesian theory explicitly trades off model complexity, as determined by prior probabilities, against the (probabilistic) fit to the data. This trade-off is a direct consequence of Bayes' theorem, requiring no additional assumptions. Bayesian theory was given a firmer foundation in a proof by Cox (1961). This proof starts from some elementary principles that one would expect a rational agent to use in computing quantitative beliefs in propositions given incomplete or uncertain information. These principles lead directly to probability theory, and to Bayes' theorem in particular, as the logically sound way of making inferences under uncertainty.

The Bayesian process of model discovery is outlined in the next two sections. We will see that this approach not only allows the comparison of models of the same complexity, such as alternative class descriptions with the same number of classes, but also allows the comparison of models with *different* complexity, such as class descriptions with different numbers of classes. In other words, Bayesian methods provide a solution to the problem of trading model complexity against the fit to the data from first principles, with no ad hoc assumptions. The implications of this approach are explored below.

2.   Bayesian Inference and Information Theory

Bayesian theory can be cast into an intuitively appealing form called the minimum message length (MML) criterion, in which the most probable theory has the shortest encoding of the theory and data combined[2] (Georgeff & Wallace, 1984; Rissanen, 1978; Wallace & Freeman, 1987). The basic idea behind MML is that new information (data) should not be stored as received, as is done in most database systems, but should instead be modeled, based on what is already known. By describing a new example as an instance of an existing abstraction, a shorter message *may* result.

For example, the Identi-Kit contains pictures of facial parts that are used to construct facial images. A small number of "standard" chins, mouths, and eyes allow enormous data compression as compared with an original photograph. The standard components are models abstracted from previous examples (the collection of facial features). If a new example (such as a dog's face) is sufficiently different from all existing examples, then there may be no compact encoding in the Idendi-Kit system. It may be simpler to just describe the new example directly. In such cases, MML indicates that the new example is very surprising relative to abstractions over the previous examples.

This is how the Bayesian (MML) method distinguishes signal from noise—if no encoding of the new example can be found that is shorter than a direct description of the example, then the new example is random as far as the existing theory is concerned. Such entirely new cases may form the basis of new abstractions if there are some common features among the cases. When a new example cannot be compactly described in terms of abstractions of previous data, it means that the previous data have very little *predictive value* for that example.

Let us explore the connection between Bayes' theorem and MML. Assume that the user has selected a set of discrete mutually exclusive and exhaustive hypotheses $\{H_0, H_1, \ldots, H_n\}$ and has assigned *prior* probabilities $p(H_i \mid c)$, where c is the given general context. If the hypotheses (theories or models) are sufficiently concrete, they give the probability of observing any data set D—in other words, $p(D \mid H_i)$ is a consequence of the definition of H_i. This latter probability is often referred to as the *likelihood*. Typically, you have some data D and want to know the (posterior) probability of a hypothesis given those data, $p(H_i \mid D)$.

---

2. For further information on the minimum message length criterion, see Wallace and Freeman (1987).

Often, you want the particular H_i that gives the maximum posterior probability, as defined by Bayes' theorem:

$$p(H_i \mid D, c) = \frac{p(H_i \mid c)p(D \mid H_i, c)}{\sum_i p(H_i \mid c)p(D \mid H_i, c)}. \tag{1}$$

Taking the logarithm[3] of this expression, turning products into sums, gives

$$-\log p(H_i \mid D, c) = -\log p(H_i \mid c) + -\log p(D \mid H_i, c) + \text{constant}. \tag{2}$$

If you are only interested in the *relative* probability of different hypotheses, then the last (constant) term in equation (2) can be ignored. From information theory, $-\log p(\text{outcome})$ is the (theoretical) minimum message length to encode the particular outcome. This quantity can be thought of as a measure of "surprise" for the outcome. If the base of the log is 2, then the message length is in bits, so surprising outcomes take many bits to encode.

The minimum message length is the sum of two terms. The first term is the information to describe the hypothesis, which is greater for more complex (less probable) hypotheses. The second term is the information required to encode the data, given the hypothesis, and this decreases for suitably selected more complex hypotheses. From equation (2), it is clear that the maximum posterior probability criterion is equivalent to MML, which can be expressed symbolically as:

Total message length   =   Message length to describe the model +
                            Message length to describe the data,
                            given the model.

The trade-off between these two terms gives a quantitative version of Ockham's razor. Although the user may have no interest in the likelihood term $p(D \mid H)$ per se, its evaluation is an essential part of finding the most probable model. Note that the Bayesian distinction between

---

3. Since probabilities are ≤ 1, $-\log$ is used instead of log to ensure positive quantities.

signal and noise is not absolute—the difference in message length between any two hypotheses is proportional to the relative probability of the hypotheses.[4] The next section outlines the practical application of this theory.

3. The Bayesian Inductive Procedure

The basic Bayesian approach to induction for a given set of data is:

1. Define the model space. This is the space that you think contains the "true" model. Finding this model is the goal of inductive learning.

2. Use domain knowledge to assign *prior probabilities*[5] to the models and probability distributions on the values of model parameters.

3. Use Bayes' theorem to obtain the (possibly parameterized) posterior probability of models, given the data. To compare models of different complexity, it is necessary to marginalize[6] all the parameters of the competing models. These marginalized probabilities are the probability of the particular model, given the data; they give the relative probability of models, regardless of model complexity.

4. Use a search algorithm that efficiently searches the space of possible models for the (locally) maximum posterior probability model.

5. Stop the search when the most probable model is found or when it is no longer worth searching to find more probable models. Typically, the stopping criterion will involve a trade-off between model optimality and search time. For most interesting inductive search problems, this occurs because the goal of finding *the* optimal model is computationally too expensive, so heuristic search methods are mandatory.

The following examples should clarify the above inductive steps and their relevance to scientific discovery.

---

4. Specifically, if ΔM is the difference in message length between the two hypotheses (in bits), then their relative probability is given by $2^{\Delta M}$. It does not take many bits of difference in message length before one hypothesis has negligible relative probability. For a more detailed description of the relationship between Bayesian probability and MML, see Wallace (1987).

5. Prior probabilities are also commonly referred to as "priors."

6. This is a procedure from probability theory in which variables are eliminated by summing over all the probabilities of the possible values of the variable to be eliminated.

3.1 Example 1: Curve Fitting

In curve fitting, the goal is to find a curve that summarizes all the actual information in a set of measurements. For example, you may model a set of noisy measurements of variable-star luminosity as a function of time using a curve defined by an orthogonal series, such as a Fourier series or Legendre polynomials. Such a curve describes the mean value of luminosity as a function of time, where the individual measurements deviate from the mean value mainly due to measurement error. This "noise" is often modeled by a Gaussian (normal) distribution because the error is typically the result of many small independent errors. Choosing a particular series representation defines the model space (step 1).

The fundamental problem is that the mean curve can always be made closer to the measurements by adding more coefficients to the series representation. If there are as many coefficients as there are measurements, an exact match can be found—clearly, at some point, the curve started fitting the noise. The basic question is, how do you tell when the curve is fitting the noise? Classical statistics has provided at least thirty different answers to this question.

In principle, the Bayesian answer to the question is simple. It takes a certain amount of information (usually measured in bits) to describe the coefficients of a particular curve—the more coefficients, the longer this model description. With additional coefficients, the information needed to describe the data decreases because the curve can be chosen to be closer to the measured values. The information required to describe a coefficient can be shown to depend only on the ratio of its prior standard deviation to its posterior standard deviation, as Gull (1988) describes in detail.

Maximum likelihood methods (the classical statistical approach) are concerned only with fitting the data to the model (the likelihood term), and these methods always favor adding more complexity to the model. In contrast, the Bayesian approach minimizes the *sum* of two competing quantities: the information required to describe the model and the information required to describe the data, given the model (step 3). The goal in Bayesian (or MML) search (steps 4 and 5) is to find the model (or models) that require the minimum *total* information.

The Bayesian criterion has some important consequences. It does *not* state how much structure there is in the domain, only how much structure can be justified by the given data and the assumed model space. For example, in solar intensity measurements, there may be a very weak periodic variation that is buried in the measurement noise. In this case, the Bayesian approach says only that the existence of such a weak periodic signal can be more compactly described by calling it noise. This conclusion may be reversed with more data. However, if the total message length is greatly reduced by assuming the model, this is overwhelming evidence that the modeled structure exists, even if it is not the whole story.

It is always possible that the model space does not contain the correct model, and in such cases the nearest approximation to the actual system in the given model space is found. For example, if the frequency of system fluctuations is not constant, then modeling them with a Fourier series will find the nearest constant approximation. In other words, the discovery of some structure by asking one question does not mean that there is no more structure to be found by asking a better question. Deciding when the current model space is inadequate and choosing a better one is a higher level problem that may also be amenable to Bayesian analysis, but one that we will not pursue here.

3.2   Example 2: Unsupervised Classification

In unsupervised classification, a set of objects can be modeled by first describing a set of classes ("prototypes"), then describing the objects using these prototypical class descriptions.[7] Each description gives the probabilities of the observable features, assuming that the object belongs to the class. The class descriptions are chosen so that the information required to describe objects in the class is greatly reduced because they are "close" to the class prototype. This information reduction arises because only the differences between the observed and expected (prototypical) values need to be described. It takes a certain amount of information (in bits) to describe a set of classes (i.e., the information required to state the probabilities of the features, given that an object belongs to the class). However, these probabilities reduce the

---

7. Many view the problem of classification as partitioning the set objects according to some criterion and generating class descriptions from these partitions. This *conceptual clustering* approach (Michalski & Stepp, 1983) requires separate criteria for partitioning and for extraction of class definitions. Since the end result of conceptual clustering is a set of class descriptions, the Bayesian approach has the advantage of working directly in class description space and so avoids the need for ad hoc partitioning criteria.

information needed to describe the objects by first describing each object's class membership, then describing how each object differs from the prototype.

The Bayesian approach involves finding the set of classes that minimizes the *total* information (class descriptions + objects, given class descriptions). If you observe a randomly chosen set of objects, it is very unlikely that class descriptions can be found for which the total information is less than that used to describe each object individually. Even though by chance some objects will tend to form clusters, the clustering will be so weak that the information required to describe the class will not compensate for the reduced information required to describe the objects. This Bayesian/MML classification criterion has a long history (Cheeseman et al., 1988a, 1988b; Wallace & Boulton, 1968). An important consequence of the Bayesian/MML approach to automatic (unsupervised) classification is that the assignment of objects to classes is necessarily probabilistic (Cheeseman et al., 1988b, 1988c). Such probabilistic class memberships capture the intuitions behind "family resemblances" described by philosophers (Wittgenstein, 1953) and the "fuzzy set" advocates (Cheeseman, 1986; Zadeh, 1981). In the Bayesian approach, the probabilistic class assignments occur as a direct consequence of Bayes' theorem. A class description assigns probabilities to observable attributes (conditional probabilities of the observations, given that the object belongs to the class) so that, when Bayes' theorem is used to invert this probability (the conditional probability of the class, given the observations of the object), the resulting class probabilities are not unity. In other words, if you cannot say for certain what will be observed when an object belongs to a particular class, then you cannot determine an object's class membership with certainty. In practice, if the classes are well separated, the probability of a particular object belonging to the nearest class is so close to unity that little is lost by assuming it belongs to that class.

3.3   Discussion of the Bayesian Procedure

Clearly, choosing the model space is a very important step—if an inappropriate choice is made, very poor results may be obtained. For example, using a polynomial representation to fit discontinuous curves is a poor choice. However, in a new domain, there is often little prior knowledge to guide the choice of model space. Fortunately, using a very general model will often turn up systematic effects, even if that model is poor. These systematic effects are a statistical shadow of the real effects that are operating. Alternatively, one could choose the most general model possible, namely, the set of Turing computable functions. This is the choice used in Chaitin (1966) complexity to establish basic asymptotic results. However, this choice is impractical for inductive inference because the halting problem makes most functions in this space uncomputable in finite time.

Also, choice of model space is a way of introducing potentially important prior knowledge. If you choose the wrong model space, the Bayesian approach still gives the most probable answer in the space selected. For example, polynomials will fit the overall shape of a discontinuous curve but will smooth out any discontinuities. If the curve is violently discontinuous, as it is for line spectra, the polynomials will probably not discover anything useful. Poor results are often a signal that a new model space should be tried. Typically, models include an explicit provision for noise, and the data that cannot be attributed to the model are automatically attributed to noise. This means that information captured in one model may be called noise by another model.

A model definition essentially gives the probability of the possible outcomes (observables), assuming the model is true. These probabilities are symbolically defined by the model, although their value may be estimated from the data. For example, a class definition for "cat" assigns an expected body-weight distribution, expected number of ears, and so on, letting one calculate the probability that a given entity is in this class. The "cat" model does not give necessary and sufficient conditions for membership; it only assigns strong probabilities to particular observables if the entity is a cat.

The requirement that prior probabilities be assigned to the model space is often cited as an insurmountable difficulty, but the practical difficulties of assigning priors over a model space have been greatly exaggerated. Another objection is that prior probabilities introduce an undesirable degree of "subjectivity" into what should be an "objective" problem. This philosophical point has been discussed by Cheeseman et al. (1988a) and Denning (1989), who have shown that the subjective components of the so-called objective approach have been hidden, partly by calling them "assumptions." The view taken here is that in any real inductive problem there is always cogent prior knowledge, and failure to use it creates an illusion of objectivity that ignores potentially important information. The Bayesian approach makes incorporation of prior information relatively easy and explicit.

Another advantage of assigning prior probabilities to models is that it does not limit the method to finite sets of theories. To prevent over-fitting, machine learning and maximum likelihood approaches often arbitrarily limit the complexity of models they consider. However, this restriction is artificial and may exclude the best answer. For example, in polynomial curve fitting, the set of possible models is indefinitely extensible by adding more and more coefficients to the polynomial representation. Instead of arbitrarily placing an upper bound on the number of coefficients, the Bayesian method adds as many coefficients as are justified by the data. In this approach, it is the amount of data and their structure that limit the complexity of induced models. Note that, in the case of polynomials, priors are required for the number of possible coefficients as well as for parameters that describe the prior probability distribution for each coefficient.

Most interesting inductive problems involve search over a space of possible models, and finding the maximum posterior probability is the goal in such an optimization search. Standard numerical search algorithms, such as steepest descent, conjugate gradient methods, and the Newton-Raphson technique are possible choices of search procedures. Which algorithm is the best for a given problem depends on the amount of data and the properties of the search space. Unfortunately, most interesting inductive tasks have many local minima in a huge search space, making search for *the* optimal solution out of the question. In such cases, it is appropriate to employ search methods that attempt to find good local minima, such as simulated annealing (Kirkpatrick, Gelatt, & Vecchi, 1983) or the EM algorithm (Dempster, Laird, & Rubin, 1977). There is no global method that is best for all problems or domains.

3.4   Overview of the Inductive Process

The Bayesian inductive process described in detail in the previous subsection is only part of the whole inductive inference task. One way of viewing the whole task, including the evaluation of models, is shown schematically in Figure 1.

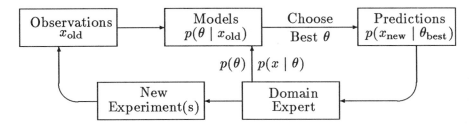

Figure 1. The hypothesize-and-test cycle for Bayesian induction.

In this cycle, the domain expert first suggests a broad class of possible models, θ, which give a probability of observations x; that is, $p(x \mid \theta)$. For example, $p(\text{weight} \mid \text{mean, variance})$, is the probability of a particular weight observation as a function of the model parameters "mean" and "variance." Model selection corresponds to step 1 of the Bayesian procedure described earlier. The domain expert also provides prior probabilities over the possible models, $p(\theta)$—step 2 of the procedure. For example, the expert may provide a prior distribution for the unknown parameters "mean" and "variance." The product of $p(\theta)$ and $p(x \mid \theta)$ is proportional to the model probability, given the data—$p(\theta \mid x)$. The goal of the Bayesian component of the inductive task is to maximize this posterior model probability, as shown on the top line of Figure 1. This search for the maximum posterior probability model corresponds to steps 3 and 4 of the Bayesian procedure. In summary, Bayesian model discovery uses the (usually) weak prior information about which model is correct for a particular domain, $p(\theta)$, the data (x_{old}), and the probability of data given the (parameterized) model, $p(x \mid \theta)$, to define the posterior probability of the model given the data, $p(\theta \mid x)$. The Bayesian inductive procedure then searches for the maximum posterior model, or the closest approximation the search procedure can find.

Although the domain expert plays a part in the overall process through model selection, prior probability assignment, and data selection, a more important role is in the evaluation of the current cycle (the bottom path illustrated in Figure 1). Typically, the domain experts pick the most likely model based on the current data and use their domain knowledge to suggest new experiments that would generate more data relevant to supporting or denying the current most probable model. If the data fail to strongly confirm any particular model, this may imply that the original model framework should be modified or expanded ($\theta \rightarrow \theta'$). The Bayesian analysis does not give direct information about this kind of model shift, but it does allow the relative evaluation of models within a given framework based on the given data. There is no guarantee that the hypothesize-and-test cycle shown in Figure 1 will terminate, but the current most probable model makes the strongest predictions that are possible by any single model, given the available data and prior knowledge.

3.5   Consequences of the Bayesian Approach

In the Bayesian approach, once the model space, priors, and data are chosen, all the degrees of freedom have been used up—there is only one answer to the question, what is the most probable model? However, finding it may be difficult. This approach has a number of consequences when applied to inductive learning and discovery that differ from much of the current practice, as described below.

3.6   Language Invariance

A basic principle in inductive learning is that the answer to the question, which model best fits the data? should *not* depend on the language used to define the model if the alternative languages are expressively equivalent.[8] For example, the discovery of similar 3-D shapes should depend on the intrinsic shapes of the objects and not on whether you use Cartesian or polar coordinates to describe them, or on the particular viewpoint. However, the use of syntactic "biases," such as minimizing the number of disjuncts in a disjunctive normal form (DNF) language or minimizing the number of rules in a particular rule syntax, violates this basic requirement.

Consider the problem of finding the "best" model for a particular set of training examples using either a conjunctive normal form (CNF) or DNF language. A model in one language can be logically mapped into a corresponding model in the other language and vice versa (that is, the languages are expressively equivalent). However, the "complexity" (bias) of the model description in the two languages can be very different. For example, describing a concept as "red or blue" is logically

---

8. This principle was originally proposed in Wallace and Freeman (1987).

equivalent to describing the same concept as "not (yellow, green, purple, ...)," but these descriptions have very different syntactic complexities and so could lead to different choices for the "best" model.

Also consider the "grue/bleen" paradox (Goodman, 1954). This paradox compares two alternative descriptions of an emerald and a sapphire. In one description, they are described as green and blue, respectively; in the other they are described as "grue" and "bleen," where "grue" is defined to be green until the year 2000, then blue after, whereas "bleen" is blue now and green after the year 2000. Both descriptions are well-defined colors, so why do we automatically prefer the standard names? From symmetry alone, neither description can be preferred over the other since colors in one language can be fully described in the other. However, different descriptions can have very different prior probabilities. In this case, descriptions that leave *observable* quantities invariant over time are a priori much more probable than those that do not, unless there is reason to expect temporal changes.

The Bayesian approach has the required property of language invariance. This is because the prior probabilities assigned to the components of one language are transformed correctly if the problem is mapped into another language. For example, a high prior probability of straight lines in a Cartesian representation is mapped into an equivalent high prior probability for straight lines in a polar representation, even though the representation of straight lines as equations in polar coordinates is more complex than for circles. Simple syntactic measures of complexity are generally not a good idea—the measure of model complexity (bias) should reflect the prior probabilities of the domain, not the language used to describe it. Note that there is a tendency for natural languages to evolve so that more probable cases are syntactically simpler. This means that syntactic measures of model complexity are often approximately correct. The tendency for common words to be short is one example of this trend, but it does not license syntactic "biases" in general.[9]

---

9. The requirement for invariance of model selection under changes of language does not hold if the alternative languages are not expressively equivalent (e.g., a 2-D versus a 3-D representation).

3.7 Models and Search

The Bayesian approach splits the problem of finding the most probable model into two components: defining the *goal criterion* (MML) for possible models and *searching* the space to find the best model under that criterion. This decomposition of the problem differs from that taken in most previous work, as outlined below.

By failing to specify their model space, some researchers implicitly question the need for explicit modeling or its appropriateness for a particular domain. For example, CNF forms, decision trees, and rule sets have all been used in machine learning, but there is usually no discussion as to why they provide suitable models (languages) for their intended applications. Similarly, neural nets implicitly model domains, but the standard language of hidden units, connection weights, and activation functions is difficult to relate to prior knowledge of a particular domain. It is easy to find domains for which a standard language is totally inappropriate—not because it cannot represent the concept to be learned (although this can also be true) but because it requires a very complex description for common examples. For example, a one-dimensional vector of pixel values can represent any image, but it fails to compactly represent two-dimensional concepts, such as line, edge, and region. Because such two-dimensional concepts have very complex descriptions in a one-dimensional language, they are unlikely to be discovered by feeding a neural net raw one-dimensional pixel data.

Model space selection should reflect domain knowledge; it is not an arbitrary choice. For example, using a classification model to capture time-dependent behavior is not generally appropriate, whereas a Markov model or time series may be. Similarly, a Fourier series is appropriate for representing periodic models, whereas a polynomial series is not. Decision trees, such as those constructed by Quinlan's ID3 (1987), implicitly assume the existence of a small number of highly relevant attributes instead of a large number weakly informative attributes. Rules relating domain properties to choice of model space, and methods for revising initial model spaces in the light of experience, would make good topics for machine learning research. Most of the literature on machine learning implicitly assumes that completely domain-independent model-free learning is possible, but Bayesian analysis denies this possibility.

Programs that do not clearly separate the search method, representation language, and model assumptions make it difficult to discern the effect of each component on the results. For example, many existing automatic classification programs (e.g., Fisher, 1987) assume that hierarchical classifications (trees) are desirable. The reasons a program generates a particular hierarchy rather than another is the result of a complex interaction between the program and the particular data set. There is little suggestion that the hierarchy discovered is an intrinsic property of the domain rather than an artifact of the program. However, when biologists, linguists, or other scientists construct hierarchies, they believe that the hierarchies they construct reflect the way the system under study actually evolved or that they reflect cascaded causal influences. A good test of any induction program is to see if it finds structure in randomly generated data. If structure is found where none exists, it is strong evidence that the program's procedure is overfitting the data. Similarly, failure to find structure where it is known to exist is also a cause for alarm.

In contrast, the Bayesian approach makes a clear separation between the criterion used to determine the "best" model and the search procedure used to find it. The properties of the search space, such as the density of local minima and their shape, are defined by the evaluation function (criterion) and the particular data set. Choosing the appropriate search algorithm then depends mainly on the properties of the search space, the available resources, and whether or not the search is incremental. Programs that mix the search procedure with the evaluation function make consideration of alternative algorithms difficult, and the choice of the particular search procedure is hard to justify in terms of the properties of the search space. For example, our studies in unsupervised class discovery (Cheeseman et al., 1988b, 1988c) suggest that gradient (hill climbing) methods fail badly in search spaces with high dimensionality but that reestimation algorithms (Dempster, Laird, & Rubin, 1977) are very robust, with good initial convergence properties.

3.8 Deterministic Models

Deterministic models with perfect data are a special case of probabilistic induction where the probabilities are zero or one. In this case, the likelihood of a particular datum, given the model, is either one (consistent) or zero (inconsistent), so only the prior probability remains to give a preference among all the consistent models. These prior probabilities are a form of "bias," although the term *bias* is now so overused that it has come to mean any kind of preference.

Logical inference methods, such as "version spaces" (Mitchell, 1982), are also appropriate for this special case, but logical methods can fail dramatically in the case where the data include noise. This can occur because logical methods may throw out the actual solution due to incompatibility with the data. The often-expressed hope that logic-based inductive methods can be extended to handle noisy data is incorrect for worst case behavior, although their typical behavior seems to be satisfactory (Quinlan, 1986).

4. Why Find the Best Model?

It might seem obvious that finding the most probable model is the goal of scientific discovery, but this assumption is questionable. The usual scientific method bases all predictions on the most probable model alone, as shown in Figure 1. However, if the goal is to make as accurate predictions as possible about future data, then taking a weighted average over possible models is optimal by the Bayesian criterion (Self & Cheeseman, 1987). Basing all predictions on a single model is jumping to conclusions, and this is especially dangerous when alternative high-probability models make radically different predictions. Given the many hotly debated models of human evolution based on the current fossil evidence, only a brave person would bet on one as correct. By taking a weighted average of the predictions from the different models, a more reliable prediction can be made (hedging your bets). This weighted average maps old data onto new via a weighted set of possible models, as shown in Figure 2.

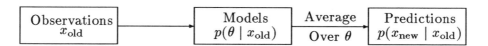

Figure 2. Optimal model-based prediction.

If the goal is understanding of the domain, and the most probable model is far more likely than the alternatives, then finding this model and using it for prediction is reasonable. This approach also reduces the storage requirements and the computational effort of prediction. Although minimal computational effort may be important for humans, it is much less important for computers. Consequently, for *automated* theory discovery, it is reasonable to keep all the high probability models simultaneously and to base predictions on a weighted combination. Retaining many alternatives lets new data change the relative weights of the competing models instead of backtracking to previously abandoned alternatives, as required for the single best-model approach.

As shown in Figure 2, the use of a weighted combination of models to make predictions adds support to the philosophical approach called "instrumentalism" (Bridgman, 1927; Cheeseman & Self, 1987). In this approach, theories (models) have no intrinsic value but are just convenient instruments for making predictions from previous experience. Bayesian theory confirms this view if predictive accuracy is the sole criterion. Even in this case, the prediction is influenced by the *class* of models considered. For example, in series prediction, if the models are restricted to the first-order Markov assumption,[10] the results may differ from predictions extracted from models that depend on all the previous data.

However, this downgrading of the importance of theories ignores the fact that they are rarely generated in a vacuum, so that any theory generated must be evaluated in the context of existing theories in the same domain. For example, even if I discover a weak statistical correlation between rainfall in the Sahara and car accidents in Los Angeles, I will interpret this as a statistical fluke. In contrast, an identical correlation between rainfall and car accidents in Los Angeles would be interpreted as wet road conditions causing accidents. Why are two identical statistical results interpreted so differently? Clearly, in one case there is an obvious link to the domain, whereas in the other case there is no plausible causal explanation.

This example suggests that in domains where there is already a strong theoretical framework (but one that is not strong enough to make specific predictions), the most probable model is *not* just dependent on the data but also on the prior probability of the model, as demanded by the Bayesian analysis. In practice, the "prior" probability of a model is more conveniently evaluated after the data have suggested strong candidate models. If such a candidate can be plausibly "explained" in the context, this makes the model much more probable, but a failure to find

---

10. The first-order Markov assumption is that the next value depends only on the present value and not on the history.

a plausible explanation reduces the model's probability. If the statistical probability of a model from the data is overwhelming, but there is no plausible explanation, this may indicate a problem with the basic theory. For instance, even though I have a very strong prior disbelief in ghosts, I would change my mind if presented with extremely strong evidence.

5.   Models and Intrinsic Properties

A central philosophical question underlies the entire process of scientific model discovery: Does the most probable model result from the way a scientist partitions the world, or does it result from the intrinsic structure of the world? Strong arguments can be made on either side of this question. On the subjective side, it is clear that the class of foods that I do not like is a result of imposing *my* tastes on the set of potential foods and is not an intrinsic property of the foods themselves.[11] However, I can teach others my likes and dislikes to the point where they can predict with high accuracy how I would react to new foods. Thus, even though the class of foods I dislike is decided by my tastes alone, the class definition can be expressed in terms of quantities observable by anyone (an "objective" expression of a subjective class). Many researchers in the machine learning community accept that classes are the result of the user's purpose (Michalski & Stepp, 1983).

The opposite extreme claims that the discovery of highly probable models in a previously unknown domain is an intrinsic property of that domain and is not a property of the observer at all.[12] For example, Cheeseman et al. (1988c) reported results with AUTOCLASS, an automatic unsupervised classification program that has discovered previously unsuspected fine structure in the infrared spectra of stars (Cheeseman et al., 1989). This program had no understanding of spectra or stars—the classes of spectra that were found could only be a property of the domain that was recovered from the data.

I believe that the answer lies somewhere between these extreme positions, depending on the particular problem. In approaches that require a model space before learning is possible, there is always a subjective

---

11. Langley, Simon, Bradshaw, and Zytkow (1987) use the term *intrinsic property* in a somewhat different sense, referring to unobserved but unvarying properties of particular objects or classes of objects, such as mass or specific heat.
12. However, quantum mechanics makes the observer an intrinsic part of the system.

element in the selection of this space. In other words, when you ask a question of nature, the answer you get depends on both the question asked and what is actually there. One reason that Figure 1 is circular is to lessen the dependence of the model discovered on the model space by selecting new spaces to be explored, based on the successes and failures of the current models. Similarly, the data used for finding the best model depends on the data available to the user, and this introduces another unavoidable subjective element into the model discovery process sampling biases. Other subjective elements that affect the model selection process are the available search time and the prior probabilities of models.

Despite the subjectivity mentioned above, the Bayesian approach will converge on the same model, given sufficient data, regardless of prior beliefs. Given incorrect priors, the nearest approximation to reality will be found in the given model space. This is the sense in which it is reasonable to claim that learning systems discover the intrinsic structure of the domain. However, one can never know for certain that there is no more structure to be found, given more data or more search. The history of science shows that those who think they have found the "right answer" have almost always been wrong or incomplete upon further examination. Models of the world seem to get more complicated. The apparent simplicity of a model is due to a failure of imagination and limited data, unless the domain really is simple. If the world were really random, chemistry, cooking, and credit would not be possible, so our models cannot be entirely figments of our imagination.

Often in scientific inquiry, we look for *any* kind of pattern in the data to try to understand the domain better. However, induction for a particular purpose (e.g., learning the probability of heart attack) reduces the possible model space drastically. Goal-directed induction is blind to patterns in the data unless they are predictive of the target variable, and this is typical in decision making, where the decision to be made directs the inductive search. Decision making and prediction of particular outcomes have a subjective component in the selection of the target prediction, whereas pure scientific inquiry searches for intrinsic patterns, regardless of the use to which they might be put.

6.   Conclusions

Figure 1 shows the basic hypothesize-and-test cycle that governs human theory discovery. This cycle can be partially formalized so that the model evaluation part of the cycle (finding the most probable model) and predictions based on the model can be automated. The theory for model selection and probabilistic prediction is Bayesian inference, which also provides the necessary trade-off between model complexity and the corresponding fit to the data. The problem of overfitting the data is well known, but many researchers seem unaware of the Bayesian response. This solution can briefly be described as trying to find the model that minimizes the total message length (MML), where this is defined as the sum of the information needed to describe the model and the data, given the model.

The model with the minimum total message length is also the model with the maximum posterior probability, by Bayes' theorem. Roughly speaking, a more complex model requires more bits to describe, but because this model can fit the data better, the message length to describe the data is correspondingly reduced. The basic question is whether the cost (in bits) of added model complexity is paid for by improved prediction (also in bits). The goal is to find the model with the minimum *total* message length. For most interesting problems, achieving this goal is NP-hard, so heuristic methods are necessary that do not guarantee to find the MML model. Even when the best model is found, there is no guarantee that it is the "true" model, as more data may make a more complex model more probable.

The MML model is guaranteed to be the best *single* model predictor (in the given model space) for new data, but even better predictions can be made by taking an average of the predictions of all the models weighted by their respective posterior probability. This calls into question the utility of always selecting the best model and acting as if it is true, especially if prediction is the goal. Since a computer can entertain many hypotheses simultaneously, it is not good practice to always jump to conclusions unless the evidence overwhelmingly supports a particular hypothesis.

A major feature of the Bayesian/MML criterion for model selection is that it is independent of the language used to express the model (for expressively equivalent languages). This is possible because the prior probabilities over the model space are transformed accordingly if a different language is used. For example, a high prior probability of straight lines remains unaltered, no matter what perspective is used. Methods that rely on syntactic "biases" do not have this language independence property.

An essential message of the Bayesian analysis is that inductive learning is inherently model dependent and that prior beliefs about the possible models affect the evaluation of the best (most probable) model. However, sufficient data will overcome prior beliefs if they do not fit the data well and new models can be tried if the current models are found to be inadequate. Domain-independent or general (assumption-free) learning is not possible—every learning system implicitly or explicitly makes assumptions. The Bayesian approach makes the use of prior knowledge explicit so that these beliefs can be checked against the data.

References

Bridgman, P. W. (1927). *The logic of modern physics*. New York: Macmillan.

Chaitin, G. J. (1966). On the length of programs for computing finite sequences. *Communications of the ACM, 13*, 547–549.

Cheeseman, P. (1986). Probabilistic versus fuzzy reasoning. In L. N. Kanal & J. F. Lemmer (Eds.), *Uncertainty in artificial intelligence*. Amsterdam: Elsevier.

Cheeseman, P. (1988). Inquiry into computer understanding + discussions and rebuttal. *Computational Intelligence, 4*, 57–142.

Cheeseman, P., Freeman, D., Kelly, J., Self, M., Stutz, J., & Taylor, W. (1988a). Autoclass: a Bayesian classificiation system. *Proceedings of the Fifth International Conference on Machine Learning* (pp. 54–64). Ann Arbor, MI: Morgan Kaufmann.

Cheeseman, P., Freeman, D., Kelly, J., Self, M., Stutz, J., & Taylor, W. (1988b). Bayesian classification. *Proceedings of the Seventh National Conference on Artifical Intelligence* (pp. 607–611). St. Paul, MN: Morgan Kaufmann.

Cheeseman, P., Goebel, J., Self, M., Stutz, M., Volk, K., Taylor, W., & Walker, H. (1989). *Automatic classification of the spectra from the infrared astronomical satellite (IRAS)*. (Reference Publication 1217). Washington, DC: National Aeronautics and Space Administration.

Cox, R. T. (1961). *The algebra of probable inference*. Baltimore: Johns Hopkins University Press.

Dempster, A. P., Laird, N. M., & Rubin, D. B. (1977). Maximum likelihood from incomplete data via the em algorithm. *Journal of the Royal Statistical Society, Series B, 39*, 1–38.

Denning, P. J. (1989). Bayesian learning. *American Scientist, 77*, 216–218.

Fisher, D. (1987). Conceptual clustering, learning from examples, and inference. *Proceedings of the Fourth International Workshop on Machine Learning* (pp. 38–49). Irvine, CA: Morgan Kaufmann.

Georgeff, M. P., & Wallace, C. S. (1984). A general selection criterion for induction inference. *Proceedings of the European Conference on Artificial Intelligence* (pp. 473–482). Amsterdam: Elsevier.

Goodman, N. (1954). *Fact, fiction and forecast*. London: Athlone Press.

Iba, W., Langley, P., & Wogulis, J. (1988). Trading off simplicity and coverage in incremental learning. *Proceedings of the Fifth International Conference on Machine Learning* (pp. 73–79). Ann Arbor, MI: Morgan Kauffman.

Jeffreys, H. (1961). *Theory of probability*. Oxford: Clarendon Press.

Kirkpatrick, S., Gelatt, C. D., & Vecchi, M. P. (1983). Optimization by simulated annealing. *Science, 220*, 671–680.

Langley, P., Simon, H. A., Bradshaw, G. L., & Zytkow, J. M. (1987). *Scientific discovery: Computational explorations of the creative processes*. Cambridge, MA: MIT Press.

Michalski, R. S., & Stepp, R. (1983). Learning from observation: Conceptual clustering. In R. S. Michalski, J. G. Carbonell, & T. M. Mitchell (Eds.), *Machine learning: An artificial intelligence approach* (Vol. 1). San Mateo, CA: Morgan Kaufmann.

Mitchell, T. M. (1980). Generalization as search. *Artificial Intelligence, 18*, 203–226.

Quinlan, J. R. (1986). The effect of noise on concept learning. In R. S. Michalski, J. G. Carbonell, & T. M. Mitchell (Eds.), *Machine learning: An artificial intelligence approach* (Vol. 2). San Mateo, CA: Morgan Kaufmann.

Quinlan, J. R. (1987). Decision trees as probabilistic classifiers. *Proceedings of the Fourth International Workshop on Machine Learning* (pp. 31–37). Irvine, CA: Morgan Kaufmann.

Rissanen, J. (1978). Modeling by shortest data description. *Automatica, 14*, 445–471.

Self, M., & Cheeseman, P. (1987). Bayesian prediction for artificial intelligence. *Proceedings of the Workshop on Uncertainty in Artificial Intelligence* (pp. 61–69). Seattle, WA.

Wallace, C. S., & Boulton, D. M. (1968). An information measure for classification. *Computer Journal, 11*, 185–195.

Wallace, C. S., & Freeman, P. R. (1987). Estimation and inference by compact encoding. *Journal of the Royal Statistical Society, Series B, 49*, 233–265.

Wittgenstein, L. (1953). *Philosophical investigations.* Oxford: Basil Blackwell.

Zadeh, L. A. (1981). Possibility theory and soft data analysis. In L. Cobb & R. M. Thrall (Eds.), *Mathematical frontiers of the social and policy sciences.*

CHAPTER 4

An Integrated Approach to Empirical Discovery

BERND NORDHAUSEN

PAT LANGLEY

1. Introduction

Philosophers of science distinguish between two forms of discovery—the generation of *empirical laws* and the formation of *theories* (Thagard, 1988). The first activity involves descriptive generalizations that summarize observations, such as Ohm's law and the ideal gas law. The second concerns the explanation of phenomena, which often involves postulating unobserved structures or processes. Examples of scientific theories include the fluid model of electricity and the kinetic theory of gases. Of course, science also involves many other components, including the design of experiments and measuring instruments, but it is often useful to focus one's attention on a limited set of phenomena. In this paper, we focus on the discovery of empirical laws.

In recent years, researchers in machine learning have investigated three main aspects of empirical discovery. The first relates to the process of taxonomy formation. Before one can formulate laws, one must first establish the basic concepts or categories that one hopes to relate. For instance, one might group certain substances together as acids, alkalis, or salts, depending on their tastes. Research on *conceptual clustering* (Fisher, 1987; Lebowitz, 1987; Michalski & Stepp, 1983) addresses this problem, even though it has seldom been cast as relevant to scientific discovery. This task involves organizing a set of observations into a conceptual hierarchy, which can then be used to classify new observations. Fisher and Langley (1985) review work on conceptual clustering and its relations to statistics, whereas Gennari, Langley, and Fisher (in press) examine incremental clustering methods.

Another facet of empirical discovery concerns the generation of qualitative laws. In this case, the goal is to uncover qualitative relations that hold for a set of observations. Thus, one might note that acids tend to react with alkalis and that the result is always some salt. Researchers who have addressed this problem include Brown (1973), Lenat (1977), Emde, Habel, and Rollinger (1983), Langley, Zytkow, Simon, and Bradshaw (1986), Jones (1986), and Wrobel (1988).

Finally, empirical discovery can involve the production of quantitative laws. Here the goal is to find mathematical relations between numeric variables. For instance, one might determine the amount of hydrochloric acid that combines with a unit amount of sodium hydroxide and note the amount of sodium chloride that results from this reaction. Researchers have developed a number of AI systems that have rediscovered a variety of numeric laws from physics and chemistry (Falkenhainer & Michalski, 1986; Kokar, 1986; Langley, Simon, Bradshaw, & Zytkow, 1987; Zytkow, 1987). In some cases, these systems find not only quantitative relations but also qualitative conditions on the laws.

Although each of these systems is successful at its specific task, no system to date has attempted to integrate these different aspects of science. The goal of our research is to develop a framework that unifies all three components of empirical discovery. To this end, we have developed IDS, an integrated discovery system that incorporates mechanisms for taxonomy formation, qualitative discovery, and numeric discovery. In this essay we describe the IDS system in detail. The next section describes the representation and organization of knowledge, and the following section presents the main discovery components and their relation to each other. Both sections contain examples to clarify the system's structures and processes. We close the chapter by describing the status of IDS, along with our plans for evaluating and extending the system.

2. Representation and Organization in IDS

IDS' representation draws upon recent work in qualitative physics, describing its observations as sequences of qualitative states. The system also uses this notation in stating its taxonomy and laws, which require some organization of memory. On this dimension, IDS borrows from recent work on incremental approaches to conceptual clustering. In this section, we give the details of representation and organization, first dealing with IDS' inputs and then with its outputs.

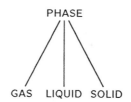

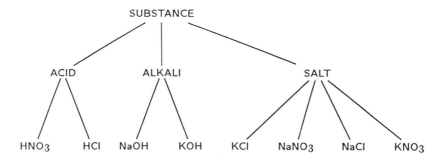

Figure 1. A simple object hierarchy for chemical substances.

2.1 Inputs to IDS

Most discovery systems start with some background knowledge, whether this bias is made explicit or not. In IDS, this takes the form of a simple background knowledge that describes classes of objects the system may encounter. This knowledge is represented as an *is-a* hierarchy, similar to what Michalski (1983) calls a "structured descriptor" and what Mitchell, Utgoff, and Banerji (1983) call a "concept description grammar." Figure 1 presents an example of such a hierarchy for certain chemical substances. Although this example involves a disjoint hierarchy, the system can also handle nondisjoint structures in which nodes can have multiple parents. For instance, one might know that the substance HCl comes in two different colors, green or blue.

An empirical discovery system also requires some data or observations on which to base its generalizations. In IDS, this information is represented as *histories* (Hayes, 1979), which are sequences of qualitative states that the system observes in an incremental fashion. Each "state"

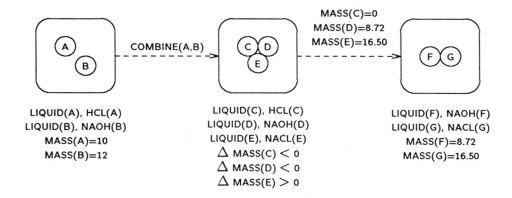

Figure 2. The sequence of qualitative states observed during a chemical reaction.

represents an interval of time during which objects exhibit "constant" behavior; this representation borrows heavily from Forbus' (1985) qualitative process theory.

Figure 2 shows the history for a simple chemical reaction with three distinct qualitative states. The initial state contains two separate objects, liquid HCl and liquid NaOH. When these substances are combined by an external agent, a new state begins that contains three objects— the two original reactants and a new product.[1] During this state, the masses of the reactants are decreasing, whereas the mass of the product is increasing. This is a form of constant behavior since the signs of the derivatives remain unchanged.

A qualitative state ends and a new one begins whenever the sign of any derivative changes—when any increase or decrease of a variable starts or stops.[2] A boundary between states also results when any structural change takes place. Both occur in Figure 2 when the mass of the HCl in state 2 reaches zero, for at this point one of the reactants disappears and the mass changes halt in the remaining objects. As a result, the history enters a new state in which these two objects are in contact but in which

---

1. We have omitted the second product, liquid H_2O, for the sake of simplicity. Also, note that the objects are labeled by pattern-match variables that are different from state to state.

2. These state boundaries correspond to *limit points* in Forbus' qualitative process theory.

their masses remain constant. Although IDS is given these boundaries, one can imagine a system that finds them on its own (Kuipers, 1985; Weld, 1986).

IDS represents each qualitative state as a frame with four slots. The *object description* slot describes the objects present in the state using the object hierarchy. For example, object C in Figure 2 is described as `liquid(C)`, `HCl(C)`. A state also includes a *structural description*, such as `touches(C, D)`. The *changes* slot contains a list of zero or more changes occurring in the state. As in Forbus' theory, we express changes in terms of derivatives. For instance, a decrease in the `mass` of C is expressed as Δ `mass(C)` < 0. Finally, the *quantity* slot describes numerical attributes that remain constant during a state.

Histories are inherently sequential, and IDS represents the successor of a given state using a *successor* link. These links can be labeled by a *transition condition*, which identifies the condition under which the current state ends and its successor begins. These transition conditions may involve either quantitative descriptions, such as `mass(B) = 12`, or the actions of an external agent, such as `combine(A, B)`. For instance, the melting and boiling points of a substance are commonplace transition conditions.

IDS processes qualitative states one at a time, in the order they occur in a historical sequence. The system also receives the information on the temporal order of these states. For example, it would first be given the initial state in Figure 2. After processing is complete, IDS is then presented with the second state, along with the fact that this state is the successor of state 1. The third state is then given in the same fashion, with the information that it is the final state in the history. The system is next presented with the first state of the next history, and so on. As we will see, IDS uses these states and the temporal relations between them to incrementally form a hierarchy of qualitative states and to discover empirical laws.

2.2 Outputs of IDS

The incremental nature of IDS means the system has no explicit outputs since it continues processing states as long as they are available. However, the system produces a knowledge structure after each experience, and one can view these structures as its "output." We shall focus on three aspects of this output—the taxonomy, qualitative laws, and numeric laws.

IDS organizes the qualitative states it observes into a taxonomic hierarchy, with specific states as terminal nodes and with abstract states as internal nodes. This hierarchy takes the form of a tree, so that no specific state can belong to more than one abstract category. IDS does not make a distinction between instances and abstract states, thus the abstractions have the same slots as particular states—a description of the structure, the objects involved, the changes occurring during the state, and the constant quantities. The structure of the IDS hierarchy is similar in form to those generated by UNIMEM (Lebowitz, 1987) and COBWEB (Fisher, 1987), though these systems do not cluster qualitative states.

Figure 3 presents the top levels of a taxonomy that summarizes qualitative states involving various acids, alkalis, and salts, with solid lines standing for *is-a* links. For example, node 4 describes cases in which a liquid alkali and a liquid salt are in contact with each other, whereas node 5 specifies cases in which acidic and salty liquids occur. Node 3 is an abstraction of these cases in which the second substance is not specified. No changes are occurring in any of these three states, though changes are present in node 2.

The IDS taxonomy connects states at varying levels of abstraction through *is-a* links, but histories also contain temporal information. Thus, the system also specifies *successor* links between nodes, indicating that one class of qualitative states follows another in time. Figure 3 also shows examples of *successor* links, using dashed arrows to indicate these temporal connections. For example, the link between node 1 and node 2 specifies that instances of node 2 occur directly after instances of node 1.

Successor links may also specify the conditions under which the transition occurs. For instance, the label on the link between nodes 1 and 2—combine(A, B)—indicates that this transition occurs when the two objects in node 1 are physically combined. Taken together, nodes and successor links represent qualitative laws similar in content to those found by GLAUBER (Langley et al., 1987). Thus, node 1, node 2, and the link between them asserts that, when a liquid acid is combined with a liquid alkali, the two substances react to form a salty liquid. This

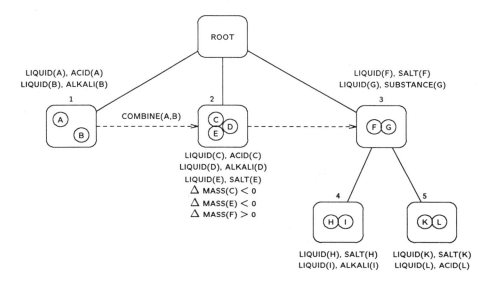

Figure 3. A taxonomy for acids, alkalis, and salts, augmented with *successor* links and transition conditions.

can be viewed as a restatement of the qualitative law[3] "acids react with alkalis to form salts."

Figure 4 shows the third aspect of IDS' output—the augmentation of nodes and successor links with numeric laws. One type of law describes the conditions for transitions between states. For instance, the successor link between node 2 and node 3 in this figure indicates that, when the mass of the liquid HCl reaches zero, the reaction state (node 2) ends and the final state (node 3) begins. Nonzero values, such as the boiling point for a substance, may be stored as well. IDS also stores numeric laws with individual states that relate attributes within that state; the ideal gas law is one example of such a relationship.

In addition, the system forms numeric laws that relate attributes in different states in the same sequence. It stores these laws on *quantity relation* links that connect the states containing the related attributes. Figure 4 shows an example of such a cross-state law. This relation specifies that liquid HCl reacts with liquid NaOH in constant proportion

---

3. We will not argue that this approach can represent *all* forms of qualitative laws, but we do feel that temporal relations among abstract qualitative states constitute an important subset of such laws.

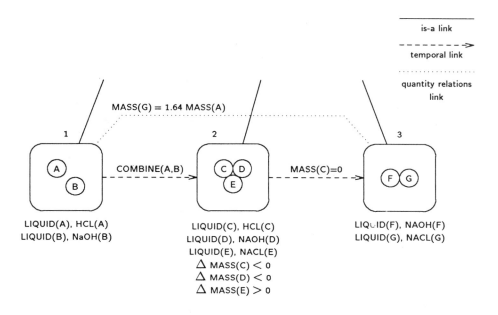

Figure 4. A numeric law relating attributes across states.

to form liquid NaCl. The mass of the resulting NaCl is 1.64 times the initial mass of the HCl; this corresponds to the chemical concept of the *definite proportions* of the reaction.

The numeric laws generated by IDS are similar to those found by BACON (Langley, Bradshaw, & Simon, 1983), ABACUS (Falkenhainer & Michalski, 1986), and FAHRENHEIT (Zytkow, 1987), but there is an important difference. These earlier systems found numeric laws and conditions for them, but their statement of the laws contained no information about the structural or physical context in which they occurred. Even as simple a relation as the ideal gas law actually involves a set of structurally related objects that change over time. This is precisely the function of the taxonomy of qualitative states and the successor links in IDS. Unlike earlier approaches to numeric discovery, IDS describes a qualitative *context* for its quantitative laws.

3.   Discovery and Prediction in IDS

Now that we have discussed IDS in terms of its inputs and outputs, we can present the mechanisms it uses for empirical discovery. Like much of the recent work in machine learning, the system can be characterized as using an incremental hill-climbing approach (Langley, Gennari, & Iba, 1987). Such systems process one instance at a time and do not reprocess substantial numbers of previous instances. Furthermore, these systems hold only one structure in memory and retain no information about their learning steps, so explicit backtracking cannot occur. Following this general strategy, IDS processes one qualitative state at a time, sorting the state through its current taxonomy, incorporating it into the hierarchy, and creating new nodes as needed. When a new node is created, IDS determines the successor of that node. If a new child does not obey a numeric law of its parent node, the system formulates an improved law. This cycle continues as long as the system receives new observations.

3.1   Forming a Taxonomic Hierarchy

Table 1 summarizes the IDS clustering algorithm, which has been heavily influenced by Lebowitz' (1987) work on UNIMEM and Fisher's (1987) work on COBWEB. When IDS receives a new qualitative state, it sorts this state through its hierarchy. Starting at the root node, the system computes the similarity score between the instance and each child of the current node. IDS measures these differences using a *lexicographic evaluation function* (Michalski, 1983). The total value of this function is computed from scores of similarity between the slots of the two states.[4]

The system then sorts the instance to the child that matched it most closely. If the match is sufficiently high (if the score is above a user-specified threshold), the selected child becomes the current node, and the sorting continues recursively. If the match with the selected child is high enough but that child does not cover the instance, the child is generalized so it covers the instance completely. If the match is not high enough, the instance is added as a new child of the current node.

---

4. The structural description is treated as most important, followed by the description of changes, then by the object descriptions, and finally the descriptions of numeric attributes.

Table 1. The IDS clustering algorithm.

```
Variables: N and P are nodes in the hierarchy.
           I is an instance (a very specific node).
           X is a similarity score between two nodes.

Cluster(N, I)
    For each child C of N
        compute the similarity score between C and I.
    Let P be the node with the highest score.
    Let X be the similarity score between I and a P.
    If X is sufficiently high,
        Then if P does not cover I,
                Then generalize P to cover I.
             Cluster(P, I).
        Else add I as a child of N.
             Merge-children(N, I).
Note:
    The similarity measure is lexicographic, using states'
        slots to calculate the similarity between two states.
    The object hierarchy is used to generalize and merge states.
```

Let us consider another example from the domain of alkalis and acids. Figure 5 shows a portion of an IDS taxonomy that describes the final states in a reaction sequence, after one of the initial substances has been completely used. The system receives a new instance (labeled 4 in the figure), which it sorts to node 1. At this point, IDS computes the similarity between the instance and the children of node 1. In this case, node 3 matches the instance more closely than node 2. However, the score for placing the instance as a descendent of node 3 is not above threshold, so the algorithm adds the instance as a new child of node 1, as seen in Figure 6.

As we noted earlier, IDS identifies the objects in each state using state-specific variables. In order to determine the best match between two states, the matcher generates all possible bindings between the variables in the two states. Using the evaluation function, the matcher then calculates a similarity score between the two states for each set of bind-

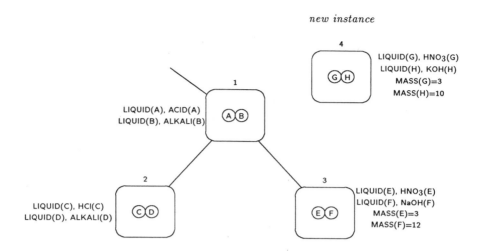

new instance

LIQUID(G), HNO₃(G)
LIQUID(H), KOH(H)
MASS(G)=3
MASS(H)=10

LIQUID(A), ACID(A)
LIQUID(B), ALKALI(B)

LIQUID(C), HCl(C)
LIQUID(D), ALKALI(D)

LIQUID(E), HNO₃(E)
LIQUID(F), NaOH(F)
MASS(E)=3
MASS(F)=12

Figure 5. A taxonomy before incorporation of a new qualitative state.

ings and selects the one with the highest score. For example, there are two possible sets of bindings for the variables between nodes 4 and 3 in Figure 7: {(G, E) (H, F)} and {(G, F) (H, E)}. Because the first binding set receives a higher score, the matcher concludes that G and H of node 4 correspond to E and F of node 3, respectively.

Whenever IDS adds a new instance as the child for a node in the hierarchy, it considers two ways to merge the node's children. First the system finds the two siblings that match the new child the closest. It then considers merging the new child with its closest sibling as well as merging these two siblings.[5] IDS computes a score for each option using its evaluation function and merges the pair that produces the higher score, creating a merged (generalized) node that subsumes the pair. The system then stores the siblings as children of the merged node, which in turn is stored as a child of the original parent node. Merges of two nodes that result in a node identical to the original parent are not executed.

In our example, node 4 (the new child) has only two siblings. Thus, the system considers two actions—merging nodes 2 and 3 and merging

---

5. For an optimal solution, the system would have to consider merging all possible pairs. However, initial experiments have shown that considering all possible pairs usually does not produce a better taxonomy.

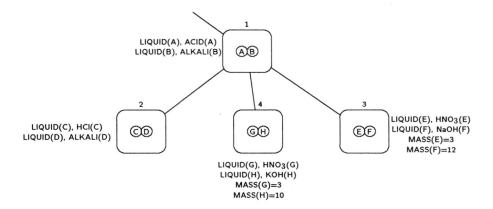

LIQUID(A), ACID(A)
LIQUID(B), ALKALI(B)

LIQUID(C), HCI(C)
LIQUID(D), ALKALI(D)

LIQUID(E), HNO₃(E)
LIQUID(F), NaOH(F)
MASS(E)=3
MASS(F)=12

LIQUID(G), HNO₃(G)
LIQUID(H), KOH(H)
MASS(G)=3
MASS(H)=10

Figure 6. A taxonomy before two children have been merged.

nodes 4 and 3. The second option receives the higher score; since merging the two nodes does not produce a node that is identical to node 1, the merge is carried out. The merged node (node 5) is added as a child of node 1, and nodes 4 and 3 are added as children of node 5. Figure 7 shows the modified taxonomy after merging has occurred, and Table 2 presents the algorithm for merging children.

The process for creating a generalized node is straightforward. IDS uses the matcher to find a correspondence between the variables in the merging nodes and the variables in the merged node, which it then uses to fill the slots of the new state. In general, each slot value in a new node is the intersection of that slot's values in the merging nodes. For example, given the situation in Figure 7, the matcher determines that variable I of node 5 corresponds to G in node 4 and E in node 3 and that variable K in node 5 corresponds to H and F. The quantity slot of node 3 has a value of mass(E) = 3, mass(F) = 12, and the quantity slot of node 4 has a value of mass(G) = 3, mass(H) = 10. Using the variable correspondence as a constraint, the intersection of these two sets is mass(I) = 3, which becomes the value for the quantity slot in node 5.

IDS computes the object descriptions of a merged state in a different manner, using the object hierarchy to determine the values of this slot. As described above, for each variable in the merged node, the matcher determines the corresponding variables in the merging nodes. The system then collects the components of the object description for

Table 2. The algorithm for merging children.

```
Variables: N, P, Q, and R are nodes in the hierarchy.
           I is the newest child.
           X and Y are similarity scores.

Merge-children(N, I)

    For each child C of N except I
        Compute the similarity score between C and I.
    Let P be the node with the highest score.
    Let R be the node with the second highest score.
    Let X be the score of merging I and P.
    Let Y be the score of merging P and R.
    If X is the best score,
        Then let Q be the resulting node of merging P and I.
            If Q is not equal to N,
                Then place Q as a child of N.
                    Remove P and I as children of N.
                    Place P and I as children of Q.
    Else if Y is the best score,
            Then let Q be the resulting node of merging P and R.
            If Q is not equal to N,
                Then place Q as a child of N.
                    Remove P and R as children of N.
                    Place P and R as children of Q.
```

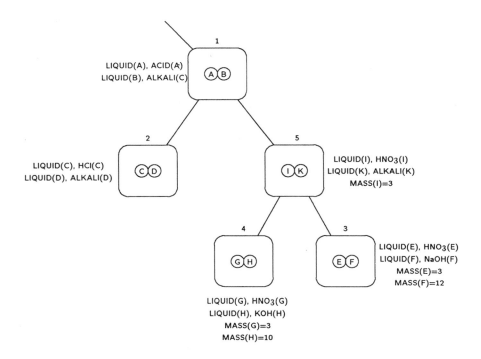

Figure 7. A taxonomy after two children have been merged.

each pair of corresponding variables. Next, IDS determines all closest common ancestors of each pair of components in the object hierarchy. These common ancestors become the components of the new object description for the variable in the merged node.

We can clarify this procedure with an example. Given the situation in Figure 7, the matcher determines that variable H of node 3 and F of node 4 correspond to K in node 5. The description components of H are `liquid` and KOH, whereas the components F are `liquid` and NaOH. The closest common ancestor of `liquid` and `liquid` (given the object hierarchy in Figure 1) is `liquid`, and the closest common ancestor of KOH and NaOH is `alkali`. The terms `liquid` and KOH have no common ancestor, nor do NaOH and `liquid`. Hence, the resulting description components for K are set to `liquid` and `alkali`.

As we noted above, IDS' clustering component has been influenced by Lebowitz' (1987) UNIMEM and Fisher's (1987) COBWEB, but there are some important differences. For instance, IDS and COBWEB form only disjoint taxonomies in which each node has a single parent. In contrast, UNIMEM can sort an instance down multiple paths, producing a nondisjoint hierarchy. IDS differs from both earlier systems in that it includes no counts or probabilities on its features; each description in the taxonomy is categorical. Our system is probably most akin to COBWEB, though it uses a different evaluation function and lacks the latter's splitting operator (the inverse of the merge operator).

Lenat's (1982) AM system also organizes its concepts into a hierarchy and dynamically extends that hierarchy over time. However, AM begins its existence with 250 heuristics and over 100 initial concepts, whereas IDS begins with a simple algorithm, a small background knowledge, and an empty hierarchy. More important, Lenat's system generates new concepts by "mutating" the definitions of existing ones and then testing them; we might call this an *exploratory* approach to discovery. In contrast, IDS (like UNIMEM and COBWEB) generates new concepts in direct response to observations, using a *data-driven* approach to discovery.

3.2 Discovering Qualitative Laws

We have seen that IDS represents qualitative laws in terms of abstract qualitative states and the successor links connecting them. With the exception of the final state in a history, every node in the taxonomy must have some successor node. This temporal information is given as part of the input, and this input specifies the links for terminal nodes in the hierarchy, but the system must induce the links between abstract nodes.

Whenever IDS forms a new abstract qualitative state (a nonterminal node in the hierarchy), it determines the successor for this new node. This is a simple process that involves finding the closest common ancestor of the successors of the new node's children.[6] For an example, consider the partial taxonomy shown in Figure 8. Node 1 has two children, which are labeled nodes 3 and 4. These nodes have as successors

---

6. This requires that all the successors of that node's children have been incorporated into the state hierarchy. Therefore, finding the closest common ancestor of a node is delayed until the successors of all of its children have been clustered. Every set of nodes has at least one common ancestor, the root node. However, a successor link from a node to the root node in the hierarchy does not form a qualitative law with any useful content.

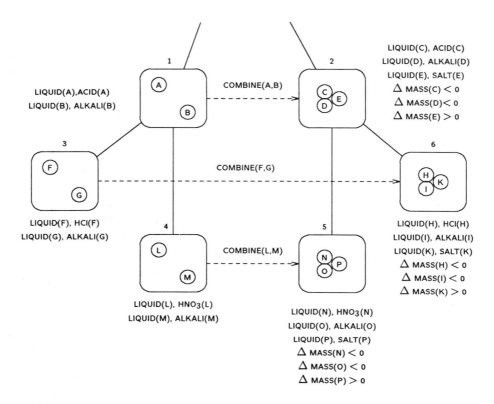

LIQUID(A),ACID(A)
LIQUID(B), ALKALI(B)

COMBINE(A,B)

LIQUID(C), ACID(C)
LIQUID(D), ALKALI(D)
LIQUID(E), SALT(E)
Δ MASS(C) < 0
Δ MASS(D) < 0
Δ MASS(E) > 0

COMBINE(F,G)

LIQUID(F), HCl(F)
LIQUID(G), ALKALI(G)

LIQUID(H), HCl(H)
LIQUID(I), ALKALI(I)
LIQUID(K), SALT(K)
Δ MASS(H) < 0
Δ MASS(I) < 0
Δ MASS(K) > 0

COMBINE(L,M)

LIQUID(L), HNO$_3$(L)
LIQUID(M), ALKALI(M)

LIQUID(N), HNO$_3$(N)
LIQUID(O), ALKALI(O)
LIQUID(P), SALT(P)
Δ MASS(N) < 0
Δ MASS(O) < 0
Δ MASS(P) > 0

Figure 8. A qualitative law that results from the merging of two successor links.

nodes 5 and 6, respectively. In this case, IDS determines that the closest common ancestor of nodes 5 and 6 is node 2, and it asserts this node as the successor of node 1.

In addition, IDS attempts to attach transition conditions to the new successor link, which may take the form of some external action or some quantity relation. The system determines these conditions in the same way that it forms merged nodes, by finding the structure common to the two links. For example, the transition formed between node 1 and node 2 is labeled `combine(A, B)` because this action is stored on the successor link connecting nodes 3 and 6 as well as that connecting nodes 4 and 5. The act of adding this condition is equivalent to inducing a law that states, "if a liquid acid is combined with a liquid alkali, they react to form a liquid salt."

If IDS finds only some common actions in the children's successor links, it includes only the shared structure in the abstract link. If it can find no common structure, the system creates the successor link but specifies no transition conditions. In other cases, the conditions on the specific links involve numeric relations, such as reaching zero mass or achieving boiling point. In this situation, IDS attempts to find a numeric relation that covers the specific cases, using the algorithm described in the next section.

As we noted earlier, IDS formulates qualitative laws with similar content to those found by Langley et al.'s (1986) GLAUBER. However, the two systems arrive at these laws in very different manners. As we saw in the acid-alkali example, qualitative discovery in IDS is a simple process of finding two nodes' closest common ancestor. In contrast, GLAUBER spent considerable effort in finding classes of objects with common attribute values and playing similar roles.

3.3 Finding Numeric Laws

The third major component of IDS focuses on discovering numeric laws. As we saw earlier, these relations augment the qualitative descriptions, and they may specify the conditions for moving from one state to another, a relation between numeric attributes within a given state, or a quantitative relation between states. Each of these cases involves storing a law at a node or link in the taxonomy that summarizes information in the children of that node or link. IDS uses a single procedure to find all three forms of numeric law. Briefly, whenever the system adds a new child to an existing node in the hierarchy, it checks to see if the child obeys the laws currently stored at the parent. If it does not, IDS searches for new laws that cover the added child and its siblings.

For a given data set, the system conducts a beam search through the space of numeric terms to find a law that covers these data. More precisely, the search task can be stated as:

- *Given:* a set of base terms $a, b, c, \ldots$, along with one designated term (a) from that set;
- *Find:* a term $x = a^{n_0} \cdot b^{n_1} \cdot c^{n_2} \ldots$ such that a linear relation of the form $a = mx + n$ holds.

IDS searches from simple terms to more complex ones, using correlation analysis (Freund & Walpole, 1980) to direct the search process. As in

BACON, the basic operators involve defining new terms as products and ratios of existing terms. The system initially examines correlations between the designated term and observable attributes, uses these to select promising products and ratios, and then recurses if it cannot find a law with the existing terms.

This search technique has a semi-incremental flavor. In cases where IDS has rejected an existing law, there is no need to reconsider the term in that law and those leading to the law. Thus, it uses the old term as the starting point for the new search, saving considerable effort over an approach that starts from scratch. However, this method does require one to store and reprocess all the data that led to the rejected law. As a result, it does not quite fit with our description of IDS as an incremental learning system, though we hope to modify this in future versions.

Table 3 presents the basic algorithm for finding numeric laws. The top-level function find-law is given three arguments—the designated term D, the set of base terms S, and a set of current terms C. If IDS is attempting to revise an existing law, C is the term occurring in the right-hand side of that law. If the system is searching for a new law, C is the set of observable terms S.

At each point in the search, IDS defines the products and ratios between the terms in the set S and those in C, but it retains only those terms having the highest correlations with the designated term D. These new terms become the current set C, and the function find-law is called recursively, with the designated term D and the base terms S remaining the same. If any term in C has a sufficiently high correlation with D, IDS ends the search and uses a regression technique to find the slope and intercept of the line relating them. The system continues along these lines until it finds such a linear relation or until it exceeds the maximum search depth. If the search fails, IDS assumes that no law covers all the observed data.

As an example, let us consider how IDS rediscovers the ideal gas law. The system receives data in the form of states with gaseous objects at different temperatures, pressures, and volumes. Figure 9 shows the hierarchy after the system has processed three states, with all instances stored under a common parent node. Given these data, IDS finds a law relating the temperature and the pressure since one can express the temperature as a linear function of the pressure. Now the system observes a fourth instance, which it adds as a child of node 1 because

Table 3. The algorithm for finding numeric laws.

```
Variables: S is the set of base terms.
           D is the designated term.
           C is the set of current terms.
           P and Q are sets of terms.
           A is a defined term.


Find-law(D, S, C)
    Let A be the term in C that has the highest
        correlation with D.
    If the correlation between D and A is high enough,
        Then call linear regression on D and A
                to find the slope and intercept.
            Return (A, slope, and intercept).
    Else if the maximum search depth is reached,
            Then return NIL.
    Else let C be Find-best-terms(D, S, C).
        Find-law(D, S, C).


Find-best-terms(D, S, C)
    Let Q be the quotients of the terms of S and C.
    For each term A in the union of P and Q,
        Compute the correlation between D and A.
    Return the terms with the N highest correlations.


Parameters:
    Width of the beam (memory size).
    Threshold of the correlation (accuracy).
    Maximum power of terms (law complexity).
    Maximum depth of the search tree (when to halt).
```

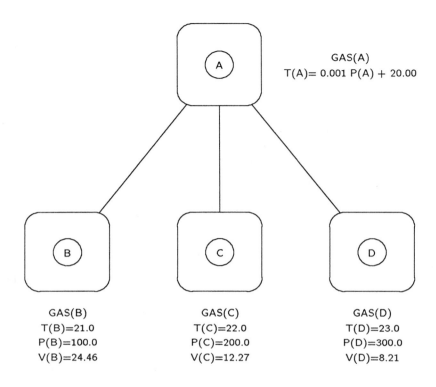

GAS(A)
T(A)= 0.001 P(A) + 20.00

GAS(B)	GAS(C)	GAS(D)
T(B)=21.0	T(C)=22.0	T(D)=23.0
P(B)=100.0	P(C)=200.0	P(D)=300.0
V(B)=24.46	V(C)=12.27	V(D)=8.21

Figure 9. A spurious relation found during discovery of the ideal gas law.

it matches the parent completely.[7] However, this new instance violates the numeric law stored at the parent node, causing IDS to search for a new relation that covers all four instances.

Since the term P was used in the rejected law, IDS calls the function **find-law** with {P} as the current set C, T as the designated term, and {P, V, T} as the base terms S. In other words, IDS uses the term P as the entry point in the search space, starting by combining P with the terms in S to form products and ratios such as PV, P^2, P/T, and P/V. Of these new terms, PV has a high enough correlation to end the search. Regression produces the numeric law T= 0.12·P V − 273; this version is equivalent to the standard form of the law, PV= 8.32(T + 273). Figure 10 shows the hierarchy that emerges after this revision is complete. As

---

7. Recall that the system does not consider whether instances satisfy numeric laws during the clustering process.

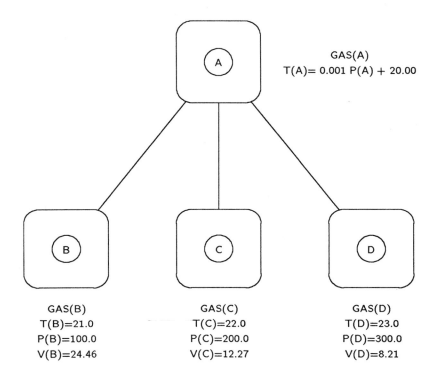

GAS(A)
T(A)= 0.001 P(A) + 20.00

GAS(B)
T(B)=21.0
P(B)=100.0
V(B)=24.46

GAS(C)
T(C)=22.0
P(C)=200.0
V(C)=12.27

GAS(D)
T(D)=23.0
P(D)=300.0
V(D)=8.21

Figure 10. A correct version of the ideal gas law, found after rejection of the spurious version in Figure 9.

the system processes more instances and stores them under the parent node, it finds that they obey this new law, so **find-law** is not called again.

Four parameters control IDS' search for numeric laws. The size of the set of current terms C determines the beam width of the search. The level of correlation used as a termination criterion influences the accuracy of the laws and the system's tolerance of noise. Finally, the maximum power of terms and the maximum depth of the search tree limit the amount of search. Although we have not undertaken a careful study of this algorithm, preliminary results suggest that it is efficient and robust.

As noted in Section 2.2, IDS finds numeric laws similar in form to those produced by BACON (Langley et al., 1983) and ABACUS (Falkenhainer & Michalski, 1986). Moreover, these systems use similar methods to control their search for useful numeric terms, using simple correlations to focus attention. However, the systems differ in the details of their search control. BACON uses a recency-based scheme, focusing on more recently defined terms in preference to older ones. ABACUS creates a "proportionality graph" to determine promising combinations of terms, then uses a modified beam search to find the best combinations. IDS also carries out a beam search through the space of numeric terms, but this search is not as sophisticated as that used in ABACUS. The main novelty of IDS' search scheme is the reuse of existing terms, which decreases the amount of reprocessing needed when new observations are made.

3.4   The Process of Prediction

Now that we have discussed IDS at an algorithmic level, let us consider the implications of the knowledge it acquires. The system inductively constructs a hierarchy of abstract qualitative states, augmented with qualitative and quantitative laws at different levels of abstraction. The taxonomy and its associated laws describe the observations that have been made. Embedded in the hierarchy are qualitative laws, such as the reactive behavior of acids and alkalis, and numeric laws, such as laws of combining weights, that summarize the histories given to the system.

However, these data structures also have predictive power. After IDS has observed a number of qualitative states, it can use its taxonomic hierarchy to predict unobserved states. In addition, once it has classified a novel state, it can predict the possible successors of that state (sometimes many steps ahead) and when they will occur. Finally, once the system has stored cross-state numeric laws, it can use them to predict the values of numeric attributes in as yet unobserved but predicted states. Moreover, IDS' incremental nature permits it to make these predictions at any point in the discovery process. In fact, the prediction process can be viewed as an integrated part of its discovery method.

IDS' three-tiered approach to discovery is also robust in that it can profit from partial understanding of a domain. In cases where IDS' numeric component cannot discover quantitative relations, it may still be able to form qualitative laws and use them for qualitative prediction. For example, the program can describe the qualitative behavior of reacting substances even if other factors make the combining weights difficult to determine. Similarly, in domains where temporal information is not available or not highly predictive, the system can still find within-state quantitative laws without the need to form qualitative laws. For instance, IDS needed no qualitative relations to formulate the variant of the ideal gas law shown in Figure 10. Finally, the system can construct taxonomies that organize qualitative states and thus can make simple predictions, even when no temporal information is present and no numeric laws can be found.

4. Discussion

In this section we discuss some general issues concerning our work on IDS. We begin by describing the status of the system and our ideas for near-term extensions. After this, we consider some approaches to evaluating IDS, including experiments on both historical and artificial domains. Finally, we discuss our longer-term plans for more extensive changes to the system. Although IDS integrates some important aspects of discovery, we think it holds the potential for supporting much more of the scientific process.

4.1 The Status of IDS

The three major parts of IDS—the clustering algorithm, the method for finding successor links, and the component for numeric discovery—have all been implemented. Initial experiments show that the system is able to form taxonomies and to find qualitative and quantitative laws. For instance, given histories of chemical reactions between acids and alkalis, IDS has successfully characterized the reactive behavior of these substances, as described earlier in the chapter. It has also summarized the qualitative behavior of heat exchange, along with a simple version of Black's law that does not involve specific heat. We have tested IDS on all the laws that BACON could handle which do not involve inferring intrinsic properties. The system found Ohm's law, Kepler's third law, and Coulomb's law with no difficulty.

The system has also successfully found numeric laws within states and on transition conditions. However, we have yet to implement the algorithm for inferring laws that relate attributes across states, as shown in Figure 4. We envision using a forward propagation strategy to find these relations. If no law between a state and its immediate successor can be found, the system will look for a numeric relation between the state and the successor of the successor, continuing this chain until it finds a relation or it reaches a final state. We will use the numeric discovery method from Table 3 to actually find these laws.

As Langley et al. (1983) have noted, *intrinsic properties* play an important role in empirical discovery, occurring in many numeric laws. An intrinsic property is some attribute of an object or class of objects that remains constant over time; often this attribute is not directly observable. For example, mass is an intrinsic property associated with particular objects, whereas density, specific heat, and boiling point are intrinsic properties associated with classes of objects. We are currently extending IDS to infer intrinsic properties, based on the parameters found in numeric laws within states, across states, and on transition conditions.

4.2   Evaluating IDS

Our work on integrated discovery is still in progress, and we have not yet carried out any systematic evaluation of IDS. We hypothesize that the system can discover a wide range of empirical laws, provided that: (1) the qualitative states can be organized in a disjoint concept hierarchy; (2) the qualitative laws can be described as deterministic finite-state machines; and (3) the numeric relations can be stated as products of exponentiated terms. Many examples from the history of physics and chemistry satisfy these constraints, suggesting that IDS will do well in such domains. However, we need to more formally specify the space of laws searched by our algorithms so that we can identify the limits of the framework. We also need to carry out careful experimental studies of the system's behavior, along the lines proposed by Kibler and Langley (1988).

In the near future, we plan to evaluate IDS along two dimensions. As with earlier discovery systems like Bacon, Abacus, and Glauber, we will test IDS' ability to rediscover laws from the history of science. There are many physical and chemical relations that should fall within the system's abilities. The laws of chemical reaction that we have used throughout the chapter are obvious candidates, and we plan to borrow test cases from the earlier work on machine discovery. However, we plan to present IDS with both qualitative and quantitative data for each of these cases. This will simultaneously test the entire system, rather than its components, in its ability to discover empirical laws of the type actually found by scientists. It may also provide more plausible historical accounts of these discoveries than earlier AI systems, although this is not our main goal.

We also plan to test IDS' ability to make predictions about unseen data. As we noted in Section 3.4, the system should be able to predict the qualitative behavior of unobserved attributes in a given state, predict the nature of succeeding states, and predict the values of numeric attributes. As IDS processes more data and its knowledge about the world improves, the accuracy of these predictions should increase. For this study we will use artificial domains, which will let us vary such factors as the structure of the taxonomy, the complexity of the qualitative laws, and the amount of noise in numeric data. We also plan to vary aspects of the system itself, such as the parameter settings used in the numeric component. Experiments of this type will provide information about the robustness of IDS' various discovery methods and suggest ideas for improving them.

4.3 Directions for Future Research

Our long-term plans call for extending IDS in a variety of more challenging directions. These include improving the clustering method, hierarchy designing experiments, and constructing new measuring instruments. We discuss each of these below.

4.3.1 Improving the Clustering Algorithm

In IDS, the discovery of qualitative and numeric laws relies on the formation of appropriate taxonomies, making the clustering process central to the overall system. The current algorithm has some important limitations, which we hope to remedy in future work.

Our experience suggests that the clustering method is sensitive to instance order, forming different hierarchies depending on the order in which it encounters qualitative states. This feature is not so important for experimental data since these can be presented in a careful order with one attribute varied at a time. However, order effects can have a major impact on the structure of a taxonomy formed from observational data and thus on the laws the system finds. Of course, any incremental hill-climbing system will have some sensitivity to order, but we would like to minimize this effect.

Fisher (1987) has argued that including additional learning operators can reduce the effect of instance order. Within the context of his COB-WEB system, he describes a *split* operator that deletes a parent node and elevates its children; he also describes a *merge* operator that is similar to the one used in IDS. These give the effect of backtracking through the space of taxonomies without the need for memory of previous learning steps. Gennari, Langley, and Fisher (in press) present empirical evidence that these operators make systems like COBWEB and IDS less order dependent, and we plan to augment the IDS clustering algorithm with a split operator. This will also require the system to update its laws as the structure of the hierarchy changes.

A second drawback of the current system is that it uses a somewhat ad hoc evaluation function to sort instances and merge nodes. In future versions of IDS, we plan to adapt Gluck and Corter's (1985) *category utility* measure, which has a theoretical grounding in information theory. Fisher's (1987) COBWEB incorporates this as an evaluation function, but his system is limited to attribute-value representations. We plan to extend the function to handle qualitative state descriptions that include multiple objects and structural relations. This approach assumes a probabilistic representation of knowledge, which should permit IDS to represent information about the likelihood of various abstractions.

Another issue involves the object hierarchy of substances that IDS is currently given by the programmer. This takes the form of a hierarchy, and there is no reason in principle why the system could not acquire this knowledge on its own, clustering objects with similar structural features. However, this would require IDS to construct two interleaved taxonomies, one that organizes qualitative states and another that organizes objects appearing in those states. This raises issues of updating the state hierarchy when changes occur in the object hierarchy.

A final problem concerns the assumption that the taxonomy is disjoint, with each observed state being sorted down a single path. This is a clear oversimplification for many domains, and we plan to alter the IDS clustering algorithm to form nondisjoint hierarchies, in which each node may have multiple parents. This simplest approach involves modifying the sorting process at each level of the taxonomy. In addition to considering the placement of the instance in each sibling, one can also consider placing it in the two best-matched siblings, the three best, and so forth. One then simply selects the option giving the highest evaluation score. Only experimentation will tell how well this heuristic approach works.[8]

The extension to nondisjoint taxonomies promises another benefit. Figure 11 shows a partial taxonomy in which the highest-level nodes describe simple qualitative processes and whose children summarize states in which these processes occurred together. The extended IDS should be able to identify such primitive processes from their occurrence in more complex states, even if they are never observed in isolation. It should then be able to use these primitive processes in indexing new states that involve previously unseen combinations. For instance, suppose the system first formed nodes 1, 2, 3, 4, and 5 and only then encountered a qualitative state of the form shown in node 6. This new observation would be sorted through nodes 1 and 3, and node 6 would be stored as their joint child.

4.3.2 EXPERIMENTATION

Another central component of science is experimentation, and we also hope to extend IDS to support this process. Given a current taxonomy and its associated laws, the experimentation component will design a new experiment to be run. This will take the form of an initial state, along with an (optional) set of actions by some external agent. For example, in the domain of acids and alkalis, the experimenter will propose different initial states by varying the reactants, their quantities, and the initial conditions. The system will execute and monitor the experiment itself in a simulated world (Nordhausen & Langley, 1987) or ask the programmer for the results.

We plan to borrow heavily from recent work in this area by Kulkarni (this volume), Rajamoney (this volume), and Karp (this volume). In particular, Kulkarni's KEKADA system includes a heuristic for focusing attention on surprising phenomena. In the IDS framework, one can instantiate this notion as an unpredicted qualitative state or a mispredicted numeric attribute. KEKADA attempts to identify the scope of

---

8. We owe this idea to Doug Fisher, who originally proposed it in the context of his COBWEB system.

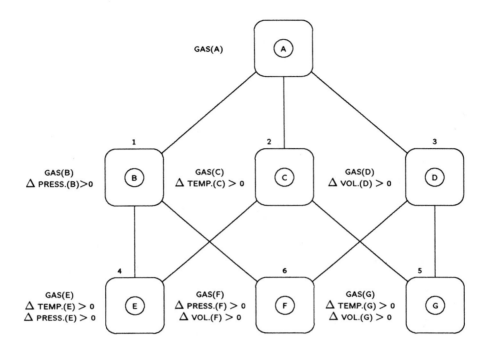

Figure 11. A nondisjoint taxonomy, with primitive processes as parents of combined processes.

the phenomenon, generating different initial conditions using a domain theory of substances much like the one in IDS.

This approach to experimentation will complement the data structures and mechanisms in the existing IDS system. Some of Kulkarni's heuristics, such as dropping factors that have no influence, emerge from IDS' methods for taxonomy and law formation. The incremental nature of the system will let the experimentation component change strategies after observing each history. Also, the systematic variation of substances, their relations, and their numeric attributes will simplify the clustering process, reducing the chances of undesirable order effects.

The IDS framework also supports the construction of new measuring "instruments." Recall that the system can store intrinsic properties, such as a substance's specific heat or boiling point. Given a sequence of abstract states containing such terms, one might place an as yet unobserved substance in the initial state and let the sequence run its course.

Data observed along the way will let one estimate the specific heat or boiling point of the new substance, effectively acting as an instrument for measuring these quantities. IDS can then use these measurement instruments in designing more sophisticated experiments.

4.4 Concluding Remarks

We have presented an integrated approach to empirical discovery that supports taxonomic hierarchies, qualitative relations, and numeric laws. Our ideas are implemented in IDS, a computational system that uses an incremental hill-climbing strategy to discover empirical laws. The system has rediscovered a number of qualitative and numeric laws from the history of physics and chemistry, and our initial experience with the system has been encouraging. However, we hope to carry out more careful experiments in the near future, using both historical and artificial domains.

The individual components of IDS borrow significantly from earlier work on empirical discovery. Each component can be improved, as we plan to do in our future work, but we believe that the overall framework is genuinely new and that it constitutes an important contribution to our understanding of scientific discovery. Moreover, the basic framework shows the potential for covering other aspects of the scientific process, including experimentation and measurement.

Science is a complex enterprise, and it is not surprising that early work on machine discovery focused on isolated aspects of the overall process. However, the field now has relatively robust mechanisms for dealing with many components of discovery, and future progress will depend on understanding the ways in which these components interact. We think that IDS is an important step in this direction, and we encourage other researchers to join us in developing integrated frameworks for discovery.

Acknowledgements

The ideas in this paper have resulted from discussions with members of the UCI machine learning group. We would particularly like to thank Randy Jones and Don Rose for their contributions, as well as Jeff Shrager, who gave valuable comments on drafts of this paper. This research was supported in part by Contract N00014-84-K-0345 from the Information Science Division, Office of Naval Research. The first author was also supported by a Regent's Dissertation Fellowship from the University of California.

References

Brown, J. S. (1973). Steps toward automatic theory formation. *Proceedings of the Third International Joint Conference on Artificial Intelligence* (pp. 121–129). Stanford, CA: Morgan Kaufmann.

Emde, W., Habel, C., & Rollinger, C. (1983). The discovery of the equator or concept driven learning. *Proceedings of the Eighth International Joint Conference on Artificial Intelligence* (pp. 455–458). Karlsruhe, West Germany: Morgan Kaufmann.

Falkenhainer, B. C., & Michalski, R. S. (1986). Integrating quantitative and qualitative discovery: The ABACUS system. *Machine Learning, 1*, 367–422.

Fisher, D. (1987). Knowledge acquisition via incremental conceptual clustering. *Machine Learning, 2*, 139–172.

Fisher, D., & Langley, P. (1985). Approaches to conceptual clustering. *Proceedings of the Ninth International Joint Conference on Artificial Intelligence* (pp. 691–697). Los Angeles, CA: Morgan Kaufmann.

Freund, J. E., & Walpole, R. E. (1980). *Mathematical statistics.* Englewood Cliffs, NJ: Prentice-Hall.

Forbus, K. D. (1985). Qualitative process theory. In D. G. Bobrow (Ed.), *Qualitative reasoning about physical systems.* Cambridge, MA: MIT Press.

Gennari, J. H., Langley, P., & Fisher, D. H. (1989). Models of incremental concept formation. *Artificial Intelligence, 40*, 11–61.

Gluck, M., & Corter, J. (1985). Information, uncertainty and the utility of categories. *Proceedings of the Seventh Annual Conference of the Cognitive Science Society* (pp. 283–287). Irvine, CA: Lawrence Erlbaum.

Hayes, P. S. (1979). The naive physics manifesto. In D. Michie (Ed.), *Expert systems in the microelectronic age.* Edinburgh: Edinburgh University Press.

Jones, R. (1986). Generating predictions to aid the scientific discovery process. *Proceedings of the Fifth National Conference on Artificial Intelligence* (pp. 513–522). Philadelphia, PA: Morgan Kaufmann.

Kibler, D., & Langley, P. (1988) Machine learning as an experimental science. *Proceedings of the Third European Working Session on Learning* (pp. 81–92). Glasgow, Scotland: Pitman.

Kokar, M. M. (1986). Determining arguments of invariant functional descriptions. *Machine Learning*, *1*, 403–422.

Kuipers, B. (1985). Commonsense reasoning about causality: Deriving behavior from structure. In D. G. Bobrow (Ed.), *Qualitative reasoning about physical systems*. Cambridge, MA: MIT Press.

Langley, P., Gennari, J., & Iba, W. (1987). Hill climbing theories of learning. *Proceedings of the Fourth International Workshop on Machine Learning* (pp. 312–323). Irvine, CA: Morgan Kaufmann.

Langley, P., Bradshaw, G. L., & Simon, H. A. (1983). Rediscovering chemistry with the BACON system. In R. S. Michalski, J. G. Carbonell, & T. M. Mitchell (Eds.), *Machine learning: An artificial intelligence approach*. San Mateo, CA: Morgan Kaufmann.

Langley, P., Simon, H. A., Bradshaw, G. L., & Zytkow, J. M. (1987). *Scientific explorations of the creative process*. Cambridge, MA: MIT Press.

Langley, P., Zytkow, J., Simon, H. A., & Bradshaw, G. L. (1986). The search for regularity: Four aspects of scientific discovery. In R. S. Michalski, J. G. Carbonell, & T. M. Mitchell (Eds.), *Machine learning: An artificial intelligence approach* (Vol. 2). San Mateo, CA: Morgan Kaufmann.

Lebowitz, M. (1987). Experiments with incremental concept formation: UNIMEM. *Machine Learning*, *2*, 103–138.

Lenat, D. B. (1977). Automated theory formation in mathematics. *Proceedings of the Fifth International Joint Conference on Artificial Intelligence* (pp. 833–841). Cambridge, MA: Morgan Kaufmann.

Lenat, D. B. (1982). AM: Discovery as heuristic search. In R. Davis and D. B. Lenat (Eds.), *Knowledge-based systems in artificial intelligence*. New York: McGraw-Hill.

Michalski, R. S. (1983). A theory and methodology of inductive learning. In R. S. Michalski, J. G. Carbonell, & T. M. Mitchell (Eds.), *Machine learning: An artificial intelligence approach.* San Mateo, CA: Morgan Kaufmann.

Michalski, R. S., & Stepp, R. E. (1983). Learning from observation: Conceptual clustering. In R. S. Michalski, J. G. Carbonell, & T. M. Mitchell (Eds.), *Machine learning: An artificial intelligence approach.* San Mateo, CA: Morgan Kaufmann.

Mitchell, T. M., Utgoff, P. E., & Banerji, R. B. (1983). Learning by experimentation: Acquiring and refining problem-solving heuristics. In R. S. Michalski, J. G. Carbonell, & T. M. Mitchell (Eds.), *Machine learning: An artificial intelligence approach.* San Mateo, CA: Morgan Kaufmann.

Nordhausen, B., & Langley, P. (1987). Towards an integrated discovery system. *Proceedings of the Tenth International Joint Conference on Artificial Intelligence* (pp. 198–200). Milan, Italy: Morgan Kaufmann.

Thagard, P. (1988). *Computational philosophy of science.* Cambridge, MA: MIT Press.

Weld, D. (1986). The use of aggregation in qualitative simulation. *Artificial Intelligence, 30,* 1–341.

Wrobel, S. (1988). Automatic representation adjustment in an observational discovery system. *Proceedings of the Third European Working Session on Learning* (pp. 253–261). Glasgow, Scotland: Pitman.

Zytkow, J. (1987). Combining many searches in the FAHRENHEIT discovery system. *Proceedings of the Fourth International Workshop on Machine Learning* (pp. 281–287). Irvine, CA: Morgan Kaufmann.

CHAPTER 5

Deriving Laws Through Analysis of Processes and Equations

JAN M. ŻYTKOW

1. Introduction

Many AI discovery systems, including BACON (Langley, 1981; Langley, Simon, Bradshaw, & Żytkow, 1987), FAHRENHEIT (Koehn & Żytkow, 1986; Żytkow, 1987), and ABACUS (Falkenhainer & Michalski, 1986), produce algebraic equations that describe numerical data. Hereafter we will call these BACON-*like* systems. Algebraic equations discovered by such systems can be interpreted as the quantitative laws of different sciences, including Ohm's law, Kepler's laws, and the law of conservation of momentum. However, scientific knowledge exceeds a set of separate laws—it applies to an infinite variety of physical situations, and different situations are usually described by different equations. BACON-like systems must discover equations for each such situation individually, but even simple physical situations can be varied in countless ways.

For example, Figure 1 illustrates three possible versions of an experiment conducted by Joseph Black in the eighteenth century, in which two liquids are combined and their temperature is measured at equilibrium. Figure 2 depicts two versions of a simple mechanical experiment, and Figure 3 shows two simple electric circuits. One can always add more wires, resistors, or transistors and can reconfigure them to produce other electric circuits. The number of modifications for each domain is clearly unbounded.

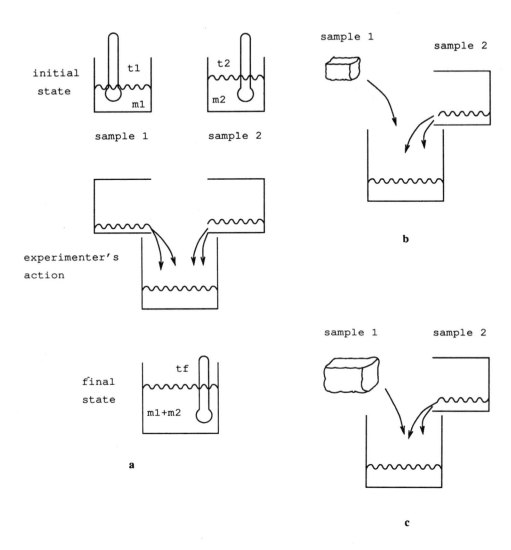

Figure 1. Thermal equilibrium for (a) two samples of water, (b) water and a piece of ice that melts completely, (c) water and a piece of ice that melts partially.

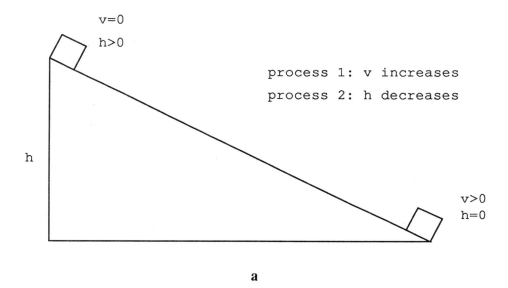

a

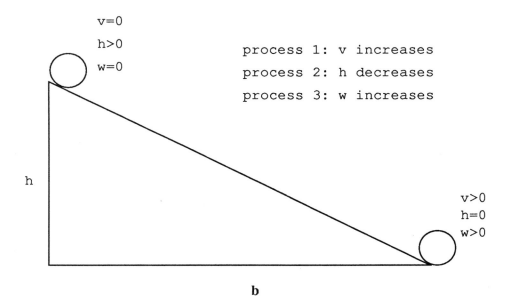

b

Figure 2. Moving down an inclined plane: (a) sliding, (b) rolling.

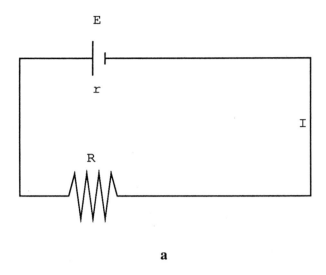

a

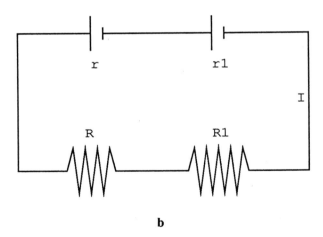

b

Figure 3. Two circuits: (a) one battery and one resistor, (b) two batteries and two resistors.

Should one apply a BACON-like system case by case to discover the corresponding equations for each different situation? This would result in an unbounded set of equations rather than in the parsimonious theories that we associate with modern physics and chemistry. Furthermore, if the discovered laws are limited to individual experimental situations, they will not allow for transfer of knowledge. One might try to reduce the number of laws discovered to a manageable size by discovering equations experimentally for simple situations and then deducing equations for complex situations. However, in most cases the equations that describe different situations are either independent or mutually inconsistent since they describe different phenomena and predict different behavior.

BACON-like systems are successful in generating equations for individual physical situations, but they do not capture the way in which science deals with a multitude of physical configurations to produce simple, finite theories that can be applied to an infinite number of situations. Science deals with the complexity of physical situations by a combination of two steps. In the first step, regularities are decomposed into simpler expressions, each of which is associated with a particularly simple situation or process. In the second step, the simple expressions are recombined to form equations that model complex situations.

This chapter describes GALILEO, a discovery system that transforms equations generated by a BACON-like system into a form compatible with the structure of physical processes they describe, so that equations can be decomposed into pieces useful in model generation. The first part of this chapter discusses the decomposability of equations and the relationship between equations and process models of the situation. This sets the stage for a detailed discussion of the GALILEO system in the following section. The fourth section evaluates the system by giving examples of its operation and describing some possible extensions of the method.

2. Equations and Situations

To understand the decomposition of quantitative regularities into simple expressions and their recombination into models, one must first understand the relationship between equations and physical situations. This section describes that relationship.

2.1   Correspondence Between Situations and Equations

Let us examine some examples in which we compare physical situations and the corresponding equations. Consider a process in which two samples of water at different initial temperatures are combined and eventually reach thermal equilibrium (Figure 1a). The experimental space is spanned by four independent variables under the experimenter's control: two initial temperatures, t_1 and t_2, and two masses, m_1 and m_2. The outcome is the final temperature of equilibrium t_f. Black's equation for this process can be discovered by a BACON-like system, usually in the form

$$t_f = (m_1 t_1 + m_2 t_2)/(m_1 + m_2) \tag{1a}$$

Now we replace the water in the first sample by a small piece of ice at the melting temperature t_1. The ice melts entirely in the water from the second sample (Figure 1b). This process is described by the equation

$$t_f = (m_1 t_1 + m_2 t_2 - m_1 c_l)/(m_1 + m_2) \tag{1b}$$

where c_l is the latent heat of melting ice. Still another experiment, with a larger piece of ice that melts only partially (Figure 1c), leads to the pair of equations

$$m_{1f} = m_1 + (m_2 t_1 - m_2 t_2)/c_l \text{ and } t_f = t_1, \tag{1c}$$

where m_{1f} is the final mass of ice. Many similar experiments are possible, in which ice is dropped into a glass of scotch, a piece of potassium is dropped into water, water is poured over dry ice, and so forth. Each time the outcome is described by a different equation.

Consider equation 1b. We tend to say that both Black's law and the latent heat law apply here, but Black's law in the form of equation 1a must undergo a suitable change before it can be applied to produce equation 1b; the same holds for the law of latent heat. Let us consider the problem in more detail.

An elementary process of melting can be described by the expression $c_l m$, where m is the melted mass and c_l is the latent heat of the phase change. A process in which the temperature increases without a phase change is described by $cm(t_2 - t_1)$, where c is the specific heat, m is the mass, and t_1 and t_2 are the initial and final temperatures. Equations 1a, 1b, and 1c can be assembled from those pieces. In general, the equations describing a complex process, P, can be assembled from pieces that describe the component processes occurring in P. The pieces themselves are not equations. We call them *expressions*, although logicians would prefer to call them "terms."

Other phenomena also can be varied in uncountable ways, but the same principle of composability applies. Consider a body sliding down an inclined plane, as shown in Figure 2a. If the initial velocity is zero and the difference in height is h, then the equation that describes a class of such experiments is

$$v^2 = c_1 h \qquad\qquad (2a)$$

where v is the final velocity and c_1 is a constant. Now suppose that another body is on the downward path but that this time it rolls without slippage, as in Figure 2b. The equation here will differ, assuming the form

$$v^2 = c_2 h \qquad\qquad (2b)$$

in which c_2 is another constant, $c_2 = 5c_1/7$. Here it is less clear what new expressions are added to equation 2a, but when we rewrite 2a and 2b as

$$mgh = mv^2/2, \qquad\qquad (2a')$$

and

$$mgh = mv^2/2 + mv^2/5, \qquad\qquad (2b')$$

we can attribute the expression mgh to the change of altitude, $mv^2/2$ to the process of sliding, and $mv^2/5$ to the process of rolling.

The correspondence of physical components and equation parts applies not only to processes but also to state descriptions, as in the following example of electric circuits. Consider the simple circuit shown in Figure 3a, which is described by the equation

$$E = IR + Ir, \qquad\qquad (3a)$$

where E is the electromotive power of the battery, r is the internal resistance of the battery, and R characterizes the resistor. By adding more elements to the circuit, as depicted in Figure 3b, the equation becomes

$$E + E_1 = IR + Ir + IR_1 + Ir_1 \qquad\qquad (3b)$$

where additional expressions E_1 and Ir_1 correspond to an additional battery, and IR_1 corresponds to the added resistor.

We have seen in these simple examples how science deals with the complexity of physical situations by combining analysis with synthesis. Scientists decompose laws into simple expressions that correspond to and are interpreted by the generic physical components of the situation. These expressions are elementary units of recombination, that is, building blocks from which complexes of equations that describe complex physical situations are recombined into models. This process allows the scientist to *transfer knowledge* extracted by analyzing simple situations to synthesize descriptions of complex situations. In most real-world situations, one must apply a combination of basic laws to generate an adequate description. However, the number of elementary components is small as compared with the total set of physical situations, and a small number of elementary expressions lets one build a limitless number of models.

2.2   The Task of Process-Directed Decomposition

The decomposition of quantitative laws into interpreted expressions requires the use of a scientific semantics of process structure. If the data represent measurements in a particular physical situation, S, then the equation discovered by the BACON-like system can be interpreted as a law that applies to S. However, this interpretation is external to the BACON-like system, in that it neither requires nor makes use of knowledge about the physical structure of S. For a BACON-like system, scientific laws are meaningful just in reference to collections of variables, whereas in science they are meaningful in reference to the structure of physical situations, including their component states, objects, and processes.

Formalisms such as Forbus' (1984) qualitative process theory provide representations for physical processes, but such descriptions will not suffice for applications in which the exact quantity of the change is important. For those applications, we need the qualitative representation of the physical situation, the corresponding equation, and the relationship between elements of the equation and elements of the physical situation. The current task is to produce such a mapping.

3.   The GALILEO System

Now we can describe GALILEO, a system that decomposes equations like the law of equation 1a into simpler expressions. The system's input consists of an equation that describes a particular physical process P, along with a description of P in which the measured variables are attached to the appropriate elements in P. Galileo transforms equations until they can be interpreted by fitting a given equation and the corresponding process description. Under a satisfactory fit, all expressions that occur in the final form of the equation are assigned to the corresponding elements of the process description. In the remainder of this section, we describe the system's representations and computational methods.

3.1 Representing Processes and Equations

The thermal process depicted in Figure 1a can be represented in a process diagram (Figure 4a) as a combination of the processes of cooling and heating, one process for each sample, coupled by energy transfer between both processes. The process diagram in Figure 4a represents a particular understanding of the actual process. Each elementary process is associated with a particular change in a particular object. For someone who believes that a particular substance (like caloric) is exchanged in this process, the process diagram will be different, as depicted in Figure 4a*. Now consider Figure 4b, which shows a process diagram for Figure 1b. One of the initial objects is a piece of ice at melting temperature, and the other is a sample of warm water at t_2. The water cools to the temperature of equilibrium t_f. As the piece of ice melts, the water produced from the ice warms to the temperature of equilibrium t_f.

Process diagrams can be generated automatically by extending qualitative process analysis (Forbus, 1984; Nordhausen & Langley, this volume). In our method, we store process diagrams in a knowledge base indexed by pointers from five "universes" for each process: a process universe, a state universe, an object universe, a space universe, and a time universe. These universes correspond to the basic categories in which we describe the world. Figure 5 depicts all the states and processes for the process depicted in Figure 4a.

3.2 Input and Output

Figure 8 reviews Galileo's input and basic algorithm. In preparation for the main search procedure, the system represents the process diagram by a process decomposition tree like those in Figures 6a and 6b, and it represents the equation by a parse tree like the one in Figure 7.

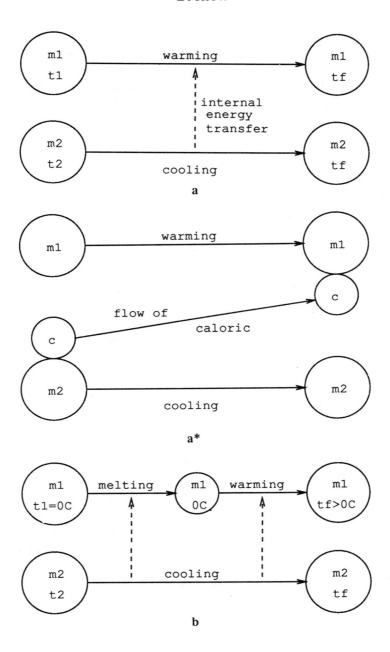

Figure 4. Process diagrams (a) for Figure 1a, (a*) for Figure 1a in the caloric theory, and (b) for Figure 1b.

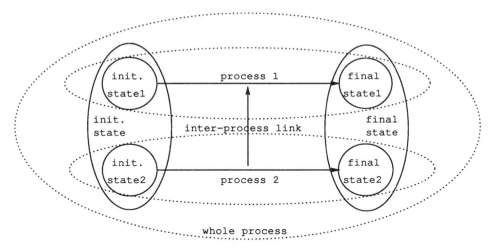

Figure 5. Conceptual framework of the process diagram.

A process decomposition tree offers a simplified representation of the process diagram sufficient to guide equation transformation.

One process diagram may be represented by several process decomposition trees, each providing a different perspective on the same process. For instance, Figures 6a and 6b depict two different trees for the process diagram of Figure 5. In Figure 6a, the process is decomposed into the initial and final states, which in turn decompose into elementary states. In Figure 6b, the same process is decomposed into two elementary subprocesses, which in turn decompose into their initial and final states. When several process decomposition trees are possible, GALILEO can operate with any of them, resulting in different possible equation decompositions. The generation of parse trees from equations and process decomposition trees from process diagrams are simple tasks that are not described here.

GALILEO's fitting procedure produces a mapping between the subtrees in the process decomposition tree and subtrees (expressions) in the equation parse tree. Each expression is then attached to the matching element in the process diagram. If the system is given the process decomposition tree in Figure 6a, then it augments the process diagram in Figure 4a by the boldface expressions in Figure 9a. For the process decomposition tree illustrated in Figure 6b, the results are the boldface expressions in Figure 9b.

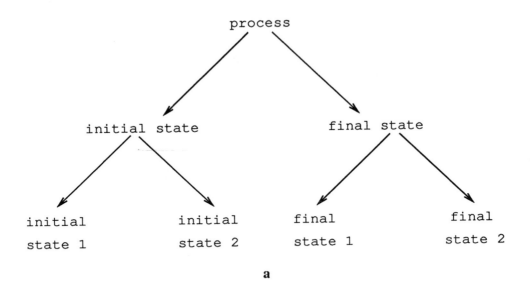

a

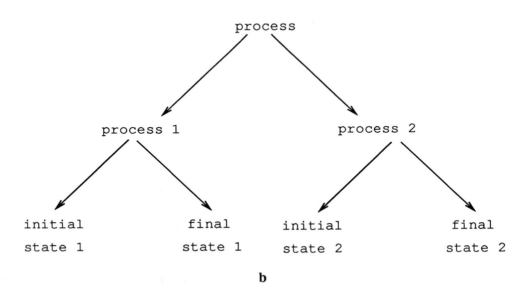

b

Figure 6. Two process decomposition trees for the process depicted in Figure 5: (a) decomposition into initial and final state, (b) decomposition into two subprocesses.

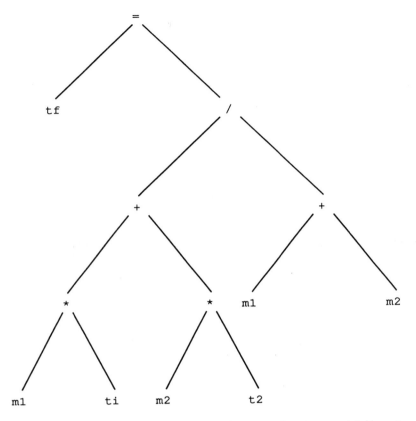

Figure 7. Equation parse tree for the equation $t_f = (m_1 t_1 + m_2 t_2)/(m_1 + m_2)$.

Before the search for a fit, the system is told what quantities represent measurements in each state. For instance, it is told that t_1 and m_1 describe the initial state 1, but it does not know that their product $m_1 t_1$ is a meaningful state description. Many expressions that are built from m_1 and t_1 are potential candidates, such as t_1, m_1, t_1/m_1, and $t_1 m_1^2$. As a result of the fitting procedure, GALILEO is able to identify the most appropriate expression, namely, $m_1 t_1$, without explicit search through the space of candidate expressions. As a result of the matching procedure, the system also infers that, for the internal nodes "initial state" and "final state" in Figure 6a, the concatenation of complex thermal states is reflected by addition of the quantities that describe elementary states, giving the expression $m_1 t_1 + m_2 t_2$. Using the tree

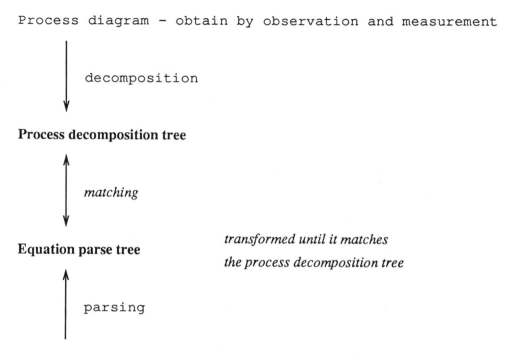

Figure 8. Overview of GALILEO's input (bold) and processing (italics).

of Figure 6b leads GALILEO to determine expressions corresponding to the elementary processes of heating and cooling: $m_1 t_f - m_1 t_1$ and $m_2 t_2 - m_2 t_f$.

It is interesting to consider this difference historically. From the perspective of modern physics and chemistry, only the expressions that describe elementary processes of heating and cooling, such as $m_1(t_f - t_1)$, are valid and useful in modeling thermal phenomena. In contrast, the product of mass and temperature is misleading as a representation of thermal states, having only limited applicability. The product of mass and temperature (multiplied by specific heat, which we disregard for the sake of simplicity) was historically interpreted as the amount of heat in physical bodies. This interpretation was particularly important in the caloric theory at the end of the eighteenth century and at the beginning of the nineteenth century. Later, the notion of the amount of heat in a body was abandoned.

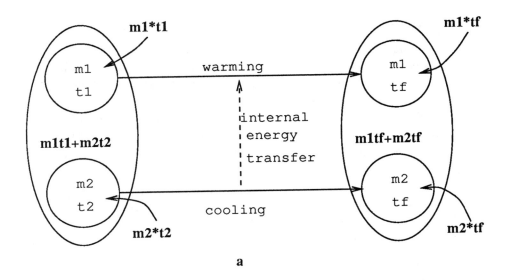

a

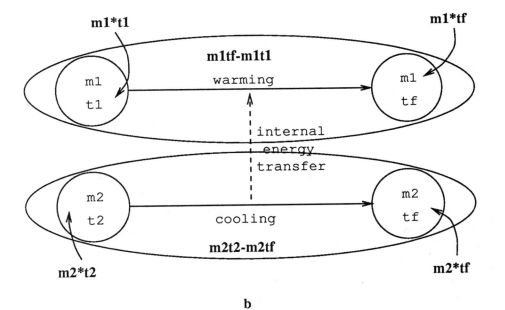

b

Figure 9. GALILEO's output (in bold) (a) for process diagram in Figure 4a, (b) for process diagram in Figure 4b.

3.3   Equation Transformation Search

GALILEO employs a combination of three search methods to transform the input equation to an interpretable form. The principal role is played by the equation transformation (Transform) search method, whose goal is to change the equation to a form in which it matches the process decomposition tree. Since matching is most natural between entities of the same type, the system tries to establish a correspondence between the parse tree of the input equation and the process decomposition tree, as described in Figure 8.

GALILEO uses a variety of transformation grammar rules (operators) to modify the parse tree. Most of these are typical operations on algebraic expressions, known from high school algebra, which are represented in if-then form, where both sides are parse trees. Figure 10 illustrates two such rules that can be recognized as standard algebraic transformations $(a + b)/c \longrightarrow a/c + b/c$ and $a = b/c \longrightarrow ac = b$. If the current equation or one of its parts matches the IF part of a transformation rule, it can be transformed into the structure shown in the THEN part. The second rule in Figure 10 applies to the whole equation, whereas the first rule applies to terms.

Equation transformation search proceeds by applying transformation rules, which act as search operators, until it establishes a complete match with the process decomposition tree or when no improvement can be made to a partial match. It is implemented as depth-first search with backtracking, guided by the degree of match.[1] Two other search methods are called as subroutines. These are described in more detail below. The first is the Match method, which tries to determine the degree of match between a given form of the equation and the process decomposition tree. A complete match terminates the search because it indicates a complete interpretation of the equation, whereas a partial match guides the next equation transformation step. The second submethod, Operator Selection, uses the guidance provided by a partial match to select operators that should be applied by Transform. Operators are heuristically selected to reduce the mismatch between the equation and the process decomposition tree. As a result, the transformation search progresses through increasingly complete partial matches.

---

1. Depth-first search works satisfactorily in many applications that involve simple equations.

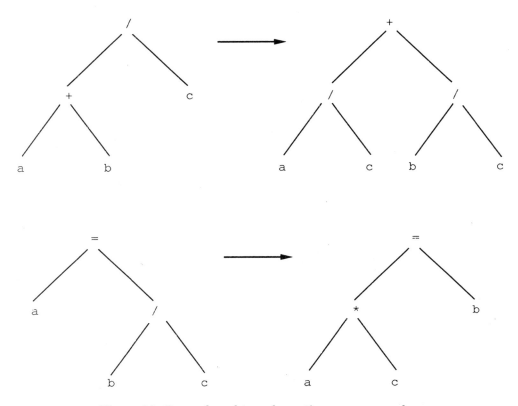

Figure 10. Examples of transformation grammar rules.

Figure 11 depicts a simple example of the Transform search. The topmost equation was first transformed by the rule depicted in Figure 10a to the form in the leftmost leaf. This was unsatisfactory, and so another path was taken (depicted along the right branch of Figure 11) by first applying the rule of Figure 10b, followed by the rule for multiplication analogous to the rule of Figure 10a. The basic approach is very similar to the method of means-ends analysis (Newell & Simon, 1972).

3.4   Tree Matching Search

The Match method searches for a match between an equation parse tree and a process decomposition tree. The matching procedure uses the equation parse tree as its search tree, implementing a depth-first traversal of that tree with backpropagation of results. Most of the action

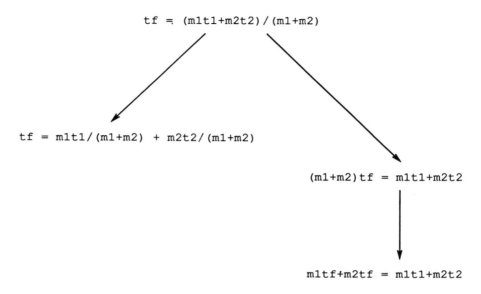

Figure 11. Part of the equation transformation search tree.

occurs during the backpropagation phase. The goals of this search are twofold. First, it decides whether a complete match has been reached between the equation parse tree and the process decomposition tree. This would indicate that the goal of the equation transformation search has been achieved. Second, if the match is not complete, the Match search returns a partial match, like that depicted in Figure 12a. This aids in the selection of operators for the next step in the Transform search that are most likely to make a transformation that improves the match.

Matching starts from the leaves of the equation parse tree, where either single variables or constants are stored. A variable matches all the leaves in the process decomposition tree that are described by that variable. For instance, m_1 matches initial state 1 and final state 1 because m_1 describes each of them, but m_1 matches neither initial state 2 nor final state 2 because m_1 does not belong to the description of these states. As a result, matching at the leaf m_1 returns the list of two elements, including both initial state 1 and final state 1.

After all children of a particular node are considered, the results are backpropagated to the parent node. Internal nodes of the equation parse tree represent arithmetic operations. Depending on the arithmetic

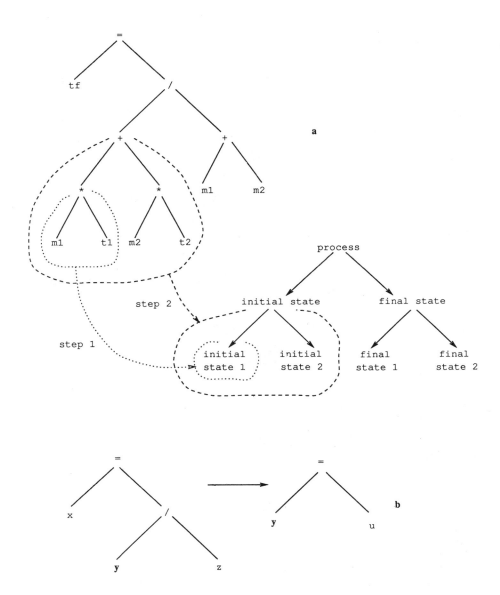

Figure 12. Results of matching: (a) partial match; (b) mismatch reduction schema for (a).

operation, one of two different list operations is applied to the lists
returned from the children of a given internal node to produce the result
at that node. For addition, subtraction, and equality, the union is
applied to the results obtained for children, whereas for multiplication
and division it is their intersection. Backpropagation stops and returns
failure of the match at any node at which an empty set has been reached.
When the backpropagation reaches the root of the equation, a list of
physical states has been attached to each node of the equation tree that
are possible candidates for the interpretation of that node.

In the next phase, GALILEO uses these lists to make detailed hypotheses about the matching subtrees. Each hypothesis about the match between two nodes in both trees is considered separately and must pass
the following tests:

1. A variable that is not relevant to a given node in the process decomposition tree must not occur in the matching subtree.
2. One vertex in the equation parse tree cannot match more than one
   vertex in the process decomposition tree; for example, t_f cannot
   match both final state 1 and final state 2 at the same time.
3. Descriptions of entities of the same type should be homogeneous; for
   instance, if $m_1 t_1$ matches initial state 1, then t_f alone cannot match
   final state 1.

The next cycle of backpropagation extends the partial matches as far
as possible. If the root is reached, a complete match has been detected;
otherwise the backpropagation stops at a particular node when no match
holds for the subtree starting at this node. If the search cannot find a
total match, it uses a consistent partial match to guide operator selection
for the next step of equation transformations.

3.5   Selecting Operators

As described above, a partial match between an equation and a process
decomposition tree helps GALILEO select an appropriate operator for
continued equation transformation. As a result of matching the equation in Figure 7 and the tree in Figure 6a, GALILEO detects a partial
match between the initial state and the expression $m_1 t_1 + m_2 t_2$, as illustrated in Figure 12a. Other elements of the equation in Figure 7 do
not match the tree in Figure 6a. This partial match lets the system
schematically characterize the desired transformation of the equation

presented in Figure 12b. This transformation schema, called a *mismatch reduction schema*, is input to the search method for selecting the relevant operator transformations. This method tries to identify the rules in the transformational grammar, such as those depicted in Figure 10, that satisfy the mismatch reduction schema.

This search proceeds in two steps. The antecedent of each rule is matched against the left side of the mismatch reduction schema; then the right sides are compared to determine whether the operator offers progress in the desired direction. For instance, the second rule in Figure 10 applies to reduce the mismatch in Figure 12b. The list of all successful matches is returned to the Transform search. They are applied in the order they were detected, and the Match search determines how well the new equations match the process decomposition tree, as compared with the equations before transformation, and selects the best matching equations for continued transformation.

3.6 Controlling the Search Process

The three interrelated searches described in the previous sections (Transform, Match, and Operator Selection) are organized in a *multisearch* framework (Jankowski & Żytkow, 1988; Żytkow, 1987; Żytkow & Jankowski, 1989). The main objective of the multisearch approach is to reduce the conceptual complexity of programming that is inevitable in an AI system that must conduct a number of searches before reaching its goal. The complexity is reduced by a uniform representation of each search, along with a driver that executes all searches. The multisearch approach enforces a clear, structured representation of each search and provides a template that can be used to enhance the code development, while the driver mechanism eliminates the necessity for writing separate control algorithms for each search.

The left side of Figure 13 depicts the top-level Transform search. The right side of the figure focuses on the structure of a single node, depicting the way in which three searches are combined. At each node of the Transform search, as soon as the node is created, the Match method is called to check whether Transform has reached its goal. If this is the case, the multisearch halts; if not, the Match search returns a mismatch reduction schema, and that schema is passed on to the Operator Selection search. Operator Selection uses the mismatch information to find and instantiate the operators that may improve the match. It returns

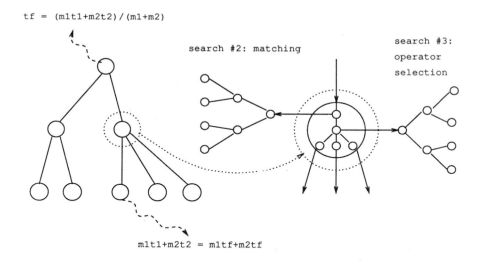

tf = (m1t1+m2t2)/(m1+m2)

search #2: matching

search #3:
operator
selection

m1t1+m2t2 = m1tf+m2tf

search #1: equation transformation

Figure 13. Equation transformation as multisearch.

the selected operator instances to the node N of Transform, which uses them to expand N. The children of node N that improve the match are retained.

4.   Evaluation of GALILEO

The GALILEO system uses an equation transformation search to decompose equations discovered by a BACON-like system into expressions that describe the constituent processes in physical situations to which those equations apply. The resulting expressions can be recombined into equations to describe new situations. We have already seen several examples of GALILEO's operation, and in this section we provide more examples of the system's performance in order to demonstrate its strengths and weaknesses. We also discuss several possible extensions to the system.

4.1   Operation of the System

GALILEO can be applied to different equations and different process decomposition trees. It usually takes few steps to bring an equation typical of the output of a BACON-like system into a form that matches the corresponding process decomposition tree.[2] Above, we used equation 1a and the process decomposition tree depicted in Figure 6a to explain GALILEO's operation. An additional example will help to understand the process as a whole.

Consider equation 1c, for which the corresponding experiment is depicted in Figure 1c. Figure 6a can be used to illustrate the process decomposition tree for this experiment. First, the leaf nodes in Figure 6a must be associated with magnitudes that describe melting of ice in water. Initial state 1, which corresponds to the piece of ice before it is dropped into water, is described by m_1, t_1, and c_l. Final state 1 is described by m_{1f}, that is, the remaining mass of ice, by t_1 and c_l.

GALILEO's operators transform the initial equation

$$m_{1f} = m_1 + (m_2 t_1 - m_2 t_2)/c_l$$

into

$$m_1 + (m_2 t_1 - m_2 t_2)/c_l = m_{1f} \tag{3.1}$$

$$m_{1f} - m_1 = (m_2 t_1 - m_2 t_2)/c_l$$

and

$$m_{1f} c_l = m_1 c_l + (m_2 t_1 - m_2 t_2) \tag{3.2}$$

from which the matching search selects equations 3.1 and 3.2, as they have the same high degree of match. Matching recognizes that m_1 and m_{1f} represent initial state 1 and final state 1, that they are at the right sides of the equation, and that they satisfy the homogeneity criterion. Matching also recognizes that $m_2 t_1 - m_2 t_2$ represents the initial and final states 2 and that this must be raised to the higher level in the equation tree. In this particular case, the matcher cannot find a satisfactory overall interpretation for the expression $(m_2 t_1 - m_2 t_2)/c_l$.

Working on equation 3.2, the matching search recognizes that the final state 2 is on the appropriate side of the equation, that homogeneity is satisfied separately for the states of ice ($m_{1x} c_l$) and the states of water ($m_2 t_i$), and that the number of states is equal to the number of states in the process decomposition tree, with both matches having the same score.[3]

---

2. GALILEO is a deductive reasoner; all its operators transform equations to equivalent form, preserving equality, that is, preserving truth.

3. If the scoring rules differed, one of these expressions might be preferred over the other, but the final result would most likely be the same. Experiments with

Both equations 3.1 and 3.2 are transformed into several new forms. The one that receives the highest score is obtained from equation 3.1 by multiplying both sides by c_l, and from equation 3.2 by swapping both sides, resulting in this equation:

$$m_1 c_l + (m_2 t_1 - m_2 t_2) = m_{1f} c_l$$

In the next step, an even higher score is obtained from this equation:

$$(m_1 c_l + m_2 t_1) - m_2 t_2 = m_{1f} c_l$$

Finally, the equation

$$(m_1 c_l + m_2 t_1) = m_{1f} c_l + m_2 t_2$$

receives the highest score. Homogeneity is not fully satisfied, but nothing better can be produced because melting and temperature change are described by expressions that are only partly homogeneous.

It is important to note that equations must include separate expressions that correspond to all elementary processes. If we transform equation 1c into the form

$$m_{1m} = m_2(t_2 - t_1)/c_l,$$

where $m_{1m} = m_1 - m_{1f}$, and present it as GALILEO's input, then, missing an item compatible to the initial state 1, the system is unable to reach a satisfactory solution.

4.2   Possible Extensions

There are several interesting possible extensions that would improve GALILEO's scope. Some of these methods have been suggested by actual scientific practice.

GALILEO terminates successfully when it detects a complete match. If a complete match cannot be achieved, the search will eventually halt, lacking candidate equations for further improvement. If the final mismatch is minor, one might hypothesize that the process diagram is not complete or that it includes irrelevant detail. Such situations are common in science. For instance, if a substance that disengages in a reaction is overlooked, a branch will be missing from the process decomposition tree. Similarly, if a process that one thought to influence a given phenomenon does not actually have a measurable impact, there will be a redundant branch in the process decomposition tree. GALILEO could be

---

GALILEO suggest that the algorithm is stable in regard to various ways in which the matching results can be combined.

extended to consider such hypotheses in order to bring the match closer to completion.

In many cases, the equation is simpler than the process decomposition tree because a number of states or processes are represented by a single expression. For example, two elementary processes occur in equation 2a—the body increases its velocity while decreasing its altitude. In the first process, the kinetic energy increases, and in the second the potential energy decreases. Both of the process decomposition trees depicted in Figure 6 adequately describe this process. However, equation 2a does not match either of the trees in the figure in a straightforward way. The equation would match the process decomposition tree in Figure 6b if we transformed equation 2a to the form:

$$mgh_1 - mgh_2 = mv_2^2/2 + mv_1^2/2$$

Special transformation rules are needed to accomplish this task. Some of these rules add and subtract the same expression or multiply and divide by the same expression:

$$a \longrightarrow a + (x - x) \longrightarrow (a - x) + x$$
$$a \longrightarrow ax/x$$

These rules would permit GALILEO to introduce components that are missing from the initial equation. In cases like equation 2a and 2b, additional terms must be introduced to the initial equation to match the initial and final states in the processes in Figures 2a and 2b. These rules can be instantiated for any value of x, except for zero in division. However, practically speaking, they should be applied when the system is able to make restricted guesses about the form of x; otherwise, the search space is infinite.

The current version of GALILEO is a deductive reasoner, but on many occasions scientists use approximate reasoning. In order to model approximate inference, we must augment the set of operators to include ones that, under specific conditions, produce approximately true conclusions from true premises. For example, the operator $a + \varepsilon \longrightarrow a$ applies when ε is much smaller than a.

Another limitation in the current system is that the Match method assumes a physical meaning for different arithmetic operations. For instance, multiplication and division apply meaningfully to pairs of variables that describe the same state. To satisfy this constraint, the match method intersects the sets of candidate states for both variables. It should be possible to discover the meaning of such arithmetic operations through a more extensive search. Currently, the set intersection operation applied in matching reduces the number of possibilities, but the system should be able to proceed without that guidance by considering more cases.

Another interesting extension would take into account finite differences that describe elementary processes. From finite differences, it is easy to make a step toward components of differential equations. Consider the finite differences $m_1(t_f - t_1)$ and $m_2(t_2 - t_f)$ in Figure 9b. These can be replaced by a finite difference $m\Delta t$ or by the infinitesimal difference mdt. When infinitesimal differences are used in model construction, we obtain differential equations.

Finally, the most important extension would consist of implementing a model-building system that would actually build equations to fit specific physical situations. GALILEO's output would be the input for such a system. In simple cases, when physical situation S is represented by a process decomposition tree, we can replace the leaves of that tree by the corresponding mathematical expressions obtained by GALILEO, and the internal nodes by the corresponding mathematical operations, to obtain an equation that models situation S.

Model construction is simple if the modeled process is simple and we can correctly decompose it into elements. In general, model construction is a complex process, guided by many sources: different theories, knowledge of particular types of structures, results of initial experimentation, previous models, precision of intended measurements, and so forth. Many models are usually considered before one is found that fits the constraints. The model construction procedure oscillates between adequacy of description and solvability of the produced equations (Lewenstam & Żytkow, 1989). At the core of this complex procedure is the knowledge of elementary building blocks and ways in which they can be put together. The GALILEO system creates those building blocks.

5. Conclusion

In order to obtain a theoretical description of a given real-world situation, S, scientists synthesize complexes of equations called *models*, which they assemble from expressions that correspond to elementary states and processes recognized in S. GALILEO is a system that decomposes empirical laws into such expressions, guided by a model of the objects and processes involved in the experimental situation in which the laws were discovered. Knowledge decomposition of this sort is a necessary step toward transfer of knowledge to new situations. Such transfer would be accomplished by model construction, using the units and association provided by GALILEO.

The approach embodied in GALILEO incorporates ideas on qualitative processes representations (Forbus, 1984; Nordhausen & Langley, this volume), quantitative discovery (Langley, 1981; Langley et al., 1987; Żytkow, 1987), and the transformation of mathematical formulas. The system's knowledge representation can be thought of as standing "perpendicular" to that of Langley and Nordhausen's IDS, which induces a network of process descriptions. In contrast, GALILEO analyzes a given equation and, by detailed comparison with a single process, reaches a finer-grained representation by assigning mathematical expressions to elementary subprocesses that describe a simple change in the properties of one object.

Unlike systems such as BACON and FAHRENHEIT, GALILEO produces simple expressions that can be applied in contexts far beyond that in which the initial empirical equation was discovered. Thus, it represents a considerable generalization of scientific knowledge, enabling an integrated scientific engine to build mathematical models for complex situations, given a description in terms of elementary objects and processes.

Acknowledgements

Special thanks to Jeff Shrager and Pat Langley for very valuable suggestions and corrections. The work described in this chapter was supported by the Office of Naval Research under Contract No. N00014-88-K-0226.

References

Falkenhainer, B. C., & Michalski, R. S. (1986). Integrating quantitative and qualitative discovery: The ABACUS system. *Machine Learning*, *1*, 367–401.

Forbus, K. D. (1984). Qualitative process theory. In D. G. Bobrow (Ed.), *Qualitative reasoning about physical systems*. Cambridge, MA: MIT Press.

Jankowski, A., & Żytkow, J. M. (1988). A methodology of multisearch systems. In Z. Ras & L. Saitta (Eds.), *Methodologies for intelligent systems* (Vol. 3). New York: North-Holland.

Koehn, B., & Żytkow, J. M. (1986). Experimenting and theorizing in theory formation. *Proceedings of the International Symposium on Methodologies for Intelligent Systems* (pp. 296–307). ACM SIGART Press.

Langley, P. (1981). Data-driven discovery of physical laws. *Cognitive Science, 5*, 31–54.

Langley, P., Simon, H. A., Bradshaw, G. L., & Żytkow J. M. (1987). *Scientific discovery: An account of the creative processes.* Cambridge, MA: MIT Press.

Lewenstam, A., & Żytkow, J. M. (1989). Model-based science of ion-selective electrodes. In E. Pungor (Ed.), *Ion-selective electrodes* (Vol. 5). Oxford, England: Pergamon Press.

Newell, A., & Simon, H. A. (1972). *Human problem solving.* Englewood Cliffs, NJ: Prentice-Hall.

Żytkow, J. M., & Jankowski A. (1989). Hierarchical control and heuristics in multisearch systems. In Z. Ras (Ed.), *Methodologies for intelligent systems* (Vol. 4). New York: Elsevier.

Żytkow, J. M. (1987). Combining many searches in the FAHRENHEIT discovery system. *Proceedings of the Fourth International Workshop on Machine Learning* (pp. 281–287). Irvine, CA: Morgan Kaufmann.

CHAPTER 6

A Unified Approach to Explanation and Theory Formation

BRIAN FALKENHAINER

1. Introduction

Deduction, abduction, and analogy are processes whose differences are normally reflected by distinct computational mechanisms. Furthermore, AI researchers typically decouple explanation and diagnosis from theory formation and discovery. Yet these tasks are intimately related and blend imperceptibly. Their integration into a unified view of explanation offers the potential for graceful degradation in the presence of an imperfect domain theory; in this approach, one provides a deductive explanation, if possible, and extends or revises the underlying theory when necessary to make explanation possible.

In this chapter, I suggest that procedural separations between deduction, abduction, and analogy are superfluous for the purpose of constructing plausible explanations of a given phenomenon. A single mechanism that proposes explanations of phenomena by their similarity to understood phenomena is sufficient, providing smoother adaptability to unanticipated or underspecified events and enabling transfer of knowledge from one domain to another. This *similarity-driven* view of explanation also lets one extend or revise imperfect theories when they fail to produce an explanation. Rather than being produced by separate processes, distinctions between the different explanation types result from the preferential ordering imposed when competing hypotheses are evaluated.

The plausibility of this conjecture is demonstrated by PHINEAS, a program that uses a single similarity-driven explanation mechanism to focus its search for explanations using its existing knowledge and to develop novel theories when its existing knowledge is insufficient. For example, when given only knowledge of liquid flow, the system is able to interpret the three situations shown in Figure 1:

   a. A beaker contains more water than a vial to which it is connected by an unknown object. Why does the water level in the beaker decrease and the water level in the vial increase?

   b. Two containers sharing a common wall of unknown substance each hold some solution. Why does one solution's level decrease and concentration increase while the other solution's level increases and concentration decreases?

   c. What causes a hot brick and cold water to change to the same median temperature when the brick is immersed in the water?

In each case, PHINEAS bases its explanation on the case's similarity to liquid flow. In the first, the phenomenon most similar to an observation of liquid flow is liquid flow itself, thus suggesting that the unknown object may be a fluid path. In this work, identicality is viewed as an extreme form of similarity. The second behavior, called *osmosis*, represents a close generalization of liquid flow when viewed as flow of solute under osmotic pressure through a selective kind of fluid path. In the final "heat flow" observation, PHINEAS draws an across-domain analogy to liquid flow phenomena and conjectures the existence of a new type of fluid that affects an object's temperature. All three interpretations are produced by a single mechanism that forms its explanations from theories about phenomena most similar to the current situation.

This chapter begins with a discussion of the relationship between explanation and analogy and suggests that they share a common core, the search for explanatory similarity. It then describes PHINEAS, along with a detailed example of its operation. The system's behavior on a variety of examples is then discussed, which indicates success in achieving adaptability and provides impetus for a number of future research themes.

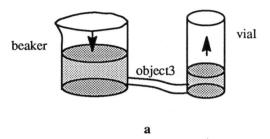

a

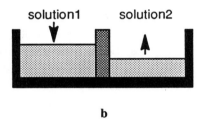

b

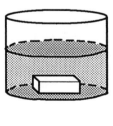

c

Figure 1. Three phenomena that PHINEAS explains by their similarity to liquid flow: (a) liquid flow, (b) osmosis, (c) heat flow.

2. Abduction as Similarity-Driven Explanation

Theory formation, explanation, and diagnosis all follow a pattern of reasoning called *abduction*, which can be defined as *inference to the best explanation*. Josephson, Chandrasekaran, Smith, and Tanner (1987) suggest that abduction is of the form:

$\mathcal{D}$ is a collection of data (facts, observations, givens);
$\mathcal{H}$ explains $\mathcal{D}$ ($\mathcal{H}$ would, if true, imply $\mathcal{D}$);
No other hypothesis explains $\mathcal{D}$ as well as $\mathcal{H}$;

Therefore, $\mathcal{H}$ is correct.

That is, if the hypothesis were true, it would explain the phenomenon.[1] There are two key phrases here. "If it were true" indicates that not all of the relevant knowledge may be available and that assumptions may be required to fill in the gaps. The process of finding the candidate hypotheses and of making assumptions along the way will be called the *interpretation-construction* task. The phrase "it would explain the phenomenon" indicates that the hypothesis would explain the phenomenon, not that it is the correct explanation. There may be other hypotheses that can also explain it. The process of deciding which hypothesis is the best explanation will be called the *interpretation-selection* task.

Abduction is traditionally characterized as using a fixed set of background theories. Assumptions needed to fill gaps due to incomplete knowledge of the situation are limited to ground atomic sentences (no new or revised rules are considered), as in

$given$ $\text{CAUSE}(\mathcal{A}, \mathcal{C}), \mathcal{C}$ $infer$ $\mathcal{A}$

These systems suffer from the *adaptability problem* (Falkenhainer, 1988): They are unable to revise or extend an imperfect domain theory to make conjectures about unanticipated events, and unable to apply knowledge of one domain to the understanding of another.

---

1. There is a distinction between the abduction process, which is normally associated with backward chaining on a set of rules, and its ultimate product, a deductive proof tree typically having at least one assumption as a leaf. Throughout this chapter, I am primarily interested in the abstract product, independent of the chaining process, in which assuming some unknown antecedent facts is required to complete the explanation.

On the other hand, theory formation typically involves making assumptions about both the situation and the incompleteness or incorrectness of current theories. It includes inferences of the form

$$given \;\; \text{CAUSE}(\mathcal{A}, \mathcal{C}) \wedge \mathcal{A} \Rightarrow \mathcal{C}, \mathcal{A}, \mathcal{C} \quad infer \;\; \text{CAUSE}(\mathcal{A}, \mathcal{C})$$

Theory formation must face the problem of generating theory-revising hypotheses and establishing a preference among a possibly infinite set of hypotheses.

These problems can be resolved by noting the strong commonalities between traditional abduction and analogy and developing a model that encompasses both. For abduction, this unified model provides the power to extend the underlying domain theory when needed. For theory formation, it enables existing knowledge, possibly of other domains, to influence hypothesis generation and evaluation, thus taking into account knowledge of the way things normally behave in the world and the way theories about those behaviors are normally expressed. This view of explanation is based on the conjecture that search for similarity between the situation being explained and some understood phenomenon suffices as the central process model for explanation tasks. Two arguments support this view.

First, consider the traditional abduction task. Simple backward-chaining models work well for explaining atomic occurrences, such as Wet(grass). However, as the complexity of the phenomenon being explained increases, the ability to backward chain to a small set of plausible candidates diminishes. One must consider the entirety of the situation and take into account all the interrelations between aspects. Hence, most abduction systems directed at complex phenomena are based on some form of macro-matching, typically in terms of schemas or frames, that seeks minimal hypothesis sets maximally fitting the data. This is true of script or schema-based models of story understanding (Charniak, 1972; DeJong, 1982; Mooney, 1987), process models for interpreting the behavior of a physical system (Forbus, 1986), and composite matching models of abduction and diagnosis (Josephson, Chandrasekaran, Smith, & Tanner, 1987; Reggia, 1983). The desire for a minimal, best match is also implicitly reflected in the Occam's razor heuristic found in simpler systems, which backward chain on one datum at a time (e.g., Pople, 1973). In other words, interpretation and explanation are a form of best match process, with the goal of matching the current situation to hypotheses that can explain it.

A second argument involves the explanation scenarios shown in Figure 2, which are summarized below:

Deduction scenario. Given phenomenon $\mathcal{P}$, where $\mathcal{P}$ represents a set of observables, a complete explanation of $\mathcal{P}$ deductively follows from existing knowledge. The only open question is whether it is *the* explanation, as there may be others. For example, suppose fluid flow is observed, and all the preconditions for fluid flow are known to hold (the source pressure is greater than the destination pressure, the fluid path is open, and so forth). Then a fluid flow explanation directly follows. Given the observed behavior and the existing preconditions, we could say that the situation is *literally similar* (Gentner, 1983) to liquid flow.

Assumption scenario. Phenomenon $\mathcal{P}$ is given, where $\mathcal{P}$ represents a set of observables. No explanation can be found using current knowledge because the status of some requisite facts is unknown. However, a complete explanation follows from the union of existing knowledge and a consistent set of assumptions about the missing facts. For example, if one observes liquid flow but does not know if the fluid path valve is open or closed, one can assume that the valve is open if there is no evidence to the contrary.

Generalization scenario. Phenomenon $\mathcal{P}$ is given, where $\mathcal{P}$ represents a set of observables. Existing knowledge indicates that candidate explanation $\mathcal{E}$ cannot apply because condition C_1 is known to be false in the current situation. However, $\mathcal{E}$ does follow if condition C_1 is replaced by the next most general relation since C_1's sibling is true in the current situation. This is a standard knowledge-base refinement scenario (e.g., Winston, Mitchell, & Buchanan, 1985) and is closely related to approaches that generalize from a set of examples (Hayes-Roth & McDermott, 1978; Winston, 1975).

Analogy scenario. Phenomenon $\mathcal{P}$ is given, where $\mathcal{P}$ represents a set of observables. No candidate explanation $\mathcal{E}$ is available directly, but explanation $\mathcal{E}_b$ is available if a series of analogical assumptions are made, that is, if the situation explained by $\mathcal{E}_b$ is assumed to be analogous to the current situation. For example, if heat flow is observed but little is known about heat phenomena, then an explanation may be constructed by analogy to liquid flow.

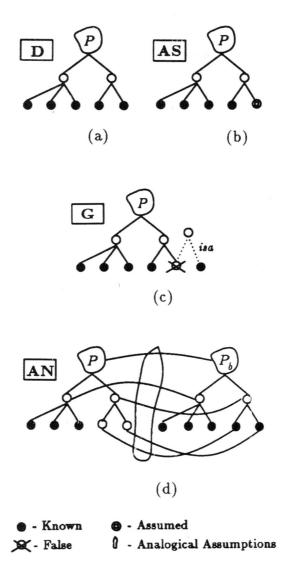

Figure 2. Four alternative explanation scenarios: (a) deduction scenario, (b) assumption scenario, (c) generalization scenario, (d) analogy scenario. In each, $\mathcal{P}$ is the phenomenon being explained and implications flow from the antecedents below to the explained consequents above.

Each scenario requires the interpretation-construction task: retrieve from memory explanatory hypotheses that match the current situation. Each also requires the interpretation-selection task: select from a set of candidate hypotheses the one that is most probable, plausible, or coherent. Importantly, each scenario represents the same process when viewed as different forms of similarity to an existing theory:

- *Deduction scenario*: complete match of identical features
- *Assumption scenario*: partial match of identical features
- *Generalization scenario*: matches between features having a close generalization
- *Analogy scenario*: a range of matches between different features and relations

A system based on this view would offer the best explanation available, ranging from application of an existing theory to distant analogy. It relies on the following conjecture:

> *Similarity conjecture*: All interpretation-construction tasks may be characterized as the search for maximal explanatory similarity between the situation being explained and some previously explained scenario. The previous situation may be drawn from an actual experience, a prototypical experience, or an imagined scenario derivable from general knowledge.

This conjecture suggests that there is no need for a strong distinction between deductive explanation processes and analogical explanation processes. The same basic process may be used in each explanation scenario, with distinctions between them emerging from how well existing knowledge supports the explanation. Deductive operations correspond to the high confidence derived from identicality matches. A corollary to the similarity conjecture is that the same basic processes are at work in both scientific theory formation and in everyday interpretation and hypothesis formation, as suggested by Leatherdale (1974).

The benefits of this view are that it suggests using a single computational architecture for explanation processes. Distinctions between explanation types influence only the weighing of evidence and the decision as to whether a new conjecture represents a revision of existing knowledge or a new separate body of knowledge. This chapter seeks to demonstrate the feasibility of this view.

3.   The PHINEAS System

The similarity-driven model of explanation discussed in the previous section is illustrated by PHINEAS, a program that offers qualitative explanations of time-varying physical behaviors. The system uses remindings of similar experiences to suggest plausible hypotheses and uses qualitative simulation as a form of *gedanken* experiment to analyze the consistency and adequacy of these hypotheses. This section begins with a discussion of the representations used in PHINEAS to describe observations and to reason about their underlying causes. It then presents an overview of the PHINEAS system and the preference criteria that gives rise to its intuitively appealing, flexible behavior.

3.1   Representation

PHINEAS' theories about the physical world and its methods for using these to generate predictions are based on research in qualitative physics (Bobrow, 1985). Given a qualitative model of a particular physical configuration, a *qualitative simulator* produces a description of the possible behaviors for the given situation, called an *envisionment*. An envisionment describes physical states and the possible transitions between them. Each state represents an interval of time during which the qualitative description of behavior does not change. A specific behavior of the system through time, either observed or predicted, may then be represented as a single path through the envisionment. I will refer to such a path as a *history*, after Hayes (1979). For example, Figure 3(a) shows a beaker connected to a vial and an observed qualitative history for this configuration.

The present work uses Forbus' (1984) *qualitative process theory* as the primary formalism to represent and reason about physical change. In QP theory, a situation is represented as a collection of objects (e.g., contained liquid), a set of relationships between them (e.g., connected), and a set of process schemas that account for all changes in the world (e.g., liquid flow). Each object has a set of continuous *quantities*, such as Temperature and Pressure. Each quantity has an amount, expressed as A[Temperature(brick)], and a derivative, expressed as D[Temperature(brick)].

Process definitions have five components: *individuals, preconditions, quantity conditions, relations*, and *influences*. The individuals specify

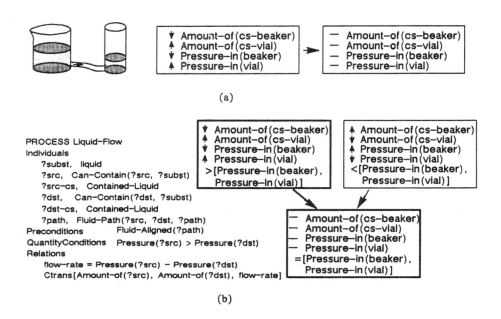

Figure 3. Qualitative physics representations: (a) qualitative observation of liquid flow from a beaker to a vial; (b) liquid flow process model and corresponding envisionment.

the objects involved in the process when it is active, the preconditions and quantity conditions indicate when the process will be active, and the relations and influences specify what relations will hold while the process is active. Figure 3(b) shows a typical QP theory definition for the liquid flow process and the envisionment it produces for the beaker-vial configuration.[2]

The explanatory consistency of a proposed model is established if there is a path through the envisionment derived from the model that corresponds to the measurements (Forbus, 1986). For example, the darkened two-state path of Figure 3(b) corresponds to the observation in Figure 3(a).

---

2. The predicate `Ctrans` refers to "continuous transfer" and is a macro for the standard QP theory pair `I-[Amount-of(?src),flow-rate]` and `I+[Amount-of(?dst),flow- rate]`. See Falkenhainer (1988) and Forbus (1984) for more details.

3.1.1 INITIAL KNOWLEDGE

PHINEAS uses three sources of knowledge during its reasoning process. These include:

1. *Initial domain theory.* Domain knowledge consists of a collection of qualitative theories about physical processes (e.g., liquid flow), entities (e.g., fluid paths), and general physical principles (e.g., mechanical coupling). This qualitative knowledge is represented using the language of Forbus' (1984) QP theory.

2. *Prior experiences.* When comparing a new observation with prior experience, PHINEAS consults a library of previously observed phenomena (structure and behavior descriptions). Focus on relevant attributes is ensured by the storage of only those aspects of a prior situation that participated in its explanation. Past reasoning traces are summarized by storing with each state in an observation the instantiated collection of theories (e.g., process definitions) that were used to explain it; for example, `Liquid-Flow(beaker1,vial8, pipe2)` might be stored with an observation of liquid flow.

   Behaviors are indexed in memory via *behavioral abstractions*, which record abstract characterizations and summaries of the phenomenon not captured by the standard QP theory representation. These correspond to graphic characterizations (e.g., `linear`, `cyclic`, `asymptotic`), movement continuity (e.g., `corpuscular`, as in a ball, or `continuous`, as in liquid flowing), and movement type (e.g., `phase-change-movement` or `invariant-form-movement`). These are arranged in generalization hierarchies, forming a forest of behavioral abstractions similar to the memory organization used by Kolodner (1984).

3. *Observation.* The final source of PHINEAS' information is the observation targeted for explanation. The system records three classes of information: the original scenario description (e.g., `Open(beaker)`), the behavior across time (e.g., `Decreasing[Amount-of(alcohol)]`), and behavioral abstractions that apply to the observation (e.g., `asymptotic`).

3.1.2 OUTPUTS

In response to a given observation, PHINEAS attempts to produce an explanatory "theory" and the envisioned behaviors it predicts. A theory consists of a set of process descriptions, entity descriptions, and atomic facts. The process and entity descriptions may be elements of the existing domain theory or new postulated theories. The system makes this distinction during hypothesis evaluation. The atomic facts are assumptions about the scenario that are required to complete the explanation.

3.2   Process Components

As depicted in Figure 4, PHINEAS operates in four stages: *access, mapping/transfer, qualitative simulation*, and *revision*. Falkenhainer (1988) provides the details of each stage, but they are briefly reviewed in this section. Throughout the discussion, the term "base" refers to a recalled analogue, and "target" refers to the current situation to be explained.

3.2.1   THE ACCESS STAGE

A new observation triggers a search in memory for understood phenomena that exhibit analogous behavior. This retrieval process involves two stages. First, behavioral abstractions of the observed situation are used to focus attention on a potentially relevant subset of memory. Second, each phenomenon in this subset is inspected more carefully by matching its detailed structural and behavioral description to the current situation. This comparison is performed by the *structure-mapping engine* (SME) (Falkenhainer, Forbus, & Gentner, 1986, 1989).[3] This partial mapping provides an indication of what objects and quantities correspond by virtue of their behavioral similarity. It serves as an important source of constraint during the mapping process.

The match also indicates where the phenomena correspond and thus what portion of the base analogue's behavior should be considered relevant. The problem of relevant theory selection is solved by retrieving only those domain theories that had been used to explain the matched portions of the base situation. Each behavioral state indicates what

---

3. SME is a flexible analogical matching system motivated by Gentner's (1983) *structure-mapping theory* of analogy. It may be configured to model a number of different theories of analogical mapping and is discussed further in Subsection 3.2.2.

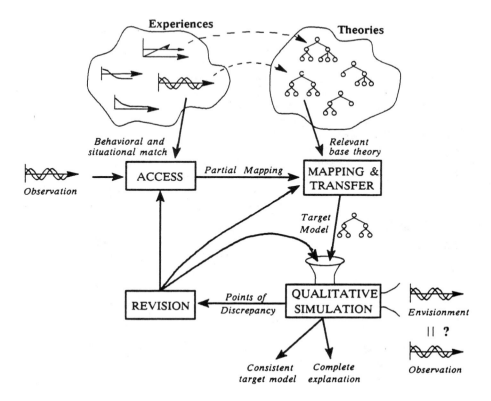

Figure 4. A functional decomposition of PHINEAS.

processes were active during that state. Thus, if the current observation matches only a subset of the states in the base observation, only those relevant process models are used.

The retrieved candidates are then ordered according to SME's evaluation score and are proposed one at a time as potential analogues on PHINEAS' global agenda.

The system's behavior can be clarified with an example. The caloric theory of heat, dominant during the eighteenth century, postulated a material heat substance called *caloric*. The temperature of an object was thought to be proportional to the amount of caloric present. Furthermore, caloric tended toward equilibrium, causing it to flow between bodies placed in contact until an equilibrium of their temperatures was

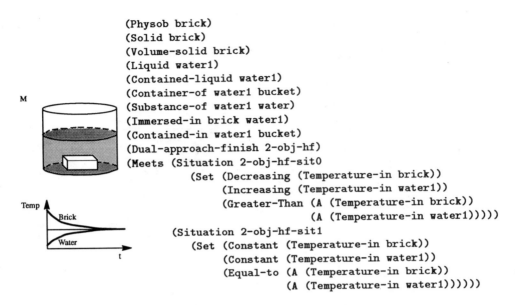

```
                (Physob brick)
                (Solid brick)
                (Volume-solid brick)
                (Liquid water1)
                (Contained-liquid water1)
    M           (Container-of water1 bucket)
                (Substance-of water1 water)
                (Immersed-in brick water1)
                (Contained-in water1 bucket)
                (Dual-approach-finish 2-obj-hf)
                (Meets (Situation 2-obj-hf-sit0
                        (Set (Decreasing (Temperature-in brick))
                             (Increasing (Temperature-in water1))
                             (Greater-Than (A (Temperature-in brick))
                                           (A (Temperature-in water1)))))
                    (Situation 2-obj-hf-sit1
                        (Set (Constant (Temperature-in brick))
                             (Constant (Temperature-in water1))
                             (Equal-to (A (Temperature-in brick))
                                       (A (Temperature-in water1))))))))
```

Figure 5. An unexplained thermal situation. When a hot brick is immersed
in cold water, the brick's temperature decreases and the water's
temperature increases. This transitions to a state in which the temperatures are constant and equal.

achieved. Let us consider how PHINEAS achieves a naive level of the
caloric view when it encounters thermal behavior for the first time.

The explanation task is illustrated in Figure 5. When a hot brick
is immersed in cold water, their temperatures asymptotically approach
each other until reaching equality. PHINEAS begins by searching memory for analogous behavior. First, the behavioral abstractions describing the observation are used to probe memory. In this case, **dual-
approach-finish** applies, which characterizes two quantities
asymptotically approaching each other and reaching equality. Only one
candidate analogue demonstrates this abstract behavior—two-container
liquid flow. This scenario describes liquid flowing from one container
(**beaker3**) to another (**vial2**), through a pipe (**pipe1**) connecting them.
Using SME to compare the current and recalled situations, PHINEAS
determines that the roles of the beaker and vial in the liquid flow description correspond to the roles of the brick and water in the thermal
situation, respectively. Additionally, it finds that **pressure** in the liquid
flow situation corresponds to **temperature** in the thermal situation.

3.2.2 THE MAPPING AND TRANSFER STAGE

The objective of the second stage is to generate an initial hypothesis about the current observation. This stage has two components, *mapping* and *transfer*.

Given a candidate analogue, PHINEAS retrieves the models used to explain analogous aspects of the recalled experience. Mapping serves to complete the initial set of correspondences (*matching*) and to propose *candidate inferences* sanctioned by those correspondences (*carryover*). The model of mapping used in this work is called *contextual structure-mapping* (Falkenhainer, 1988, 1989), a knowledge-intensive adaptation of Gentner's (1983, 1988) structure-mapping theory of analogy. It uses knowledge of the various contextual factors affecting analogical interpretation, such as the role of each element in the two analogue descriptions, to analyze the similarity between their descriptions. The mapping is constructed by SME_{CSM}, that is, by SME configured to run the rules of contextual structure-mapping.

An important observation made in contextual structure-mapping is that the correspondences and inferences proposed by the mapping stage may be incomplete. The *transfer* stage analyzes the results of the mapping stage to elaborate its correspondences and minimize unnecessary conjectures. This centers around two issues. First, if the candidate inference references an object needed in the base scenario that has no apparent target correspondent, one must find a corresponding target object or conjecture its existence. These unknown objects are represented by (:skolem *base-object*) and are called *skolem objects*.[4] Second, candidate inferences represent relevant base expressions having no apparent correspondent in the target. However, they may not be applicable to the target. Thus, the domain theory is consulted, and more detail about each candidate inference is sought. Alternate, analogous target expressions may be found or a new vocabulary (predicates) may be created. If new information is found, mapping is repeated to see how it affects the overall mapping. A *map and analyze* cycle may ensue.

Let us return to the caloric heat flow example. Upon completion of access, PHINEAS attempts to map the relevant liquid flow domain theory into the current thermal situation. First, the domain theory

---

4. The term *skolem object* derives from standard logical use of a *skolem constant* to denote the existence of an unknown object and enable removal of an existential quantifier.

Table 1. A mapping from the liquid flow process instance to the hot brick in cold water scenario, as generated by SME_{CSM}.

Match Hypotheses: pressure-in(beaker3) $\Longleftrightarrow$ temperature-in(brick)
                        pressure-in(vial2) $\Longleftrightarrow$ temperature-in(water1)
                        beaker3 $\Longleftrightarrow$ brick
                        vial2 $\Longleftrightarrow$ water1

Weight: 2.07

Candidate Inferences:

```
Contained-Fluid((:skolem cs-water-beaker), (:skolem water), brick)
   EFFECTS  Quantity[amount-of(:skolem cs-water-beaker)]
            Quantity[temperature(:skolem cs-water-beaker)]
            Qprop[temperature(:skolem cs-water-beaker),
                amount-of(:skolem cs-water-beaker)]
                •
                •
                •

Contained-Fluid((:skolem cs-water-vial), (:skolem water), water1)
   EFFECTS  Quantity[amount-of(:skolem cs-water-vial)]
            Quantity[temperature(:skolem cs-water-vial)]
            Qprop[temperature(:skolem cs-water-vial),
                amount-of(:skolem cs-water-vial)]
                •
                •
                •

Liquid-Flow((:skolem cs-water-vial), (:skolem water), water1)
   PRECONDITIONS  Liquid(:skolem water)
                  Can-contain(brick, (:skolem water))
                  Fluid-path(:skolem pipe1)
                  temperature-in(:skolem cs-water-beaker)
                    > temperature-in(:skolem cs-water-vial)
   EFFECTS  Quantity[flow-rate]
            flow-rate = temperature-in(:skolem cs-water-beaker)
                    - temperature-in(:skolem cs-water-vial)
                •
                •
                •
```

---

used to explain the two-container liquid flow experience is retrieved. This consists of the liquid flow process and two instantiations of

(Contained-Fluid *contained-fluid substance container*),

one for the beaker water and one for the vial water. SME is then invoked with knowledge of the partial mapping established during access, giving the results shown in Table 1. Its candidate inferences propose a new contained-fluid relationship, in which the temperature of the container (brick and water1) is proportional to the amount of substance it contains. This substance is currently unknown but is analogous to the water in the liquid flow situation. Additionally, a new process is proposed: When two objects of differing temperature are connected by a physical path, the unknown substance continuously flows from the object of higher temperature to the one of lower temperature, at a rate equal to their difference in temperatures.

The candidate inferences are next passed to the transfer stage. It first determines that none of the proposed expressions is inconsistent in their current state. Next, these inferences are inspected for the presence of skolem objects, and four are found: (:skolem cs-water-beaker), (:skolem cs-water-vial), (:skolem water), and (:skolem pipe1). The first two are compound objects (objects defined solely by their constituents) and are therefore ignored. The unknown (:skolem pipe1) indicates that no correspondent for the pipe connecting the beaker and vial was found. However, when PHINEAS is given the task of locating an object satisfying the conjunction

(Physical-Path brick water1 *?pipe*) ∧ (Fluid-Aligned *?pipe*) ∧
    (Fluid-path *?pipe*)

it finds that

(Physical-Path brick water1 (common-face brick water1))

and

(Fluid-Aligned (common-face brick water1))

are true in the current scenario and that the third conjunct can be assumed. Therefore, the system establishes (common-face brick water1) as the analogue for pipe. This demonstrates the utility of the transfer stage in filling out an incomplete analogical mapping. An analogy

will often evoke additional information or perspectives about the two
analogues made relevant by its consideration.

The remaining unknown, (:skolem water), indicates that no correspondent for the water flowing from beaker to vial was found. Additionally, no correspondent is found when an object satisfying the relevant
conditions is sought:

```
(Substance ?pipe) ∧ (Liquid ?pipe) ∧
(Can-Contain brick ?pipe) ∧ (Can-Contain water1 ?pipe)
```

However, when a new entity token is made for the missing water correspondent, a contradiction arises:

```
(Liquid sk-water-1) ∧ (Volume-Solid brick) ⇒
    ¬(Can-Contain brick sk-water-1)
```

As a result, (Liquid sk-water-1) is changed to (Phase-1 sk- water-1),
with Phase-1 added as a new kind of Phase. This illustrates PHINEAS'
ability to create new object tokens (sk-water-1) when it cannot resolve
a skolem object produced by mapping. Further, it is able to distinguish
between assuming the presence of an unobserved object and conjectur-
ing a theoretically novel entity. This is important information that it
can use in theory evaluation and selection.

At this point, the transfer task is completed, resulting in the model
shown in Table 2. This model postulates that the brick and water
each contain sk-water1-1 and that their temperatures are proportional to the amount they contain. Additionally, it proposes the new
Process-1, which might be called a *heat flow* process. This indicates
that sk-water1-1 will flow from the object of higher temperature to
the object of lower temperature. PHINEAS does not generalize beyond
replacing constants with variables, hence only the brick and water1
are believed to contain sk-water1-1.[5]

3.2.3 THE VERIFICATION STAGE

Verification-based analogical learning (Falkenhainer, 1986, 1988) depicts
analogical learning as an iterative process of hypothesis formation, verification, and revision, centered around the requirement to confirm ad-

---

5. Research on explanation-based learning has shown that this is not sufficient to
   ensure proper generalization (DeJong & Mooney, 1986; Mitchell, Keller, & Kedar-
   Cabelli, 1986). Explanation-based generalization might be performed at this
   point, but it has not been necessary so far.

Table 2. A final PHINEAS hypothesis explaining the behavior of a hot brick in cold water. This hypothesis was derived from the system's theories about liquid flowing between two containers.

```
(DEFPROCESS  (PROCESS-1 ?SUBST ?SOURCE ?SRC-CS ?DESTINATION
             ?DST-CS ?PATH)
  INDIVIDUALS  ((?SUBST :CONDITIONS (SUBSTANCE ?SUBST)
                                    (PHASE-1 ?SUBST))
               (?SOURCE :CONDITIONS (CAN-CONTAIN ?SOURCE ?SUBST))
               (?SRC-CS :CONDITIONS (CONTAINED-FLUID-1 ?SRC-CS
                                     ?SUBST ?SOURCE))
               (?DESTINATION :CONDITIONS
                   (CAN-CONTAIN ?DESTINATION ?SUBST))
               (?DST-CS :CONDITIONS (CONTAINED-FLUID-1 ?DST-CS
                                     ?SUBST ?DESTINATION))
               (?PATH :CONDITIONS (FLUID-PATH ?PATH)
                                  (PHYSICAL-PATH ?SOURCE
                                   ?DESTINATION ?PATH)))
  PRECONDITIONS  ((FLUID-ALIGNED ?PATH))
  QUANTITYCONDITIONS  ((GREATER-THAN (A (TEMPERATURE-IN ?SOURCE))
                                     (A (TEMPERATURE-IN ?DESTINATION)))
                       (GREATER-THAN (A (AMOUNT-OF ?SRC-CS)) ZERO))
  RELATIONS  ((QUANTITY (FLOW-RATE ?SELF))
              (Q= (FLOW-RATE ?SELF)
                  (- (TEMPERATURE-IN ?SOURCE)
                     (TEMPERATURE-IN ?DESTINATION)))
              (GREATER-THAN (A (FLOW-RATE ?SELF)) ZERO))
  INFLUENCES ((CTRANS (AMOUNT-OF ?SRC-CS) (AMOUNT-OF ?DST-CS)
                      (A (FLOW-RATE ?SELF)))))

(DEFENTITY  (CONTAINED-FLUID-1 ?V-1 ?V-2 ?V-3)
    (CONTAINER-OF ?V-1 ?V-3)
    (SUBSTANCE-OF ?V-1 ?V-2)
    (QUANTITY (AMOUNT-OF ?V-1))
    (QUANTITY (TEMPERATURE-IN ?V-3))
    (QPROP (TEMPERATURE-IN ?V-3) (AMOUNT-OF ?V-1)))

(ASSUME (SUBSTANCE SK-WATER-1))
(ASSUME (PHASE-1 SK-WATER-1))
(ASSUME (CAN-CONTAIN BRICK SK-WATER-1))
(ASSUME (CONTAINED-FLUID-1 SK-CS-WATER-BEAKER-1 SK-WATER-1 BRICK))
(ASSUME (CAN-CONTAIN WATER1 SK-WATER-1))
(ASSUME (CONTAINED-FLUID-1 SK-CS-WATER-VIAL-1 SK-WATER-1 WATER1))
(ASSUME (FLUID-PATH (COMMON-FACE BRICK WATER1)))
```

equacy of use in explaining a given phenomenon. In PHINEAS, it sanctions the use of gedanken experiments in the form of qualitative simulations to analyze the adequacy of proposed models. Specifically, the predictions of a proposed model are compared against the observed behavior, enabling the system to test the validity of the analogy and sanction refinements where the analogy is incorrect. The system generates an envisionment of the scenario, which it then compares with the original observation. If the envisionment is consistent and complete with respect to the observation, then the explanation is considered successful. If it is inconsistent or fails to provide complete coverage, then revision is aimed at the points of discrepancy.

PHINEAS produces envisonments of predicted behavior using Forbus' (1988) *qualitative process engine* (QPE). The process of comparing and identifying points of discrepancy between the predicted and observed behaviors is performed by DeCoste's (1989) *dynamic across-time measurement interpretation* system, DATMI.

Only one test remains in the ongoing caloric example, to verify the adequacy of the model in explaining the original observation. As shown in Figure 6, the model produces a five-state envisionment, with state S2 transitioning to state S0, demonstrating that the model is able to predict the observed temperature changes. In state S2, `Process-1` is active, the substance `sk-water-1` is flowing from the brick to the water, and the temperature of the brick is decreasing while the temperature of the water is increasing, each at a rate equal to the difference in their temperature. In state S0, the brick and water temperatures are equal, and all quantities are constant.

3.2.4   THE REVISION STAGE

If PHINEAS' initial hypothesis is inadequate, an attempt should be made to adapt it around points of inaccuracy. We advocate, but have not fully implemented, a model of revision that relies on past experiences to guide the formation and selection of revision hypotheses (Falkenhainer, 1988). It considers behavior analogous to the current anomaly and considers differences between the current anomalous situation and the prior situations that were consistently explained. This is the only component of PHINEAS that is not fully implemented.

Quantity	S2-I	S2	S0	S1-I	S1
Ds[FLOW-RATE(PI0)]	–	–	–	-1	-1
Ds[FLOW-RATE(PI1)]	-1	-1	–	–	–
Ds[AMOUNT-OF(SK-CS-WATER-BEAKER-1)]	-1	-1	0	1	1
Ds[AMOUNT-OF(SK-CS-WATER-VIAL-1)]	1	1	0	-1	-1
Ds[PRESSURE(SK-CS-WATER-BEAKER-1)]	-1	-1	0	1	1
Ds[PRESSURE(SK-CS-WATER-VIAL-1)]	1	1	0	-1	-1
Ds[TEMPERATURE-IN(BRICK)]	-1	-1	0	1	1
Ds[TEMPERATURE-IN(WATER1)]	1	1	0	-1	-1
A[AMOUNT-OF(SK-CS-WATER-BEAKER-1)]	>0	>0	>0	>0	=0
A[AMOUNT-OF(SK-CS-WATER-VIAL-1)]	=0	>0	>0	>0	>0
A[TEMPERATURE-IN(BRICK)] A[TEMPERATURE-IN(WATER1)]	>	>	=	<	<
ACTIVE(PI0)	F	F	F	T	T
ACTIVE(PI1)	T	T	F	F	F

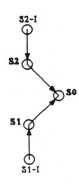

Processes:
```
  PI0:  PROCESS-1(SK-WATER-1 WATER1 SK-CS-WATER-VIAL-1 BRICK
SK-CS-WATER-BEAKER-1 (COMMON-FACE BRICK WATER1))
  PI1:  PROCESS-1(SK-WATER-1 BRICK SK-CS-WATER-BEAKER-1 WATER1
SK-CS-WATER-VIAL-1 (COMMON-FACE BRICK WATER1))
```

Figure 6. Envisionment produced by the hypothesized caloric model when applied to the brick immersed in water scenario. States are distinguished only by derivative and process values. They are split by QPE when this distinction produces a state lasting an interval of time (S2) and also lasting for an instant (S2-I).

3.3 Preference Criteria

PHINEAS is primarily concerned with *interpretation construction*—to find candidate explanations and the assumptions on which they rest. However, a system that exhaustively generated an unordered set of possible hypotheses would not be of much use. It should focus on the most promising explanations first and provide a preferential ordering on fully developed hypotheses. Correspondingly, PHINEAS incorporates two types of preference criteria; one influences the focus of problem-solving efforts and the other selects among competing, fully developed hypotheses.

The preceding sections presented PHINEAS as a sequential process concerned with the development of a single hypothesis. However, its operation is controlled by a task agenda that maintains multiple hypotheses in various stages of development. Eight task types are currently used, including access, mapping, transfer, and simulate. Repeatedly, the task–hypothesis pair at the front of the agenda is selected and executed, resulting in further development of its corresponding hypothesis. This task may in turn spawn other tasks, modify tasks waiting for execution, or signal the acceptance of a hypothesis, which halts the cycle.

The first type of preference criterion influences the ordering of tasks on PHINEAS's agenda and guides it toward developing the most promising hypotheses first. Each task is given a priority level, which induces a roughly depth-first behavior. In addition, the `mapping` and `transfer` tasks have an auxiliary score for sorting tasks within the same priority level. This auxiliary score is SME's evaluation metric for the match between the current observation and the task's associated base analogue. When determining which of two candidate analogues to consider next, PHINEAS selects the one with the higher similarity score. This metric supports the *similarity conjecture*—interpretation-construction tasks may be characterized as the search for maximal explanatory similarity between the situation being explained and some explainable scenario.

Once the system has formed a complete hypothesis (the output of the transfer task), it uses a second type of preference criterion. This criterion considers the characteristics of the hypothesis itself and enables selection among competing hypotheses. A complete account of theory selection requires consideration of many complex factors, such as a theory's plausibility, coherence, effect on prior beliefs, simplicity, and specificity in accounting for the phenomenon. Unfortunately, these are significant open research problems in their own right and are certainly beyond the scope of this chapter. However, a number of important, more specific preference criteria are readily available and have been found useful in PHINEAS for establishing preference between competing hypotheses. These are:

C_{CE} *Conjectured entities.* Does the hypothesis conjecture the existence of a novel kind of entity, and, if so, how many?

C_{VE} *Vocabulary extensions.* Does the hypothesis require the creation of new predicates, and, if so, how many?

C_{CA} *Composite assumptions.* Does the hypothesis conjecture the existence of new physical processes or new knowledge structures (e.g., schemas), and, if so, how many?

C_{AE} *Assumed entities.* Does the hypothesis assume the presence of a known type of entity not mentioned in the original scenario description, and, if so, how many?

C_{AA} *Atomic assumptions.* Does the hypothesis make additional assumptions about the properties and interrelationships of objects in
the scenario, and, if so, how many?

The single-preference criterion used to evaluate a hypothesis or compare two competing hypotheses is a function of these five metrics. The
method for combining them is adapted from Michalski (1983), who describes the use of a *lexicographic evaluation functional* (LEF) for evaluating alternate inductive concept descriptions. This approach specifies
a list of elementary criterion–tolerance pairs, in which each elementary
criterion is applied sequentially to prune the space of hypotheses. In
PHINEAS, the elementary preference criteria are ordered according to
an approximate measure of decreasing "cost":

$$\text{LEF} = (C_{CE}, \ C_{VE}, \ C_{CA}, \ C_{AE}, \ C_{AA}).$$

Thus, an explanation that postulates the existence of a novel kind of
entity (C_{CE}) is at all times deemed inferior to one that does not. Each
criterion returns a number $(N \geq 0)$ as described above, where a value
of zero indicates success and a value greater than zero indicates failure.
The function is used to select the most preferable explanation(s) from
a given set as follows: First, each proposed explanation is evaluated by
criterion C_{CE}, and those that pass C_{CE} are retained. The process is
repeated with the next criterion on the set of retained hypotheses until
only a single hypothesis remains or the list of criteria is exhausted.
If at any point all hypotheses evaluated by a particular criterion fail,
the process stops, and the current set is returned in increasing order
according to their score, N, for that criterion.

This evaluative function produces an interesting property when viewed
from the perspective of the four explanation scenarios described in Section 2:

1. *Deductive scenario.* Given phenomenon $\mathcal{P}$, where $\mathcal{P}$ represents a set
   of observables, a complete explanation of $\mathcal{P}$ deductively follows from
   existing knowledge. This corresponds to explanations passing every
   criterion. It occurs when all the antecedent features of the base are
   present in the target.

2. *Assumption scenario.* No explanation can be grounded with current
   knowledge because not all the relevant facts are known. However, a
   complete explanation follows from the union of existing knowledge
   and a consistent set of assumptions about the missing facts. This

corresponds to explanations passing every criterion but one of the last two, C_{AE} and C_{AA}. It occurs when some of the antecedent features of the base have no correspondent in the target, but may be consistently assumed to hold in the target.

3. *Generalization scenario.* Existing knowledge indicates that candidate explanation $\mathcal{E}$ cannot apply because condition C_1 is known to be false in the current situation. However, $\mathcal{E}$ does follow if condition C_1 is replaced by the next most general relation since C_1's sibling is true in the current situation. This corresponds to explanations passing the first two criteria, C_{CE} and C_{VE}, but failing C_{CA}, in which a knowledge structure is viewed as "new" if it represents a modification of an existing knowledge structure.[6] It occurs when some of the antecedent or consequent features of the base match an analogous set of features in the target, thus mapping the base theory to a situation beyond its declared scope.

4. *Analogy scenario.* No candidate explanation $\mathcal{E}$ is available directly, but explanation $\mathcal{E}_b$ is available if a series of analogical assumptions are made, that is, if the situation explained by $\mathcal{E}_b$ is assumed analogous to the current situation. This corresponds to explanations failing one of the first three criteria, C_{CE}, C_{VE}, or C_{CA}. It occurs when some of the features of the base match an analogous set of features in the target or when new vocabulary must be created to complete the mapping.

All four scenarios arise as a result of the same basic mechanism. The evaluative function causes PHINEAS to propose standard, deductive explanations if any are found. In their absence, conventional abductive explanations will be preferred. If existing theories are insufficient to provide an explanation, explanations adapting knowledge of potentially analogous phenomena will be offered. By using similarity as the single source for explanation generation, PHINEAS is able to offer a "best guess" in the presence of an imperfect or incomplete domain theory.

4.   Examples of PHINEAS' Behavior

The previous section focused on PHINEAS' explanation of heat flow using a cross-domain analogy with liquid flow. This section describes

---

6. The issue of whether to actually create a new knowledge structure or modify the existing one is an important but orthogonal issue. Here we are concerned with hypothesis evaluation rather than storage of an accepted hypothesis.

examples that begin to blur the distinction between analogous phenomena and identical phenomena. In each example, PHINEAS initially begins with knowledge of nine processes—liquid flow, liquid drain (to constantly empty an ideal sink), heat flow, boiling, heat-replenish (e.g., to constantly maintain the heat of a stove), dissolve, osmosis, linear motion, and spring-applied force. The section closes with a discussion of what PHINEAS' behavior indicates about the utility of the proposed analogical model of explanation.

4.1   Oscillation

Oscillation is a common phenomenon in physical systems. PHINEAS' initial knowledge contains theories about a prototypical spring-mass system, in which a spring is anchored to a wall on one end and attached to a mobile mass on the other. If the block is pulled and then released, it will oscillate back and forth forever.[7] Drawing from this knowledge, PHINEAS is able to explain several examples of simple harmonic motion, such as an induction-capacitance (LC) circuit and a cantilever pendulum. Here we consider the behavior of a torsion oscillator.

PHINEAS is initially given a description of a disk rotating while suspended by a rubber rod, and the disk's sinusoidal behavior is represented as a cycle of eight qualitatively described temporal intervals (Figure 7). Each interval contains facts describing the derivatives and amounts of angle, and angular velocity. In addition, it is told that the disk is a rotating object and the rod is a twisting object.

When PHINEAS probes memory for prior experiences with sinusoidal oscillation, it finds the spring-mass system. A detailed comparison of this behavior and that of the rotating disk reveals a correspondence between the eight behavioral states of each system. This correspondence indicates that the compressing spring corresponds to the twisting rod and the translating block corresponds to the rotating disk. Additionally, position is mapped to angle, and velocity is mapped to angular velocity, due to their similar behavior.

With a behavioral correspondence established, PHINEAS fetches the domain theory used to explain the spring-mass system. This consists of a Force process that applies the spring's force to the attached block, a

---

7. Modeling friction and resistance in oscillators is a difficult problem in QP theory. Ideal, frictionless oscillators are discussed throughout this section.

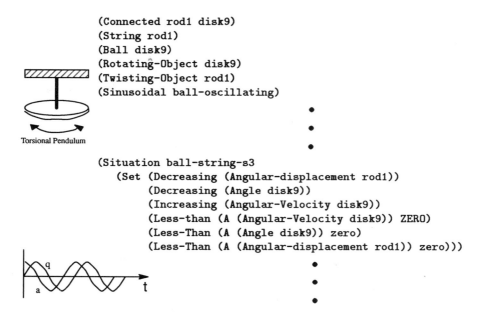

```
(Connected rod1 disk9)
(String rod1)
(Ball disk9)
(Rotating-Object disk9)
(Twisting-Object rod1)
(Sinusoidal ball-oscillating)
                                              •
                                              •
                                              •
(Situation ball-string-s3
    (Set (Decreasing (Angular-displacement rod1))
         (Decreasing (Angle disk9))
         (Increasing (Angular-Velocity disk9))
         (Less-than (A (Angular-Velocity disk9)) ZERO)
         (Less-Than (A (Angle disk9)) zero)
         (Less-Than (A (Angular-displacement rod1)) zero)))
                                              •
                                              •
                                              •
```

Torsional Pendulum

Figure 7. A torsional oscillator and its behavior when the disk is rotated and then released.

spring-mass-system object definition describing the system's total energy and the relationship between the block's position and the spring's displacement, and a **spring** object definition describing its restorative force as a function of displacement. When the spring-mass theory is mapped into the oscillating disk situation, transfer first examines each relation and finds no inconsistencies. The transfer phase next checks for skolem objects in the candidate inference and finds (:skolem sm-sys). The symbol **sm-sys** is a token that represents the spring-mass system taken as a whole. This compound object token is replaced by **sk-sm-sys-23**, which represents the newly defined rod-disk system:

```
(spring-mass-system-22 sk-sm-sys-23 rod1 disk9)
```

The proposed model of the rotating disk scenario is now usable. When the model is applied to the disk–rod pair, it produces an envisionment containing an eight-state cycle, as shown in Figure 8. When PHINEAS examines the envisionment, it finds a perfect match between the observed and predicted behavior. Thus, the model is adequate and the explanation process is completed.

•
•
•

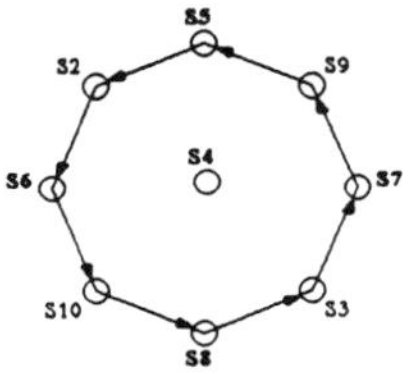

In behavioral segment 3
    (Angular-Displacement rod1) is Decreasing
    (Angle disk9) is Decreasing
    (Angular-Velocity disk9) is Increasing
    (A (angular-velocity disk9)) is Less Than zero
    (A (angle disk9)) is Less Than zero
    (A (angular-displacement rod1)) is Less Than zero

Due to the following processes being active:
FORCE-PROCESS(ROD1 RESTORATIVE-FORCE
                DISK9 ANGULAR-VELOCITY)

•
•
•

Quantity	S5	S2	S6	S10	S8	S3	S7	S9	S4
Ds[ANGLE(DISK9)]	-1	-1	-1	0	1	1	1	0	0
Ds[ANGULAR-DISPLACEMENT(ROD1)]	-1	-1	-1	0	1	1	1	0	0
Ds[ANGULAR-VELOCITY(DISK9)]	-1	0	1	1	1	0	-1	-1	0
Ds[KINETIC-ENERGY(DISK9)]	1	0	-1	0	1	0	-1	0	0
Ds[POTENTIAL-ENERGY(ROD1)]	-1	0	1	0	-1	0	1	0	0
Ds[RESTORATIVE-FORCE(ROD1)]	1	1	1	0	-1	-1	-1	0	0
ACTIVE(PI0)	T	T	T	T	T	T	T	T	T
ACTIVE(PI1)	T	T	T	T	T	T	T	T	T
A[RESTORATIVE-FORCE(ROD1)]	<0	=0	>0	>0	>0	=0	<0	<0	=0
A[ANGULAR-VELOCITY(DISK9)]	<0	<0	<0	=0	>0	>0	>0	=0	=0

Processes:

PI0: FORCE-PROCESS(ROD1,DISK9)
PI1: DERIVATIVE-PROCESS(ANGLE(DISK9),ANGULAR-
VELOCITY(DISK9))

Figure 8. Complete envisionment produced by the hypothesized torsional os-
           cillator model.

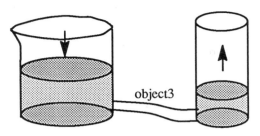

Figure 9. A beaker and a vial, each containing water, are connected by object3. What is causing the water in the beaker to decrease while the water in the vial is increasing?

4.2   Liquid Flow

Consider the scenario illustrated in Figure 9, in which a beaker and a vial, each containing water, are connected by object3. In this situation, the water in the beaker decreases while the water in the vial increases. To explain this phenomenon, PHINEAS begins by probing memory for the best set of candidate analogues. It finds four initial possibilities: `two-container-liquid-flow` (score = 28.07), `leaky-container` (score = 15.77), `dissolving` (score = 14.24), and `boiling` (score = 14.14).

Examining each of these possibilities, PHINEAS finds that not only is the `two-container-liquid-flow` scenario most similar to the current situation, it is the only candidate that produces a consistent set of predictions. Thus, the system concludes with the single assumption (`Fluid-Path object3`), which is sufficient to completely explain the observed behavior. PHINEAS' explanation is shown in Table 3. Under this assumption, the situation is viewed as a normal instance of liquid flow, with `object3` serving as the fluid path. The conclusion that this is an instance of liquid flow arises because PHINEAS' model of liquid flow mapped to the current scenario without change, rather than because the liquid flow process was instantiated by deduction.

5.   Discussion

This chapter has described a unified, similarity-driven method for explanation that seeks the best match between an observation to be explained and understood phenomena. This enables explanation from existing

Table 3.   PHINEAS' analysis of the observation that the amount of water in the beaker is decreasing and the amount of water in the vial is increasing. This analysis requires only a single assumption: `object3` is a fluid path.

```
Hypotheses for theory OBJECT3-FLOW-THEORY-5 derived from
2-CONTAINER-LF:

    (ASSUME (FLUID-PATH OBJECT3))

Analysis of OBJECT3-LF according to theory OBJECT3-FLOW-THEORY-5

    In behavioral segment 1
        (PRESSURE-IN VIAL6) is Increasing
        (AMOUNT-OF CS-WATER-VIAL1) is Increasing
        (PRESSURE-IN BEAKER6) is Decreasing
        (AMOUNT-OF CS-WATER-BEAKER1) is Decreasing
        (A (PRESSURE-IN BEAKER6)) is Greater Than
            (A (PRESSURE-IN VIAL6))

        Due to the following processes being active:
            LIQUID-FLOW(WATER BEAKER6 CS-WATER-BEAKER1 VIAL6
                CS-WATER-VIAL1 OBJECT3)

    In behavioral segment 2
        (PRESSURE-IN VIAL6) is Constant
        (AMOUNT-OF CS-WATER-VIAL1) is Constant
        (PRESSURE-IN BEAKER6) is Constant
        (AMOUNT-OF CS-WATER-BEAKER1) is Constant
        (A (PRESSURE-IN BEAKER6)) is Equal To
            (A (PRESSURE-IN VIAL6))

        There are no processes active.
```

theories if possible and theory formation or revision when necessary. Importantly, all explanations are formed with a single mechanism, with distinctions among deductive, abductive, and novel analogical explanations arising out of the evaluation process. Initial viability of the method has been demonstrated by PHINEAS on a variety of complex examples from several domains.

This work may be viewed as addressing problems in abduction. Traditional abduction systems reason from a fixed set of theories. However, most concepts, particularly when a theory is developing, do not lend themselves to precise, intensionally defined theories whose boundaries are perfectly specified by a set of necessary and sufficient conditions. Thus, in this work the underlying domain theory is assumed to be imperfect. Reasoning from similarities enables adaptation of the underlying domain theory when needed to explain observed phenomena.

This work may also be viewed as addressing problems in theory formation, in that it provides a way of constraining the set of possible revisions or extensions to existing theories. Reasoning from similarities suggests which theories are relevant and thus are candidates for revision. Also, it enables existing knowledge, possibly of other domains, to influence hypothesis generation and evaluation. It takes into account knowledge of the way things normally behave in the world and the ways theories about those behaviors are normally expressed.

This section briefly evaluates the viability of the approach, reviews related approaches to explanation, and closes the chapter with a discussion of plans for future research.

5.1   Viability of the Model

PHINEAS has been tested on over a dozen examples representing variations on a set of nine basic explanation tasks. In addition to those discussed previously, these examples include explanations of evaporation by analogy to boiling, liquid flow, and dissolving; osmosis by analogy to liquid flow; and floating of a balloon by analogy to an object floating in water.

5.1.1 SUCCESSES OF THE APPROACH

One of the goals of this work has been to show the feasibility of similarity as a single mechanism for both analogical and more traditional explanations involving deduction or abduction. In PHINEAS, the distinctions between deductive, abductive, and analogical explanations arise as an emergent result of the evaluation process. This offers an elegant, general alternative to special case solutions for theory inadequacy problems, such as generalizing a theory's preconditions or developing novel theories through cross-domain analogies. In the examples presented, PHINEAS was shown to explain a simple instance of liquid flow, apply linear oscillation concepts to an angular case, and develop a new "caloric" model of heat flow by analogy to liquid flow.

The primary region of flexibility and power corresponds to what I termed the "generalization scenario" (also classifiable as a form of within-domain analogy). PHINEAS adapts very well to situations close to, but not included in, the stated applicability boundaries of existing theories. An example of this is the mapping of liquid flow through a pipe to liquid flow through an open conduit. Informal experiments have shown that PHINEAS' behavior degrades smoothly with "analogical distance," the degree to which a distant, cross-domain analogy is required.

Due to attempts to find alternative justifications for unsupported dependencies, many of the within-domain analogies examined possess an another interesting characteristic. A new explanation may fall within the scope of existing knowledge, but the relevant explanation schema (i.e., QP theory process definition) is conservatively associated with a more restricted set of scenarios than may be applicable. This would correspond to the often observed comment "I never thought of it working for such a case, but I can see why it should." In other words, one often has the knowledge to safely extend a theory beyond its preconceived boundaries. When viewed from the perspective of explanation-based generalization (DeJong & Mooney, 1986; Mitchell, Keller, & Kedar-Cabelli, 1986), PHINEAS is able to apply a compiled explanation schema by adapting it to fit the current situation, rather than being forced to abandon the explanation schema and solve the current, similar explanation task from scratch. This view is similar in spirit to SWALE (Kass, 1986) and is examined further in Falkenhainer (1989).

5.1.2 LIMITATIONS OF THE APPROACH

PHINEAS falls short of a complete model of explanation in several ways. First, it will always produce a conjecture, no matter how weak, unless it cannot find a candidate analogue to initiate the explanation process or cannot form a hypothesis that is consistent with the observation. This is part of the intended design, and manifests itself in PHINEAS' ability and willingness to generalize a theory in response to an unanticipated observation. However, this is a two-edged sword; violations of existing knowledge are used to indicate new phenomena rather than to indicate false hypotheses. Although the preference criteria will ensure that such hypotheses are selected only if nothing better exists, an explanation system should also be able to know when it lacks knowledge. The framework needs an evaluative measure that takes into account the cost of overthrowing prior beliefs for the benefits of a more coherent belief state. Additional factors, such as plausibility and specificity in accounting for the phenomenon, are required as well.

PHINEAS also differs from more traditional abduction methods in its inability to recognize simultaneous instances of the same phenomenon. Extending the example of liquid flow between two containers in the previous section to "three-container liquid flow" demonstrates this limitation. Three containers connected in series (`can1` to `can2` by `pipe12` and `can2` to `can3` by `pipe23`) produce two simultaneous instantiations of the liquid flow process. When PHINEAS is given a description of the scenario, it finds two different analogies with the potential analogue, "two-container liquid flow." One analogy is with the `can2` to `can1` flow; the other is with the `can2` to `can3` flow. These are processed as two independent candidate explanations. Additionally, they are both adequate, since QPE fortuitously applies the proposed liquid flow model (intended for the `can2` to `can1` pair, for example) to both container pairs. Thus, the system concludes with two consistent, functionally equivalent explanations. It does not possess the knowledge that a single phenomenon was simply occurring twice. This is an important problem that research in analogy has yet to address.

Finally, PHINEAS lacks composability and flexibility to situations that do not fit prepackaged patterns. This is due in part to its inability to merge multiple analogies when forming candidate explanations, as described by Burstein (1983). It is also due to the granularity of theories considered. Schema-application approaches to explanation typically do not adapt as well as rule-chaining systems to novel configurations.

5.2   Related Approaches to Explanation

Although explanation systems differ along many dimensions, two aspects seem particularly relevant in forming class-wide comparisons with PHINEAS. First, knowledge content and use range from rule chaining and schema application to extensionally defined instance-based models. Second, explanation systems differ in how they treat lack of knowledge about the domain or lack of knowledge about the scenario to be explained. This subsection briefly reviews selected approaches to explanation along these two dimensions.

The traditional model of explanation in artificial intelligence depicts a knowledge-rich process that draws inferences from general, intensionally defined domain knowledge in the form of rules or schemas. This inference process further requires a match of antecedent information that is *complete* (all required features are present) and *exact* (each required feature is present in its prespecified form) to enable inference chaining. These characteristics lead to brittleness due to the lack of exact or complete matches to the real world and the need to anticipate all future scenarios. There have been attempts to remedy this situation. For example, Anderson's (1983) ACT system allows partial matching of antecedents using activation levels to control production-rule firing. Alternatively, probabilistic or default reasoning models enable inference in the presence of incomplete knowledge (e.g., Josephson, Chandrasekaran, Smith, & Tanner, 1987; Pearl, 1987). However, these systems still require exact matches of antecedent features and are thus insensitive to the presence of analogous but syntactically distinct features. These limitations are addressed by PHINEAS' analogy mechanism. It enables matching of analogous rather than identical features, reduces the need to have a precisely defined set of necessary and sufficient conditions for each theory, and enables knowledge of a familiar domain to aid reasoning about another domain.

This work shares much of the philosophy behind case-based reasoning, which uses similar past problem solving experiences to solve new cases (Hammond, 1989; Kolodner, Simpson, & Sycara-Cyranski, 1985). Systems in this paradigm have predominately been knowledge weak, with "match strength" used as a basis for believing that the current

and retrieved cases share common principles. More recently, attempts
have been made to include knowledge in the process, both in matching equivalent yet nonidentical features (Bareiss, Porter, & Wier, 1987)
and in subsequent testing and transformation of the initial hypothesis through examination of deeper domain knowledge (Kass, Leake, &
Owens, 1986; Simmons & Davis, 1987). Both processes are found in
PHINEAS, in which the domain and the concern with across-domain
analogies required more sophisticated representations and a more sophisticated notion of analogical similarity. Further, it required a deep
causal analysis of the consistency of a hypothesis, both internally and
with respect to the observation. Finally, neither the case-based or the
traditional knowledge-intensive models tend to address problems in theory formation, such as anticipated yet unknown objects or the creation
of new terms (e.g., postulating intrinsic properties of objects).[8]

Explanation systems also differ in their reaction to gaps in available
knowledge. As in PHINEAS, most explanation systems can offer explanations in the presence of incomplete knowledge about the scenario to
be explained. Probabilistic approaches (Buchanan & Shortliffe, 1984;
Josephson, Chandrasekaran, Smith, & Tanner, 1987; Pearl, 1987) examine a priori probabilities assigned to antecedent information. When addressing open-ended, common-sense problems about the world, having
such probabilities seems unrealistic. PHINEAS follows work in interpretation and story understanding (e.g., Charniak, 1988; DeJong, 1982),
that tends to use schema-based models and identify the assumables as
the unknown elements of a relevant and consistent schema. More work
is needed to better understand what can be assumed and when.

Explanation systems rarely address a second type of knowledge gap—
lack of applicable knowledge about the domain. However, there are a
few exceptions in addition to PHINEAS. Pazzani's (1987) OCCAM can
infer new causal rules by using knowledge of abstract patterns of causality (e.g., temporally and spatially connected events), and Rajamoney's
(this volume) COAST revises existing theories primarily through experimentation. Falkenhainer and Rajamoney (1988) show how COAST
and PHINEAS have been integrated, with similarity-driven explanation
providing focus and experimentation providing empirical testing of hypotheses. O'Rorke et al. (this volume) introduce explicit metatheoret-

---

8. See Karp (this volume), O'Rorke et al. (this volume), and Rajamoney (this
   volume) for alternative approaches to theory formation and explanation.

ical rules that elegantly give the effect of extending the basic notion of abduction to include assumption of new causal rules. However, that work is still in progress and has not yet addressed problems associated with focusing this process or relating elements of a developing theory to existing theories.

5.3   Directions for Future Research

The problem of retrieving a plausibly useful analogue from memory still stands as the least understood, most important unsolved problem in analogy. Some important progress has been made (Hammond, 1989; Kolodner, 1984), but models of access are still limited by simple representations and specialized forms of within-domain analogy. The two-stage mechanism described in this chapter (first use abstractions to focus on a candidate set, then use structural comparison to prune and order this set) sidesteps important issues. How are these abstractions formed for the stored situations? How are they recognized in the target situation? How are they organized so that an excessive number of analogues are not retrieved?

Composability is a fundamental requirement for any model of explanation. However, in its current form PHINEAS relies on a single analogous explanation structure to explain each new observation. Two capabilities are needed to address this limitation. First, the ability to draw from multiple sources of knowledge is required, as in Burstein's (1983) work on multiple analogies and their composition. Second, a theory revision ability is needed to let PHINEAS repair initial hypotheses that provide incomplete or inconsistent explanations. The two capabilities must interact, since an explanation's inadequacy may arise from an incomplete theory, which requires retrieval of additional knowledge, or a slightly incorrect theory, which requires modification of its components.

PHINEAS and most analogy systems built to date use analogy as their sole learning method. However, analogy, like any other single learning mechanism, is best viewed as a single component in a synergistic cooperation of learning methods. In addition to analogical inference, learning and explanation may be accomplished through sufficient knowledge of unexplained components (Hall, 1989) or abstract patterns of causality (Pazzani, 1987). In scientific investigation, an analogically derived hypothesis may suddenly "come to mind." However, this *flash of insight* may have been preceded by a tedious, incremental process in which data were collected and analyzed, patterns sought, and overall familiarity increased (Langley & Jones, 1988). In order to build a general investigative system, we must integrate analogy with directed experimentation, empirical learning, and analytic learning. Some work on developing a general protocol enabling such interaction has already begun (Falkenhainer & Rajamoney, 1988). However, the protocol leaves many questions unanswered, such as how to take advantage of prior problem solving and trend detection and how to integrate the results of analogy into memory.

Acknowledgements

This work has benefited from discussions with Ken Forbus, Dedre Gentner, John Collins, and Dennis DeCoste. Jeff Shrager and Pat Langley provided greatly appreciated writing assistance. Some of PHINEAS' components were written by Ken Forbus (QPE), Johan DeKleer (ATMS and portions of QPE's rule system), and Dennis DeCoste (DATMI).

This chapter describes work conducted as part of the author's doctoral disseration at the University of Illinois, Department of Computer Science. Support was provided through an IBM Graduate Fellowship and the Office of Naval Research, Contract No. N00014-85-K-0559.

References

Anderson, J. R. (1983). *The architecture of cognition.* Cambridge, MA: Harvard University Press.

Bareiss, E. R., Porter, B. W., & Wier, C. C. (1987). PROTOS: An exemplar-based learning apprentice. *Proceedings of the Fourth International Workshop on Machine Learning* (pp. 12–23). Irvine, CA: Morgan Kaufmann.

Bobrow, D. (Ed.). (1985). *Qualitative reasoning about physical systems.* Cambridge, MA: MIT Press.

Buchanan, B. G., & Shortliffe, E. H. (1984). *Rule-based expert systems: The MYCIN experiments of the Standord heuristic programming project.* Reading, MA: Addison-Wesley.

Burstein, M. (1983). Concept formation by incremental analogical reasoning and debugging. *Proceedings of the Second International Workshop on Machine Learning.* Monticello, IL. (Revised version appears in R. S. Michalski, J. G. Carbonell, & T. M. Mitchell (Eds.), *Machine learning: An artificial intelligence approach* Vol. 2. San Mateo, CA: Morgan Kaufmann, 1986).

Charniak, E. (1972). *Towards a model of children's story comprehension.* Doctoral dissertation, Laboratory for Artificial Intelligence, Massachussetts Institutue of Technology, Cambridge, MA.

Charniak, E. (1988). Motivation analysis, abductive unification, and nonmonotonic equality. *Artificial Intelligence, 34,* 275–295.

DeCoste, D. (1989). *Dynamic across-time measurement interpretation: Maintaining qualitative understandings of physical system behavior.* Master's thesis, Department of Computer Science, University of Illinois at Urbana-Champaign.

DeJong, G. (1982). An overview of the FRUMP system. In W. Lehnert & M. Ringle (Eds.), *Strategies for natural language processing.* Hillsdale, NJ: Lawrence Erlbaum.

DeJong, G., & Mooney, R. (1986). Explanation-based learning: An alternative view. *Machine Learning, 1,* 145–176.

Falkenhainer, B. (1986). *An examination of the third stage in the analogy process: Verification-based analogical learning* (Technical Report UIUCDCS-R-86-1302). Urbana: University of Illinois, Department of Computer Science. *Proceedings of the Tenth International Joint Conference on Artificial Intelligence.*

Falkenhainer, B. (1988). *Learning from physical analogies: A study in analogy and the explanation process.* Doctoral dissertation, Department of Computer Science, University of Illinois at Urbana-Champaign.

Falkenhainer, B. (1989). *Contextual structure-mapping.* SSL Technical Report, Xerox PARC, 1989.

Falkenhainer, B., Forbus, K. D., & Gentner, D. (1986). The structure-mapping engine. *Proceedings of the Fifth National Conference on Artificial Intelligence* (pp. 272–277). Philadelphia: Morgan Kaufmann.

Falkenhainer, B., Forbus, K. D., & Gentner, D. (1989). The structure-mapping engine: Algorithm and examples. *Artificial Intelligence*, *41*, 1–63.

Falkenhainer, B., & Rajamoney, S. (1988). The interdependencies of theory formation, revision, and experimentation. *Proceedings of the Fifth International Conference on Machine Learning* (pp. 353–366). Ann Arbor, MI: Morgan Kaufmann.

Forbus, K. D. (1984). Qualitative process theory. *Artificial Intelligence*, *24*, 85–168.

Forbus, K. D. (1986). Interpreting measurements of physical systems. *Proceedings of the Fith National Conference on Artificial Intelligence* (pp. 113–117). Philadelphia: Morgan Kaufman.

Forbus, K. D. (1988). The qualitative process engine, a study in assumption-based truth maintenance. *International Journal for Artificial Intelligence in Engineering*, *3*, 200–215.

Gentner, D. (1983). Structure-mapping: A theoretical framework for analogy. *Cognitive Science*, *7*, 155–170.

Gentner, D. (1988). Mechanisms of analogical learning. In S. Vosniadou & A. Ortony (Eds.), *Similarity and analogical reasoning*. London: Cambridge University Press.

Hall, R. J. (1989). Learning by failing to explain: Using partial explanations to learn in incomplete or intractable domains. *Machine Learning*, *3*, 45–77.

Hayes, P. J. (1979). The naive physics manifesto. In D. Michie (Ed.), *Expert systems in the micro-electronic age*. Edinburgh: Edinburgh University Press.

Hayes-Roth, F., & McDermott, J. (1978). An interference matching technique for inducing abstractions. *Communications of the ACM*, *21*, 401–411.

Josephson, J. R., Chandrasekaran, B., Smith, J. W., & Tanner, M. C. (1987). A mechanism for forming composite explanatory hypotheses. *IEEE Transactions on Systems, Man, and Cybernetics*, *17*, 445–454.

Jones, R., & Langley, P. (1988). A theory of scientific problem solving. *Proceedings of the Tenth Meeting of the Cognitive Science Society*, (pp. 244–250). Montreal: Lawrence Erlbaum.

Kass, A. (1986). Modifying explanations to understand stores. *Proceedings of the Eighth Meeting of the Cognitive Science Society* (pp. 691–696).

Kass, A., Leake, D., & Owens, C. (1986). SWALE, A program that explains. In R. Schank (Ed.), *Explanation patterns: Understanding mechanically and creatively.* Hillsdale, NJ: Lawrence Erlbaum.

Kolodner, J. (1984). *Retrieval and organizatinal strategies in conceptual meory: A computer model.* Hillsdale, NJ: Lawrence Erlbaum.

Kolodner, J., Simpson, R. L., & Sycara-Cyranski, K. (1985). A process model of case-based reasoning in problem solving. *Proceedings of the Ninth International Joint Conference on Artificial Intelligence.* (pp. 284–290). Los Angles, CA: Morgan Kaufmann.

Leatherdale, W. H. (1974). *The role of analogy, model and metaphor in science.* Amsterdam: North-Holland.

Michalski, R. S. (1983). A theory and methodology of inductive learning. In R. S. Michalski, J. G. Carbonell, & T. M. Mitchell (Eds.), *Machine learning: An artificial intelligence approach.* San Mateo, CA: Morgan Kaufmann.

Mitchell, T. M., Keller, R. M., & Kedar-Cabelli, S. T. (1986). Explanation-based generalization: A unifying view. *Machine Learning, 1,* 47–80.

Mooney, R. (1987). *A general explanation-based learning mechanism and its application to narrative understanding.* Doctoral dissertation, Department of Computer Science, University of Illinois at Urbana-Champaign.

Pazzani, M. J. (1987). Inducing causal and social theories: A prerequisite for explanation-based learning. *Proceedings of the Fourth International Workshop on Machine Learning* (pp. 230–241). Irvine, CA: Morgan Kaufmann.

Pearl, J. (1987). Embracing causality in formal reasoning. *Proceedings of the Sixth National Conference on Artificial Intelligence* (pp. 369–373). Seattle, WA: Morgan Kaufmann.

Pople, H. (1973). On the mechanization of abductive logic. *Proceedings of the Third International Joint Conference on Artificial Intelligence* (pp. 147–152). Stanford, CA: Morgan Kaufmann.

Reggia, J. A. (1983). Diagnostic expert systems based on a set covering model. *International Journal of Man-Machine Studies, 19,* 437–460.

Simmons, R., & Davis, R. (1987). Generate, test and debug: Combining associational rules and causal models. *Proceedings of the Tenth International Joint Conference on Artificial Intelligence* (pp. 1071–1078). Milan, Italy: Morgan Kaufmann.

Smith, R., Winston, P., Mitchell, T. M., & Buchanan, B. G. (1985). Representation and use of explicit justification for knowledge base refinement. *Proceedings of the Ninth International Joint Conference on Artificial Intelligence* (pp. 673–680). Los Angeles, CA: Morgan Kaufmann.

Winston, P. (1975). Learning structural descriptions from examples. In P. Winston (Ed.), *The psychology of computer vision.* New York: McGraw-Hill.

CHAPTER 7

Theory Formation by Abduction: A Case Study Based on the Chemical Revolution

PAUL O'RORKE

STEVEN MORRIS

DAVID SCHULENBURG

1.   World Model Revision

Much of the recent progress in AI has been characterized by the slogan *in the knowledge lies the power* (Feigenbaum, 1979). The performance of AI systems has become more impressive as the systems have become increasingly knowledge intensive. Unfortunately, knowledge-intensive systems are bound to exhibit imperfect behavior when they are based on imperfect knowledge. Existing AI systems often exhibit fragility or brittleness as a result. For this reason, revision methods are needed to correct and extend knowledge that is incorrect or incomplete.[1]

"Routine belief revision" methods already exist as a result of progress in AI research. Automated reasoning systems have used methods such as contradiction backtracing (Shapiro, 1981), dependency-directed backtracking, and truth maintenance (Doyle, 1979) in a large number of tasks requiring simple changes in systems of beliefs.

We all make simple changes in beliefs during everyday life, but dramatic changes in systems of beliefs such as occur in scientific revolutions appear to require extraordinary creative genius. This sort of "world model revision" is at the more difficult, more creative end of the spec-

---

1. In this chapter, *knowledge base, model, theory,* and *belief system* are used as roughly interchangeable terms since knowledge bases can be viewed as models, theories, or sets of beliefs.

trum of belief revision problems.[2] Great changes in our way of looking at the world represent the height of human intellectual achievement and are identified with intellectual giants such as Galileo, Newton, Lavoisier, and Einstein.

Until recently, it has not been clear how to advance toward reasoning systems capable of world model revision. The claim of this chapter is that computational methods for *theory formation by abduction* can provide a basis for world model revision. Abduction is the process of constructing explanations (Charniak, 1988; Josephson, Chandrasekaran, Smith, & Tanner, 1987; Peirce, 1931–1958; Pople, 1973; Reggia, Nau, & Wang, 1983; Schank, 1986). This chapter focuses on a theory-driven form of abduction that can be used to derive explanations of anomalous observations, given rules and facts encoding a general theory and the details of a specific situation. If a prediction of a given theory contradicts an observation, the approach to revision advocated here involves explaining the observation in terms of basic principles. We claim that the process of finding an explanation can lead by abductive inference to new hypotheses that can form crucial parts of new theories.

To support this claim, we give a detailed description of a computer simulation viewing one of Lavoisier's key insights in the chemical revolution as an example of theory formation by abduction. In Section 2 we present some background on this particular world model revision. Then in Section 3 we show how advances in qualitative physics provide a language for representing some aspects of chemical processes in the form of rules and facts. In Section 4 we discuss our overall framework for theory revision. In particular, we discuss how abduction can provide a means for theory formation. In Section 5, we illustrate how our abduction method generates qualitative chemical explanations. These sections then put us in a position to show in Section 6 how an observation in conflict with the phlogiston theory can lead via abduction to hypotheses that correspond to a key insight that contributed to Lavoisier's revolutionary shift to an oxygen theory of combustion. Section 7 discusses related work, and Section 8 provides a brief summary and conclusion.

---

2. McDermott and Doyle (1980) first made the distinction between routine belief revision and world model reorganization. They also present a modal approach to the formalization of nonmonotonic reasoning and routine belief revision.

2.   The Chemical Revolution

Conant, Nash, Roller, and Roller (1957) argue that case studies of revolutionary advances in science can facilitate the understanding of science by nonscientists.   Cognitive scientists take this one step further and argue that case studies based on the history of science can be used to achieve a deeper understanding of the cognitive processes underlying scientific discovery (Langley, Simon, Bradshaw, & Zytkow, 1987; Thagard, 1988). One immediate aim of such case studies of scientific revolutions is to develop computational models of the evolution of specific scientific theories over time. However, the ultimate goal is not so much to capture individual case histories but to improve our understanding of how theory shifts occur.

In this chapter, we present some initial results of a case study of the chemical revolution—the replacement of the phlogiston theory by the oxygen theory (1775–1789). This particular theory shift has attracted a great deal of interest, partly because it occurred in the early days of chemistry when the theories and experiments were still close to common knowledge and everyday experience. In addition, a great deal is known about the chemical revolution because of detailed records left by the scientists involved and the large number of books and papers on the subject by historians and philosophers of science (Conant, 1957; Guerlac, 1961; Ihde, 1980; Thagard, in press).

Prior to the chemical revolution, the phlogiston theory of chemistry provided the predominant explanation for the processes of combustion and calcination. Under this theory, developed by the German chemist G. E. STAHL (1660–1734), it was thought that all combustible substances contained an element called *phlogiston*. Combustion was thought of as a flow of phlogiston from the combustible substances into the surrounding air. Calcination is an alchemist's term for the process of changing things to calx or powder by applying heat. It also applies to rusting (now called *oxidation*) and was thought of as a loss of phlogiston from metals.[3] The phlogiston theory thus predicts a decrease in the weight of combusting and calcining substances.

---

3. Calx is the ashy powder left after a metal or mineral has been calcined. It is the same as rust, a coating or film formed on metals by corrosion or calcination. The reddish rust that appears on iron exposed to air and moisture is the most familiar example.

Consequently, Lavoisier, the eighteenth century French chemist who was the chief protagonist in the chemical revolution, placed great importance on the observation that the weights of some substances increase when they undergo combustion and calcination. Just after this "augmentation" effect was demonstrated conclusively, Lavoisier deposited a sealed note on November 1, 1772, with the Secretary of the French Academy of Sciences:

> About eight days ago I discovered that sulfur in burning, far from losing weight, on the contrary, gains it; it is the same with phosphorus ... This discovery, which I have established by experiments, that I regard as decisive, has led me to think that what is observed in the combustion of sulfur and phosphorus may well take place in the case of all substances that gain in weight by combustion and calcination; and I am persuaded that the increase in weight of metallic calxes is due to the same cause. (Conant, 1957)

With the help of colleagues such as Joseph Priestly, Lavoisier went on to discover that (contrary to the century-old phlogiston theory) a gas contained in the atmosphere combines with burning combustibles and calcining metals. This gas was first isolated by heating "mercurius calcinatus" (red calx of mercury, now called red oxide of mercury) until the gas in the calx was liberated. Lavoisier named the new gas *oxygen*.

3. Representing Qualitative Chemical Knowledge

In this section, we show how advances in research in qualitative physics provide a language for describing some important ideas associated with the chemical revolution. First we present a qualitative process schema for combustion according to the phlogiston theory and then consider how some of the ideas associated with qualitative physics and with phlogiston can be encoded in terms of facts and rules. As described in Sections 5 and 6, this encoding enables our abduction method to construct explanations of observations involving changes in the weights of burning and calcinating substances.

Table 1. A qualitative process description of the phlogiston theory of combustion.

Process: combustion	Individual-view: complex-stuff
Individuals:	Individuals:
combustible a complex-stuff	complex a substance
phlogiston a simple-stuff	$\{S_i \mid S_i$ a substance$\}$
air a gas	Preconditions:
Preconditions:	components($\{S_i\}$, complex)
$\sim$wet(combustible)	QuantityConditions:
component-of(phlogiston, combustible)	$\forall\, S_j \in \{S_i\}$, A[amount-of-in($S_j$, complex)] $>$ ZERO
surrounds(air, combustible)	Relations:
QuantityConditions:	There is p $\in$ piece-of-stuff
A[temp(combustible)] $\geq$ A[flashpoint(combustible)]	made-of(p, complex)
A[amount-of-in(phlogiston, combustible)] $>$ ZERO	$\Sigma_{\{S_i\}}$ amount-of-in(S_i, p) $=$ amount(p)
A[amount-of-in(phlogiston, air)] $<$	
A[capacity-of-for(air, phlogiston)]	

Relations:

  Let combustion-rate be a quantity

  combustion-rate $\propto_+$ (A[capacity(air, phlogiston)] $-$ A[amount-of-in(phlogiston, air)])

  combustion-rate $\propto_+$ amount-of-in(phlogiston, combustible)

Influences:

  I $-$ (amount-of-in(phlogiston, combustible), A[combustion-rate])

  I $+$ (amount-of-in(phlogiston, air), A[combustion-rate])

3.1 A Qualitative Process Description of Combustion

Qualitative process (QP) theory (Forbus, 1984) provides a language for describing qualitative changes due to processes acting on quantities. Table 1 shows a QP representation of a fragment of STAHL's phlogiston theory. This representation is intended to capture the phlogiston theorist's notion that combustion is similar to a "flow" of phlogiston from a combustible substance to the surrounding air.

This framework represents processes as frames or schemata called *qualitative process descriptions*. These frames contain knowledge about the objects (*individuals*) involved in the process as well as specific knowledge about the process itself. Objects are also described using additional frames known as *individual views*.

The individuals slot of a QP schema specifies the objects associated with a process or individual view. In the qualitative description of *combustion*, the individuals include a piece of some combustible substance (such as a chunk of charcoal), a piece of phlogiston, and a volume of air. In the individual view of *complex-stuff*, the relevant individuals include some complex substance and a set of component substances.

Among other things, the preconditions slot of the combustion schema captures the phlogiston theorist's belief that only complex substances can burn. According to phlogiston theory, all combustibles are compounds containing the element *phlogiston*. The quantity-conditions state that, for combustion to occur, the combustible must be "hot enough," there must be some phlogiston in the combustible, and the surrounding air must not be "saturated" by phlogiston. The relations state that the combustion rate is qualitatively proportional to the remaining capacity of the air for phlogiston and to the phlogiston content of the combustible substance. The influences state that phlogiston is leaving the combustible and "escaping" into the air and that this flow is directly influenced by the *combustion-rate*.

The individual view of *complex-stuff* states that there is a set of substances, namely the components of the *complex-stuff*, and that the sum of the amount of each of these substances equals the amount of the piece of *complex-stuff*.

3.2 Aspects of the Phlogiston Theory

The qualitative process description of the phlogiston theory of combustion sketched in the previous section is intended to capture a number of inferences and explanations made by phlogiston theorists. In this section, we simplify the theory and identify a fragment relevant to predicting and explaining why the weight of charcoal decreases as it burns.

Table 2 presents two classes of qualitative laws that capture important aspects of the phlogiston theory. The first class (GL4, GL6, GL7, and GL8) is concerned with certain basic properties of substances. Rule GL4 states that the weight of any substance is proportional to the amount of the substance. Rules GL6, GL7, and GL8 state that the amount of a complex substance is equal to the sum of the amounts of its components.

The other class of laws captures certain aspects of the phlogiston theorist's view on the nature of combustion and calcination. In this view, all combustible substances are complex substances that contain phlogiston. In our qualitative process description of combustion, rule GL5a states that combustion is a process that negatively influences the amount of phlogiston in charcoal. That is, if combustion is active, it drives down the amount of phlogiston in a partially burned piece of charcoal. Similarly, Rule GL5b states that calcination drives down the

Table 2. Some key laws of the phlogiston theory.

The weight of an object is qualitatively proportional to the amount.
    GL4:   qprop(weight(P), amount(P), pos).

Combustion is a negative influence on the amount
    GL5a: influence(combustion, amount-of-in(phlogiston, charcoal),
            neg).
Calcination is a negative influence on the phlogiston
    GL5b: influence(calcination, amount-of-in(phlogiston, m-c), neg).

The amount of a complex substance equals the sum
    GL6:   qty-eq(amount(C), qty-sum(Qs)) ←
            complex(C), is-a-set-of-amounts-of-components-of(Qs, C)
    GL7a: is-a-set-of-amounts-of-components-of([Qi | Qs], C) ←
            is-an-amount-of-a-component-of(Qi, C),
            is-a-set-of-amounts-of-components-of(Qs, C).
    GL7b: is-a-set-of-amounts-of-components-of([ ], C).
    GL8:   is-an-amount-of-a-component-of(Qi, C) ←
            complex(C),
            component(Ci, C), Qi = amount-of-in(Ci, C).
Note: 'm-c' abbreviates 'mercurius calcinatus' and stands for a piece
of partially calcinated red calx of mercury.

---

amount of phlogiston in a partially calcinated piece of mercury. The laws
in Table 2 encode a fragment of the knowledge that a phlogiston chemist
might have used in reasoning qualitatively about chemical phenomena.

3.3   General Laws of Qualitative Physics

This section presents some general laws of qualitative process theory
(see Table 3). These laws are important in commonsense reasoning
about the physical world. *The law of direct influences* (GL1) states that
a quantity may be changing because some process is directly influencing
it. The quantity increases or decreases according to whether the **sign**
is positive or negative.

The *laws of indirect influences* (GL2a and GL2b) are meant to capture the notion that a quantity may change because it is qualitatively
proportional to some other quantity. A qualitative proportionality may
be either positive or negative. If there is a positive qualitative proportionality, a change in one quantity may be accounted for by a similar

Table 3. Some general laws of qualitative physics encoded as rules.

Direct Influences:
    GL1:   `deriv-sign(Q1, Sign) ←`
              `process(Process), active(Process),`
              `influence(Process, Q1, Sign).`

Indirect Influences:
    GL2a: `deriv-sign(Q1, Sign) ←`
              `qprop(Q1, Q2, pos), deriv-sign(Q2, Sign).`
    GL2b: `deriv-sign(Q1, Sign1) ←`
              `qprop(Q1, Q2, neg), deriv-sign(Q2, Sign2),`
              `opposite(Sign1, Sign2).`

The Law of Sums:
    GL3:   `qprop(Q, Q`$_i$`, pos) ←`
              `qty-eq(Q, qty-sum(Qs)), member(Q`$_i$`, Qs).`

Note:
```
In GL1, ''deriv-sign(Q1) = Sign''
  means ''the sign of the time derivative of quantity Q1 is Sign.''
In GL2a, ''qprop(Q1, Q2, pos)'' means ''quantity Q1 is positively
  qualitatively proportional to quantity Q2.''
In GL3, ''qty-eq(Q, qty-sum(Qs))'' means
  ''Q is a quantity equal to the sum of quantities Qs,''
  where Qs is a list of quantities. Also, ''member(Qi, Qs)'' means
  ''Qi is a member of the list of Qs.''
```

---

change in some other quantity. In the case of a negative qualitative proportionality, a change in one quantity may be accounted for by an opposite change in another quantity.

The *law of sums* (GL3) states that a quantity is qualitatively proportional to a second quantity if the first quantity is equal to a sum of a number of quantities, one of which is the second quantity. For example, qprop(weight(body), Q$_i$, pos) ← qty-eq(weight(body), qty-sum([weight(lean-body-mass),weight(other-body-mass)])), member(weight(lean-body-mass), [weight(lean-body-mass),weight(other-body-mass)]).

Note that the "implications" in these "laws" are somewhat ambiguous. The implication in the "law of sums" should be interpreted as material implication, whereas the implications in the "laws of influences" should be interpreted as specifying potential causal associations. Used in backward chaining, these rules specify possible causes for events. Used in forward chaining, they predict potential consequences.

Inferences based on the "laws of influences" typically focus on one aspect of a situation under the assumption that other aspects can be safely ignored. In particular, inferences involving a change in some quantity ignore other potential influences or proportionalities involving the affected quantity. In the case of the "laws of indirect influences," a quantity may be qualitatively proportional to another quantity, and this second quantity may be changing, but this change does not necessarily completely determine what will happen to the first quantity. In the case of the "law of direct influences," an active process may be driving a quantity up or down, but that does not rule out the possibility that there are other direct or indirect influences acting in the opposite direction.

A classic example of a set of conflicting influences involves a bathtub with a faucet valve open but with the drain open as well. The water level in the tub is driven upward by the water flowing in through the faucet, but it is simultaneously driven downward by the water flowing out through the drain. Forward chaining on rules like the ones in Table 3 could be used to predict possible consequences, such as that the water level in the tub may go up (or down). Alternatively, backward chaining on these rules could be used to generate possible causes, such as explaining why the level of the water is observed to increase (or decrease).

4. Abduction, Hypothesis Formation, and Theory Revision

The previous section described the kinds of knowledge needed to explain certain observations according to the phlogiston theory. In this section, we describe how this sort of knowledge can be used to construct explanations by a process of abductive inference. Next, we explain how this form of abduction can be used to generate hypotheses, and we describe a method that uses this capability in theory revision.

According to Peirce (1931–1958), abduction is explanatory hypothesis generation. Peirce's formulation of abduction was basically: "The surprising fact, C is observed; but if A were true, C would be a matter of course, hence there is reason to suspect that A is true." An analysis of Peirce's views on abduction and hypothesis formation and evaluation may be found in Thagard (1981). Interestingly, Peirce originally intended the term *abduction* to apply only to the initial formulation of hypotheses, whereas AI researchers usually include evaluation and acceptance as part of the process. We conform to this convention and use the term *abduction* loosely as shorthand for methods for constructing and evaluating explanations.

4.1   Theory-Driven Abduction

AI researchers have cited Peirce's notion of abduction as the basis for a number of different methods and systems. In this section, we describe the particular form of abduction ("theory-driven abduction") used in our case study of the chemical revolution. This approach to abduction is related to the philosophical view of explanations as deductive arguments in which the thing to be explained follows from a set of general laws and specific facts. Hempel (1965) calls explanatory accounts of this kind "explanations by deductive subsumption under general laws or *deductive-nomological explanations.* (The root of the term *nomological* is the Greek word *nomos* for law.)" The form of abduction explored here could be considered to be a deductive-nomological form of abduction, viewing explanations as deductive proofs. The proofs show how observations follow from sets of rules and facts that encode general theories and facts that describe specific situations.

The abduction machinery used here is closely related to theorem provers, and, in particular, to the standard technique called *backward chaining* (Charniak & McDermott, 1986). In this technique, a query C? is used to generate a query A? by backward chaining on a rule $A \rightarrow C$. In technical terms, this is done by first "unifying" the query C? with the conclusion C of the rule. Unification produces a substitution θ that shows how to bind variables in C and C? so as to make them identical. This substitution is then applied to the rule's antecedent A. The result, $A\theta$, is taken as a new query A?. This query may "ground out" by unifying with known facts (statements "known to be true" and given as input to the abduction engine), or it may lead to new queries by way of additional backward chaining. In using backward chaining for abduction, observations to be explained are viewed as queries, and general theories and other observations are expressed in terms of rules and facts. Backward chaining attempts to reduce the observations to known facts by way of the rules contained in the theory.

4.2 Abduction and Hypothesis Formation

In order to see how this form of abduction may be used for hypothesis formation, it is important to distinguish between the process that constructs explanations and the resulting explanations. Explanations may be deductive, even when the process of constructing them is not deductive. A conclusion may follow deductively from a set of assumptions given the truth of those assumptions, but the process of generating the assumptions required to complete the proof may be nondeductive.

In our particular abduction engine, the process of backward chaining on an observation produces partial proof trees. The leaves of these trees may or may not correspond to known facts. In some cases, backward chaining "grounds out" so that *all* the leaves of a proof tree unify with facts given as part of the input to the abduction engine. For example, Section 5 presents a proof that explains the decrease in a piece of charcoal's weight as it burns. This explanation is derived by backward chaining on the rules given in Section 3 that describe the phlogiston theory, such as the rule that combustion drives down the amount of phlogiston in a piece of charcoal. The proof "grounds out" in statements from this theory and in statements that encode observations (for example, that the charcoal is burning).

However, when used in constructing explanations, backward chaining often fails to produce complete proof trees. In this case, the ungrounded leaves of the partial proof trees correspond to the explanatory hypotheses generated in Peirce's formulation of abduction. If the propositions corresponding to these leaves were true, the observation would follow, and so there is some reason to suspect that they are true. Yet, even if no better explanation of the observation can be found, the leap to this conclusion is a nondeductive, *abductive inference*. Section 6 shows how this sort of abductive hypothesis formation can be used to generate aspects of the oxygen theory by explaining "augmentation effects," such as the observation that a metallic calx gains weight in calcination.

4.3 Revising Theories Using Abductive Hypothesis Formation

Now we can sketch our approach to theory revision through abductive hypothesis formation. The need for revision is typically recognized when a theory is found to contradict new observations. The task is then to determine what revisions will give a new theory that is in accord with observation. Most approaches to theory revision involve direct transformations that produce the new theory from the original "old" theory. These tranformations are generally very much like "editing" or "tweaking." Two combinatorial problems occur in these transformations. One involves the identification of the erroneous subset of the original theory, and the other involves the identification of the correct changes in the erroneous parts of the original theory. In some situations, these combinatorics are likely to overwhelm editing approaches to theory revision, and there is some evidence that people employ a different approach. In Shrager and Klahr's "instructionless learning" experiments, subjects were asked to "figure out" complex programmable devices. Shrager (1987) comments:

> ... we observed that between interactions with the BigTrak, subjects changed their theory of the device. A number of empirical generalizations seem to hold about the nature of these changes ... Instead of trying to determine in detail what led to a failed prediction, subjects usually observed what (positive behavior) took place and changed their theory according to that observation ...

We believe that our approach to theory revision is compatible with Shrager's results. When a surprising observation contradicts a prediction of the original theory, our approach involves retracting questionable beliefs. However, one need not start by trying to identify an individual incorrect belief or even a small set of culprits. Instead, we assume that the initial theory has some internal structure and that more general fundamental principles can be separated from relatively specific, less basic statements. A "core" subset of the original theory—a set of basic statements having nothing to do with the anomaly—is retained while less central beliefs are suspended. Our approach involves explaining the unexpected new observation in terms of the remaining, relatively solid basic principles. As we will see in Section 6, this explanation process can generate hypotheses, suggesting extensions to the basic principles. If these hypotheses are added to the basic principles, the resulting set of rules and facts is a candidate revised version of the original theory.

This approach to theory revision is sketched in Figure 1 using Venn diagrams. In the first stage (a) of theory revision an anomaly is noted. A new observation contradicts a prediction of the old theory, as indicated by the X linking a point in the old theory and a point outside of it. In the

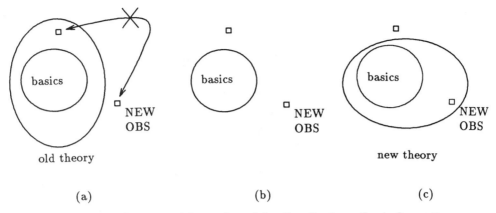

Figure 1. Theory revision using abduction for hypothesis formation.

next stage (b), the old theory is reduced to the core subset.[4] Starting from this subset, an explanation of the new observation is abduced, with hypotheses being introduced in the process. These hypotheses then form the basis for extensions to the core theory resulting in a new theory (c). This revised theory no longer makes the erroneous prediction of the old theory.

We will not explore the initial step of falling back on basic principles and shrinking the original theory in this chapter, and we will not explore the evaluation of the original theory and the candidate revision. Instead, we focus on the step from Figure 1(b) to Figure 1(c). We concentrate on the claim that the process of explaining unexpected new phenomena can lead by abductive inference to new hypotheses which can form crucial parts of new theories. We substantiate this claim by showing how an aspect of the oxygen theory of combustion and calcination can result from revising the phlogiston theory using this framework.

5.   The Phlogiston Account of Burning Charcoal

Now we are in a position to illustrate the use of the qualitative physical laws and the phlogiston theory in the construction of explanations, showing how our "abduction engine" generates explanations of an ob-

---

4. Notice that neither the prediction nor the surprising observation are included in the reduced core subset of the original theory. The circles and ellipses designate theories closed under deductive inference. The figure captures the notion that neither the prediction nor the contradictory observation should be implications of the core theory.

Table 4. The weight of charcoal decreases as it burns.

Observation:	
O1:	deriv-sign(weight(charcoal), neg).
Case facts:	
CF1:	process(combustion).
CF2:	active(combustion).
CF3:	complex(charcoal).
CF4:	component(phlogiston, charcoal).
CF5:	component(ash, charcoal).

servation by attempting to reduce it to known facts using general laws. We have implemented this approach in AbE, a PROLOG metainterpreter that uses best-first heuristic search to construct explanation trees and evaluate partial explanations. AbE's heuristic evaluation function is based on the "weighted abduction" method proposed by Hobbs, Stickel, Martin, and Edwards (1988). This section shows how the system constructs an explanation of the drop in the weight of burning charcoal in accord with the phlogiston theory.

AbE is given the observation that, upon burning, the weight of some charcoal decreases. This is expressed as a statement (labeled O1 in Table 4) that the sign of the derivative of the weight of the charcoal is negative. The system is also given some specific facts (CF1–5) that combustion is occurring and that charcoal is a complex substance containing phlogiston and ash. (This was the model of charcoal held by the phlogiston chemists.) In addition, the system is given the general laws of qualitative physics and the phlogiston theory described earlier in Tables 2 and 3. AbE is asked to explain the observation by using the given laws to connect the observation to the given facts. The output of the system is a set of proof trees like that shown in Figure 2. In the process of constructing explanations, the observation is treated as a query that gives rise to new queries by backward chaining on rules representing logical and causal laws. The remainder of this section traces the construction of the tree, visiting the nodes in the order shown by the labels in Figure 2.

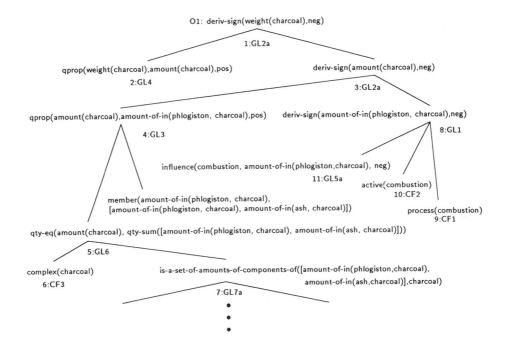

Figure 2. An explanation of the weight decrease in burning charcoal.

The initial query addressed by AbE is *why is the weight of the charcoal decreasing?* According to the laws of indirect influences (GL2), a change in some quantity may be explained by a change in some other quantity, provided those quantities are qualitatively proportional. This raises the question of whether the decrease in the weight of the charcoal may be explained in terms of a decrease in some amount that is positively qualitatively proportional to the weight of the charcoal. The question of whether there is any such quantity is answered as an instance of the general "fact" that the weight of any object is positively proportional to the amount of that object (GL4).

At this point, the question is *why is the amount of charcoal decreasing?* To explain this change, the system again attempts to find a positive qualitative proportionality between the amount of charcoal and some other decreasing quantity (using GL2). An appropriate proportionality is found using the law of sums (GL3). This states that some quantity Q is proportional to some other quantity Q_i if Q is equal to the sum of some set of quantities Q_s and Q_i is a member of that set. In this case, Q is the amount of charcoal.

The question is now *whether there is some set of quantities whose sum is equal to the amount of charcoal.* AbE answers this in terms of its knowledge about complex substances. In particular, the system knows that the overall amount of a complex substance is equal to the sum of the amounts of the components of the substance (GL6). In Figure 2, the amount of charcoal is shown as a quantity sum over a list consisting of the amount of phlogiston in the charcoal and the amount of ash in the charcoal. This list of components is actually derived through several applications of rules GL7 and GL8 (which, if shown in Figure 2, would construct a subtree beginning at node 7). In particular, the two applications of rule GL8 ground out using the case facts CF3 (charcoal is a complex substance), CF4 (phlogiston is a component of charcoal), and CF5 (ash is a component of charcoal). The result is a proof tree for node 4, which states that the overall amount of the charcoal is positively qualitatively proportional to the amount of phlogiston in the charcoal.

The question now is *whether the amount of phlogiston in the charcoal is decreasing.* The law of direct influences (GL1) can be used to explain this decrease, assuming that an active process can be found to have a negative influence on the amount of phlogiston in the charcoal. At this point, the system completes its explanation by recognizing that, according to facts of the case and a key statement in the phlogiston theory, combustion is an active process that negatively influences the amount of phlogiston in the charcoal.

AbE also generates a similar proof, in which the amount of charcoal is seen as proportional to the amount of ash. However, it cannot ground this proof at node 11 via a direct, negative influence of combustion on the amount of ash because this influence is not a fact. Consequently, in the course of its heuristic search, AbE's evaluation function ranks this incomplete explanation less favorably than the explanation presented in Figure 2.

6. Abduction of Aspects of the Oxygen Theory

In Section 3, we saw that recent progress in knowledge representation and automated reasoning makes it possible to capture key ideas contained in early chemical theories such as the phlogiston theory. In Section 4, we claimed that recent progress on automated abduction makes

Table 5. Ablation of the phlogiston theory.

The weight of an object is qualitatively proportional to the amount.
    GL4:   qprop(weight(P), amount(P), pos).

~~*Combustion is a negative influence on the amount of phlogiston in charcoal.*~~
~~   GL5a: influence(combustion, amount-of-in(phlogiston, charcoal), neg).~~
~~*Calcination is a negative influence on the phlogiston in mercurius calcinatus.*~~
~~   GL5b: influence(calcination, amount-of-in(phlogiston, m-c), neg).~~

The amount of a complex substance equals the sum of the amounts of the components.
    GL6:   qty-eq(amount(C), qty-sum(Qs)) ← complex(C), is-a-set-of-amounts-of-components-of(Qs, C)
    GL7a: is-a-set-of-amounts-of-components-of([Qi | Qs], C) ← is-an-amount-of-a-component-of(Qi, C),
                                               is-a-set-of-amounts-of-components-of(Qs, C).
    GL7b: is-a-set-of-amounts-of-components-of([ ], C).
    GL8:   is-an-amount-of-a-component-of(Qi, C) ← complex(C),
                                        component(Ci, C), Qi = amount-of-in(Ci, C).

Anomaly: The weight of red calx of mercury increases as it calcinates.
Observations:
    O1:   deriv-sign(weight(m-c), pos).
Case facts:
    CF1:   process(calcination).
    CF2:   active(calcination).

---

it possible to capture significant aspects of the reasoning that occurred in the chemical revolution. In Section 5, we showed how abduction on a simplified phlogiston theory can generate an explanation that seems to capture the phlogiston theorists' views that substances lose weight when they burn or calcine because they lose phlogiston. This explanation was constructed by reducing the observation (that a piece of charcoal lost weight) to "known facts" (given to AbE *before* it started searching for an explanation).[5] However, in Section 4 we claimed that abduction could also be used to go beyond deductive inference to form the kinds of hypotheses involved in major theory shifts.

In this section, we describe how abduction can generate a crucial aspect of the oxygen theory. Let us assume as *given* the phlogiston theory account of combustion and calcination, along with the observation and case facts shown in Table 5. Ignore for the moment that several items in Table 5 are crossed out. Recall that m-c is an abbreviation for mercurius calcinatus, which is partially calcinated mercury. According to the phlo-

---

5. PROLOG could have been easily used to construct the explanation in Section 5.

giston theory, pure metallic calxes were more primitive substances than metals. Metals were formed by heating calxes in the presence of a source of phlogiston such as charcoal; the calxes combined with the phlogiston to form the metals. On the other hand, metallic calxes resulted when phlogiston, which was viewed as a "metallizing principle," calcined out of metals.

The phlogiston theory explains and predicts a *decrease* in the weight of substances undergoing combustion or calcination. This prediction contradicts the given observation that the weight of mercurius calcinatus increases during calcination. Assume that, as a result, questionable parts of the theory and the case facts responsible for the contradiction have been identified and deleted as indicated by the offending statements crossed out in the table.[6]

Assume, then, that our abduction engine AbE is given this reduced phlogiston theory along with the observation and case facts shown in Table 5. The altered theory and observation make no mention of phlogiston, so that it is no longer considered as an essential component of combustible substances or as involved in combustion or calcination.

AbE is asked to explain, in terms of the given laws of qualitative physics and the ablated phlogiston theory, the observation that the weight of mercurius calcinatus increases (O1) during calcination (CF1 and CF2). The system does this by attempting to reduce the observation to the given facts, but if this is not possible it will propose some hypotheses in an effort to explain the observation. Figure 3 shows one explanation that AbE generates. This explanation is obviously very similar to the explanation of the decrease in the weight of charcoal according to the phlogiston theory discussed earlier. Let us examine how this explanation was constructed.

The initial query is *why is the weight of the mercurius calcinatus increasing?* The system answers this question in terms of an increase in the amount of the substance.

This leads to the question *why is the amount of mercurius calcinatus increasing?* To explain this, AbE tries to find a positive proportionality

6. Existing contradiction backtracing (Shapiro, 1981) and dependency-directed backtracking methods (Doyle, 1979) could contribute to identifying candidates for deletion or temporary suppression, but some method of evaluating plausibility will be needed in order to decide that a potential culprit should be excised. Basic principles that contribute to many explanations (such as conservation laws) should be preferentially retained.

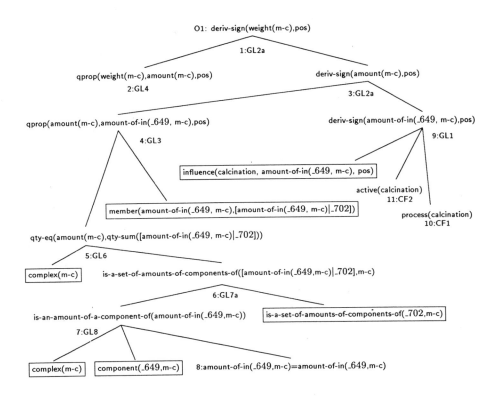

Figure 3. An explanation of the weight increase in mercurius calcinatus during calcination.

between the amount of mercurius calcinatus and some other increasing quantity. The system finds an appropriate proportionality using the law of sums (GL3), which it then uses to hypothesize that the amount of mercurius calcinatus is increasing because the amount of one of its components is increasing. Using laws about complex substances (GL6, GL7, GL8), AbE hypothesizes the existence of an unknown component of mercurius calcinatus and the existence of an unknown quantity corresponding to its amount. The system also hypothesizes that there is a set of remaining components and associated amounts, without identifying any particular elements of this set.[7] Unlike the burning charcoal

---

7. These new individuals are represented by Skolem constants generated as a natural consequence of backward chaining. As shown in Figure 3, the set of the component amounts is represented as the list [amount-of-in(_649,m-c) | _702], where

example, in the present case AbE is *not* given a case fact stating that mercurius calcinatus is complex. Instead, in using the law of sums, AbE hypothesizes that mercurius calcinatus is a complex substance.[8]

The question now is *whether the amount of the unknown component of mercurius calcinatus is increasing.* The law of direct influences (GL1) can be used to explain this increase, if one assumes an active process that has a *positive* influence on the amount of the component of the mercurius calcinatus. At this point, since calcination is known to be an active process, AbE completes its explanation by hypothesizing that calcination has a direct positive influence on the amount of the unknown component.

The construction of the above explanation used only general rules and facts, such as the rule that a quantity may have changed as a result of an active process and the fact that the weight of a substance is qualitatively proportional to its "amount." There was no use of knowledge encoding the chemical content of specific qualitative chemical theories, such as the phlogiston or oxygen theory. Instead, AbE employed the basic theory to generate hypotheses corresponding to parts of the oxygen theory by abductive inference. The hypotheses generated in this manner are enclosed in boxes in Figure 3. These abductive inferences correspond to Lavoisier's insight that something was being added during calcination.

Explaining new, surprising observations is a key step in theory revision. In the case of the chemical revolution, Lavoisier's hypothesis that something was added by calcination to the calx of mercury, in conjunction with experimental results of Priestley and others, eventually led him to posit the existence of a hitherto unknown component of air. Lavoisier called this new theoretical entity "oxygen." During the next decade, he and his colleagues worked out a new theory of combustion, calcination, and respiration that eventually displaced the phlogiston theory. This occurred because most chemists of the time were persuaded that the new theory explained the new observations (and reexplained old observations) in a more coherent manner than did modified versions of the phlogiston theory.

_649 is the hypothesized unknown component and _702 is the remaining subset of component amounts.

8. The apparently tautological node labeled "8:" in Figure 3 reflects the unification of Qi to amount-of-in(_649,m-c) in law GL8.

7. Relation to Other Work

Our work is related to, and builds on, work on commonsense reasoning about the physical world, qualitative physics, and scientific discovery. The approach fits into the four-stage theoretical framework for learning in physical domains sketched by Forbus and Gentner (1986). The learning taking place in our chemical revolution appears to fit in the third stage ("learning naive physics"). Here we briefly consider its relation to other work in machine discovery.

7.1 STAHL and STAHLp

Recent work in scientific discovery has produced two systems, STAHL (Zytkow & Simon, 1986) and STAHLp (Rose & Langley, 1986), that can automatically detect and correct errors in chemical theories. Both systems represent chemical theories in terms of *reaction and component models*. A reaction is specified by its *input* and *output* substances—the substances entering the reaction and the substances resulting from the reaction. A component model specifies the components of an individual complex substance as a list of substances.

These systems could conceivably model the shift from the phlogiston to the oxygen theory as a change from a set of reaction rules and component models involving phlogiston to a set of reaction rules and component models involving oxygen. However, in our opinion, such an account of the theory shift would be incomplete if only because the models of the phlogiston and oxygen theories would be limited to reactions and component models. For example, both the phlogiston theory and the oxygen theory explained why a flame burning in an enclosed place eventually expires—but their explanations cannot be expressed solely in terms of component models and reactions.

The inputs to STAHL and STAHLp are reactions. From these reactions, they use rules to derive new, *inferred* reactions and component models. An inferred reaction is generated from a parent reaction by two methods: (1) *reduction*: a substance that appears on both sides of the parent reaction is removed on both sides; and (2) *substitution*: a substance in the parent reaction is replaced by the components of that substance as specified by its component model. A component model of a complex substance is inferred in one way only: When an input or inferred reaction has exactly one input substance, the inference is that the outputs of the reaction specify the components of that substance. Both programs, using different techniques, carry out belief revision in an incremental fashion; beliefs are revised when inconsistent sets of reactions or component models are inferred and detected.

We now discuss some aspects of belief revision in STAHL and STAHLp. In STAHL, inconsistent inputs may lead to the inference of component models that result in infinite recursion. For example, the following two component models could be inferred:

```
mercury = calx-of-mercury + phlogiston
calx-of-mercury = mercury + oxygen
```

Substitution of the second model into the first produces a component model of mercury that is self-referential and leads to infinite recursion:

```
mercury = mercury + oxygen + phlogiston
```

STAHL solves this problem by renaming calx-of-mercury in one of the two component models as "calx-of-mercury-proper." This can be seen as a model of the historical practice of chemists casting doubt on the proposed identity of a substance in a reported reaction. This allows the introduction of new substances, *but only by way of renaming substances already mentioned in reactions.*

In STAHLp, beliefs are revised when the system infers an inconsistent reaction in which either the input or output side has no substances while the other side has one or more substances. The program corrects this situation by revising its input reactions. Each revision involves deleting or adding a substance from one side of a reaction, and an input reaction can have more than one such revision made to it. Belief revision is effected by identifying a set of revisions of the input reactions that satisfies two conditions: (1) a balanced reaction will be inferred from the revised input reactions, and (2) the number of component models that will be changed by the revision is minimal. Once this revision is identified, all beliefs (inferred reactions and component models) that depend on the revised input reactions are deleted; then the revisions to the input reactions are made; and then STAHLp generates the new reactions and component models that follow from the revised theory. The result of the theory revision can be the elimination or modification of previously held component models and inferred reactions and the addition of new component models and inferred reactions.

Although such revision in STAHLp amounts to hypothesizing the existence of unobserved substances in the input reactions (adding substances), and retracting previously believed observations of substances in the input reactions (deleting substances) all such substances must have been named in previous input reactions. The system is not capable of hypothesizing the existence of a new substance that has not previously appeared in an input. This is in contrast to AbE, which can hypothesize a new component substance on the basis of general laws concerning qualitative physics, sums of quantities, and complex substances.

7.2   COAST and PHINEAS

Falkenhainer and Rajamoney (1988) describe a closely related approach to theory revision. The PHINEAS system (Falkenhainer, 1988) extends abductive hypothesis formation to include qualitative physical analogies, and the COAST system (Rajamoney, 1988) revises qualitative physical theories involving processes such as evaporation and osmosis, using "explanation-based theory revision" to propose changes in an initial theory in response to an anomaly. The theory revision process takes an "editing" approach, in the sense that it focuses on both the prediction of the initial theory and the surprising observation that contradicts the prediction. Revision rules are used to generate ways of changing the initial theory so that the prediction is no longer made but the unexpected observation is predicted instead. An advantage of this "theory debugging" strategy is that the errors in the initial theory and their corrections are identified together. In related work, Rajamoney (1989) describes a method for using "exemplars" to guide theory revision, in which he uses qualitative process schemata for a phlogiston theory in an example. However, the revisions proposed by his method are essentially "patches" of the phlogiston theory, and the existence of a new substance (oxygen) is not hypothesized.

7.3   ECHO and PI

Thagard (1989) presents a theory of explanatory coherence and a connectionist implementation. His ECHO program is given data representing observations and the phlogiston and oxygen theories. Using activation and inhibition links between data and theoretical statements, the program attempts to determine which of the two theories best "coheres" with the data.

Thagard's ECHO focuses on the *evaluation* of *existing* theories. With regard to the question of where such theories might have come from in the first place, Thagard hints that PI (Holland, Holyoak, Nisbett, & Thagard, 1986) might be able to construct them. In another paper (Thagard, in press), he examines the conceptual changes that occurred during the overthrow of the phlogiston theory and gives a fairly detailed conceptual map of several important intermediate stages of chemical theory in the transition from the phlogiston theory to the oxygen theory. He also suggests that the mechanisms for concept formation and rule abduction present in PI can be used to form conceptual networks that can chart the conceptual changes which occurred during the chemical revolution. Our work on AbE has moved beyond this, showing a detailed example of how abduction, in concert with ideas from qualitative physics, can make some crucial inferences associated with the discovery of oxygen.

8. Conclusion

Theory revision can profitably be viewed as a process that involves hypothesis formation by abduction. When an anomaly is encountered, the best course is often to forget or suppress questionable details of the original theory and to derive an explanation of the anomalous observation based on more solid, more basic principles. In this way, the process of looking for explanations of unexpected new phenomena can lead by abductive inference to new hypotheses that can form crucial parts of a revised theory.

The main result of this chapter is that recent progress on abduction and qualitative process theory makes it possible to automate significant aspects of the reasoning that occurred in the chemical revolution. We believe that the language for describing processes and causal relationships resulting from work on qualitative physics, together with inference mechanisms such as automated abduction, will enable automation of many crucial but relatively commonsense insights associated with scientific revolutions. If this proves true, it suggests that automated abduction is a key to advancing beyond "routine theory revision" towards automated reasoning systems capable of "world model revision."

Acknowledgements

This is a revised and expanded version of an earlier paper (O'Rorke, Morris, & Schulenburg, 1989). The ideas in this chapter have evolved in discussions with many members of the machine learning community at the University of California, Irvine. Special thanks are due to Pat Langley, Deepak Kulkarni, and Don Rose for numerous discussions of scientific discovery that sparked our interest in modeling the chemical revolution. Discussions with Paul Thagard on Peirce, abduction, the chemical revolution, and scientific revolutions in general were also inspirational. Thanks to Pat Langley and Jeff Shrager for constructive criticism of an earlier draft. This chapter is based on work supported in part by an Irvine Faculty Fellowship from the University of California, Irvine, Academic Senate Committee on Research and by grant number IRI-8813048 from the National Science Foundation.

References

Charniak, E. (1988). Motivation analysis, abductive unification, and nonmonotonic equality. *Artificial Intelligence, 34*, 275–295.

Charniak, E., McDermott, D. V. (1986). *Introduction to artificial intelligence.* Reading, MA: Addison-Wesley.

Charniak, E., Riesbeck, C. K., McDermott, D. V., & Meehan, J. R. (1987). *Artificial intelligence programming* (2nd ed.). Hillsdale, NJ: Lawrence Erlbaum.

Conant, J. B. (1957). The overthrow of the phlogiston theory: The chemical revolution of 1775–1789. In J. B. Conant, L. K. Nash, D. Roller, & D. H. D. Roller (Eds.), *Harvard case histories in experimental science.* Cambridge, MA: Harvard University Press.

Conant, J. B., Nash, L. K., Roller, D., & Roller, D. H. D. (Eds.). (1957). *Harvard case histories in experimental science.* Cambridge, MA: Harvard University Press.

Doyle, J. (1979). A truth maintenance system. *Artificial Intelligence, 12*, 231–272.

Falkenhainer, B. (1988). *Learning from physical analogies: A study in analogy and the explanation process.* Doctoral dissertation, Department of Computer Science. University of Illinois, Urbana-Champaign.

Falkenhainer, B., & Rajamoney, S. A. (1988). The interdependencies of theory formation, revision, and experimentation. *Proceedings of the Fifth International Conference on Machine Learning* (pp. 353–366). Ann Arbor, MI: Morgan Kaufmann.

Feigenbaum, E. A. (1979). Themes and case studies of knowledge engineering. In D. Michie (Ed.), *Expert systems in the micro electronic age.* Edinburgh: Edinburgh University Press.

Forbus, K. D. (1984). Qualitative process theory. *Artificial Intelligence,* *24,* 85–168.

Forbus, K. D., & Gentner, D. (1986). Learning physical domains: Toward a theoretical framework. In R. S. Michalski, J. G. Carbonell, & T. M. Mitchell (Eds.), *Machine learning: An artificial intelligence approach* (Vol. 2). San Mateo, CA: Morgan Kaufmann.

Guerlac, H. (1961). *Lavoisier—the crucial year—the background and origin of his first experiments on combustion in 1772.* Ithaca, NY: Cornell University Press.

Hempel, C. G. (1965). Aspects of scientific explanation. In C. G. Hempel (Ed.), *Aspects of scientific explanation, and other essays in the philosophy of science.* New York: Macmillan.

Hobbs, J. R., Stickel, M., Martin, R., & Edwards, D. (1988). Interpretation as abduction. *Proceedings of the Twenty-Sixth Annual Meeting of the Association for Computational Linguistics* (pp. 95-103). Buffalo, NY: The Association for Computational Linguistics.

Holland, J. H., Holyoak, K. J., Nisbett, R. E., & Thagard, P. R. (1986). *Induction: Processes of inference, learning, and discovery.* Cambridge, MA: MIT Press.

Ihde, A. J. (1980). Priestley and Lavoisier. In *Joseph Priestly symposium, Wilkes-Barre, Pa., 1974.* London: Associated University Presses.

Josephson, J. R., Chandrasekaran, B., Smith, J. W., Jr., & Tanner, M. C. (1987). A mechanism for forming composite explanatory hypotheses. *IEEE Transactions on Systems, Man and Cybernetics*, *17*, 445–454.

Langley, P., Simon, H. A., Bradshaw, G. L., & Zytkow, J. M. (1987). *Scientific discovery: Computational explorations of the creative processes*. Cambridge, MA: MIT Press.

McDermott, D., & Doyle, J. (1980). Non-monotonic logic I. *Artificial Intelligence, 13*, 41–72.

O'Rorke, P., Morris, S., & Schulenburg, D. (1989). Abduction and world model revision. *Proceedings of the Eleventh Annual Conference of the Cognitive Science Society* (pp. 789–796). Ann Arbor, MI: Lawrence Erlbaum.

Peirce, C. S. S. (1931–1958). *Collected papers of Charles Sanders Peirce (1839-1914)*. C. Hartchorne, P. Weiss, & A. Burks (Eds.). Cambridge, MA: Harvard University Press.

Pople, H. E. (1973). On the mechanization of abductive logic. *Proceedings of the Third International Joint Conference on Artificial Intelligence* (pp. 147–152). Stanford, CA: Morgan Kaufmann.

Rajamoney, S. (1988). *Explanation-based theory revision: An approach to the problems of incomplete and incorrect theories*. Doctoral dissertation, Coordinated Science Laboratory, University of Illinois, Urbana-Champaign.

Rajamoney, S. (1989). Exemplar-based theory rejection: An approach to the experience consistency problem. *Proceedings of the Sixth International Workshop on Machine Learning* (pp. 284–289). Ithaca, NY: Morgan Kaufmann.

Reggia, J. A., Nau, D. S., & Wang, P. Y. (1983). Diagnostic expert systems based on a set covering model. *International Journal of Man-Machine Studies, 19*, 437–460.

Rose, D., & Langley, P. (1986). Chemical discovery as belief revision. *Machine Learning, 1*, 423–452.

Schank, R. C. (1986). *Explanation patterns: Understanding mechanically and creatively*. Hillsdale, NJ: Lawrence Erlbaum.

Shapiro, E. Y. (1981). *Inductive inference of theories from facts* (Research Report 192). New Haven, CT: Yale University, Department of Computer Science.

Shrager, J. (1987). Theory change via view application in instructionless learning. *Machine Learning, 2*, 247–276.

Thagard, P. (1981). Peirce on hypothesis and abduction. *Proceedings of the C. S. Peirce Bicentennial International Congress* (pp. 271-274). Lubbock, TX: Texas Technical University Press.

Thagard, P. (1988). *Computational philosophy of science.* Cambridge, MA: MIT Press.

Thagard, P. (1989). Explanatory coherence. *The Behavioral and Brain Sciences, 12*, 435–502.

Thagard, P. (in press). The conceptual structure of the chemical revolution. *Philosophy of Science.*

Zytkow, J. M., & Simon, H. A. (1986). A theory of historical discovery: The construction of componential models. *Machine Learning, 1*, 107–137.

CHAPTER 8

A Computational Approach
To Theory Revision

SHANKAR A. RAJAMONEY

1. Introduction

One of the principal goals of scientific discovery is the development of a
theory—a coherent body of knowledge that can be used to provide explanations and predictions for the domain under investigation. Theory
development is a complex process involving three principal activities—
theory formation, theory revision, and paradigm shifts. The process
commences with the formation of an initial theory that accounts for
a collection of known observations of a domain. Such a newly formed
theory will seldom be perfect, and its shortcomings will be exposed as
familiarity with the domain increases. Theory revision eliminates these
shortcomings by augmenting or modifying the theory. Through a series of revisions, the initial theory can evolve into one that provides
sophisticated explanations. However, the evolutionary changes due to
theory revision can frequently result in a theory that is very complex
and difficult to use, provides grossly inelegant explanations, or does not
satisfactorily account for some of the observations. In such cases, the
problems can be eliminated by a paradigm shift (Kuhn, 1970), a revolutionary change that involves a conceptual reorganization of the theory.

This chapter addresses the theory revision aspect of theory development, which extends a theory to assimilate anomalous observations. It
includes identifying problems with a theory, proposing revisions to the
theory to eliminate the problems, and developing methods to test the
proposed theories. Theory revision differs considerably from the other

two major components of theory development. Theory formation addresses the problem of constructing a theory for a domain from scratch based on an initial collection of observations. Due to the dearth of knowledge, methods for theory formation are typically data-driven, and they often involve the generation and testing of large search spaces of candidate theories. On the other hand, theory revision deals with the problem of modifying an already existing theory that has adequately dealt with many previous observations. Consequently, methods for theory revision can utilize this theory to constrain the generation and testing of candidate theories, making them primarily theory-driven methods.

The major differences between theory revision and paradigm shifts are the nature and scope of the changes to the theory. Theory revision involves small, local changes to the theory that have minimal impact beyond the assimilation of the anomalous observations. In contrast, paradigm shifts involve large, global changes to the theory that entail an entirely new perspective on the domain. The explanations provided by the new theory are substantially different from the earlier explanations for previously observed phenomena. Examples of such revolutionary changes from the history of science include the paradigm shifts from the Ptolemaic theory to the Copernican theory in astronomy, from the phlogiston theory to the oxygen theory in chemistry, and from the Newtonian theory to the theory of relativity in physics.

Theory revision serves a critical function in theory development—the gradual, incremental transformation of a rudimentary, ill-formed theory into a comprehensive, well-developed theory. The history of science contains many examples of theory revision:

- The *phlogiston* theory was initially proposed by early chemists to explain combustion. According to the theory, when a substance burns, a physical entity called *phlogiston* is expelled. This theory was gradually extended to provide qualitative explanations for a large and varied collection of observations of physical phenomena, including rusting, respiration, calcination of metals, and other chemical reactions.

- The Ptolemaic theory was an ancient theory of astronomy that predicted the changing positions of stars and planets. Starting from a simple system in which the planets and the stars revolved around the Earth in circular orbits, the Ptolemaic theory gradually evolved into a complex system of epicycles and compounded circles that could account for retrograde motion, variations in orbital speed, and discrepancies in the previously predicted positions and observed positions.

- The *caloric* theory was formulated to explain the flow of heat. The initial theory postulated that heat flow was effected by an elastic, fluid-like, material substance called *caloric* that flowed from the hotter body to the colder body. Caloric was ascribed typical characteristics of matter: it was indestructible and uncreatable, and it had appreciable weight. Conflicting observations, such as a seemingly inexhaustible supply of caloric in substances and the failure to detect appreciable changes in weight due to heat flow, necessitated revisions to the original theory.

This chapter describes a computational method called *explanation-based theory revision*. This is a knowledge-intensive approach that incorporates detecting anomalies, proposing revisions to the original theory, testing the proposed theories by designing experiments to gather additional information and by maintaining consistency with earlier observations, and ranking the remaining theories. The method has been implemented in a system called COAST. The next section describes explanation-based theory revision and the system. The third section presents a detailed example of the incremental theory revision carried out by COAST. The fourth section discusses related research, and the final section discusses some of the limitations of the method.

2.   Explanation-Based Theory Revision

Explanation-based theory revision (Rajamoney, 1989a) is an iterative process consisting of five principal stages: (1) The revision process commences when one encounters anomalous observations that violate the expectations of the theory. (2) The method proposes revisions that augment or modify the theory to assimilate the anomalous observations. (3) In general, a large number of revised theories will be proposed to account for the anomalies, many of which will be incorrect; thus, the method designs experiments to identify and eliminate the incorrect theories. (4) Before the anomalous observations were encountered, the initial theory may have successfully explained many observations from different situations. The proposed theories must also be capable of explaining these previous observations. Explanation-based theory revision collects examples of relevant previous observations and uses them to test

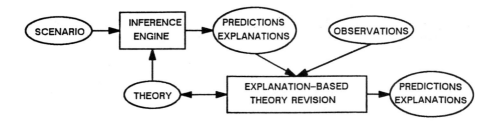

Figure 1. The architecture of the COAST system.

the consistency of the proposed theories. (5) Finally, the method rates the remaining theories and selects one for use in future prediction and explanation.

Explanation-based theory revision[1] has been implemented in a system called COAST (for **CO**rrecting and **A**ugmenting **S**cientific **T**heories). The system revises qualitative theories of the physical world, using the architecture depicted in Figure 1. COAST's inference engine analyzes given situations based on the given domain theory. When anomalous observations are encountered, it revises the theory using explanation-based theory revision. The system then uses the revised theory to explain the anomalous observations and to make predictions and construct explanations for situations encountered later.

2.1   Inputs to COAST

COAST requires the specification of three types of knowledge structures: a *domain theory*—an encoding of relevant domain knowledge required for making predictions and explaining observations; *scenar-*

---

1. The method has been named *explanation-based* because explanations play a central role. Problems with the theory are detected when it fails to provide explanations for observations; the proposed theories must provide explanations for the anomalies; the proposed theories must provide explanations for the experimentally obtained data and for previously made observations; and the proposed theories are rated according to the quality of explanations they provide. However, the method is different from traditional explanation-based learning methods (De-Jong & Mooney, 1986; Mitchell, Keller, & Kedar-Cabelli, 1986); it learns at the knowledge level (Dietterich, 1986), does not require complete, correct domain theories, and does not utilize the explanations primarily to formulate general rules.

ios—situations of interest from the domain whose behavior must be predicted or explained; and *observations*—known changes in a given scenario.

2.1.1 Domain Theories

Coast adopts Forbus' (1984) qualitative process (QP) theory for representing its knowledge of domains. In this framework, physical changes such as boiling, cooling, heating, and dissolving are intuitively viewed as effects of *processes*. A process is characterized by five pieces of information: *individuals*—a set of objects and other processes that participate in the process; *preconditions* and *quantity conditions*—a set of conditions that must be true for the process to be active; and *relations* and *influences*—a set of statements about the world that are true when the process is active. Quantity conditions specify inequality relationships and statuses of process instances, whereas preconditions specify all other conditions. Influences specify all the direct effects of a process. Dependent changes, including those due to qualitative proportionalities, are specified by relations. In QP theory, objects like the liquid in a container are described by *individual views*. Individual views are also specified by individuals, preconditions, quantity conditions, and relations. However, unlike processes, individual views do not have influences and, therefore, cannot effect changes. QP theory models the continuous properties of objects, such as the temperature or pressure of a liquid, by *quantities*, but it does not require numerical values for these quantities. Instead, it computes and reasons with inequalities between the quantities and qualitative changes to the value of a quantity.

The representation of a domain in QP theory consists of definitions for all the processes and individual views in the domain. For example, a domain theory for fluids includes process definitions for boiling, heat flow, flow of fluids, evaporation, absorption, and dissolving, and individual view definitions for liquids, solutions, and gases. Table 1 illustrates definitions for the evaporation process and the heat flow process. The definition of the evaporation process states that a liquid in an open container evaporates, and that evaporation negatively influences the amount of the liquid and positively influences the amount of the vapor at rates proportional to their area of contact. The definition of the heat flow process states that a heat flow results when two objects of different temperatures are connected by a heat conducting path. The

temperature of the object with the initially higher temperature is negatively influenced and that of the second object is positively influenced;
the rate of the flow of heat depends on the geometry of the conducting
path and the temperatures of the two objects.

Table 1.   A partial theory of fluids showing the process definitions for evaporation and heat flow.

Evaporation (?liquid ?vapor)
  Individuals
    ?liquid ?vapor
  Preconditions
    (open? (container ?liquid))
  Quantity Conditions
  Relations
    (Q+ (evaporation-rate ?self) (contact-area ?liquid ?vapor))
  Influences
    I-[(amount-of ?liquid), (A (evaporation-rate ?self))]
    I+[(amount-of ?vapor), (A (evaporation-rate ?self))]

Heat-Flow (?source ?destination ?path)
  Individuals
    ?source ?destination ?path
  Preconditions
    (heat-aligned? ?path)
  Quantity Conditions
    (greater-than (A (temperature ?source)) (A (temperature ?destination)))
  Relations
    (Q+ (heat-flow-rate ?self) (temperature ?source))
    (Q- (heat-flow-rate ?self) (temperature ?destination))
    (Q+ (heat-flow-rate ?self) (cross-sectional-area ?path))
    (Q- (heat-flow-rate ?self) (length ?path))
  Influences
    I-[(temperature ?source), (A (heat-flow-rate ?self))]
    I+[(temperature ?destination), (A (heat-flow-rate ?self))]

2.1.2 Scenarios

Coast's representation of a scenario consists of two components: (1) a *layout*, which specifies the objects in the scenario, the static physical distribution of the objects (that is, the particular manner in which the objects are organized in the situation), and the initial relationships between the quantities of the objects, and (2) a *behavior* that describes and explains the qualitative changes to the quantities. Coast's inference engine, which is a simplified version of Forbus' qualitative process engine,[2] computes the behavior of different scenarios from the layouts and the specified theory.

Figure 2 shows a scenario from the fluids domain, and Table 2 shows Coast's representation for the scenario. The individuals in the scenario are the water in the container, the vapor, the stone, the heat path connecting the water to the vapor, and the heat path connecting the stone to the water. The layout includes statements describing the scenario—the two heat paths are not heat-aligned (that is, the walls of the container and the vapor are insulators), the container is open, and the stone is at a lower temperature than water.

The behavior of the scenario is computed by Coast's inference engine using the theory shown in Table 1. The only active process is the evaporation of the water in the container. The inactive processes are a heat flow from the water to the stone and a heat flow from the water to the vapor. Both the heat flows are inactive since the respective paths are insulators. The predicted changes are a decrease in the amount of the water in the container and an increase in the amount of the vapor.

2.1.3 Observations

The observed behavior of physical quantities is input to the system by the user in the form: (?change ?quantity) where ?change represents the qualitative change in the value and is an increase, a decrease, or a constant. For example, measurements of the amount of the water and the temperature of the stone in the scenario in Figure 2 yield the observations: (decrease (amount-of water1)) and (constant (temperature stone1)).

---

2. In particular, COAST's inference engine computes only the dynamic changes within a qualitative state; it does not describe the evolution of states with time.

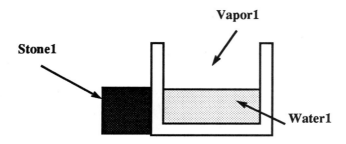

Figure 2. A scenario involving evaporation.

2.2   Detecting Anomalous Observations

The theory is routinely used to construct explanations for observations and to make predictions. Explanation-based theory revision is invoked when the theory fails to fulfill these functions properly. The failures are of three principal types (Rajamoney & DeJong, 1987): *broken explanations*—the explanations cannot be fully constructed due to missing knowledge; *contradictions*—missing or incorrect knowledge results in conflicting conclusions; and *multiple explanations*—missing knowledge or inadequate detail results in multiple, mutually inconsistent explanations for an observation. For qualitative theories, these three types of failures are detected through discrepancies in the predicted behavior and the observed behavior. The discrepancies are classified as unexpected observations— quantities predicted to remain constant are observed to change, failed predictions—quantities predicted to change are observed to remain constant, and inverse behavior—quantities are observed to change in a manner opposite to the predicted change.

Each stage of COAST's theory revision will be illustrated with an example (henceforth called the evaporation example) involving the revision of the theory shown in Table 1 when the situation depicted in Figure 2 is encountered. COAST is provided with the observation that the temperature of the water is decreasing. However, according to the behavior (Table 2) computed by the system using the given theory (Table 1), the temperature is predicted to remain constant, since the only active process (the evaporation of the water) does not influence it. Consequently, the given observation is unexpected.

Table 2. The layout and behavior of the scenario shown in Figure 2.

Layout1:
  Individuals:
    water1 vapor1 stone1 water-vapor-path water-stone-path
  Facts:
    (greater-than (A (temperature water1)) (A (temperature stone1)))
    (not (heat-aligned? water-stone-path))
    (not (heat-aligned? water-vapor-path))
      .
      .
      .

Behavior1:
  Theory: evaporation heat-flow
  Scenario: evaporation-temperature-scenario
  Active Processes:
    (evaporation water1 vapor1)
  Inactive Processes:
    (heat-flow water1 vapor1 water-vapor-path)
    (heat-flow water1 stone1 water-stone-path)
  Predicted Changes:
    (increase (amount-of vapor1))
    (decrease (amount-of water1))
  Explanations:
    (increase (amount-of vapor1))
      I+[(amount-of vapor1), (A (evaporation-rate
        (evaporation water1 vapor1)))]
        (active (evaporation water1 vapor1))
        (open? (container water1))
    (decrease (amount-of water1))
      I-[(amount-of water1), (A (evaporation-rate
        (evaporation water1 vapor1)))]
        (active (evaporation water1 vapor1))
        (open? (container water1))

2.3 Proposing Revised Theories

When a theory fails, the failure must be eliminated by identifying and fixing the deficiencies of the theory that contributed to the failure. COAST uses *theory revision operators* to modify and augment the theory. These operators add new components, delete existing components, extend or restrict the scope of components, and invert or negate the effect of components. A straightforward generate-and-test approach to theory revision would apply each of the operators to every applicable component of the theory and collect all the revised theories that eliminate the failure. However, for real-world, nontrivial domain theories, the space of generated theories will be very large and difficult to test. Consequently, constraints must be imposed on the generation of revised theories.

Explanation-based theory revision curtails the space of revised theories by imposing three types of constraints. The *scenario constraint* defines the boundaries of the theory for revision. According to this constraint, only those components of the theory that are relevant to the computation of the behavior of the scenario in which the failure occurred are considered for revision. This constraint is especially useful if only a small portion of a large, complex theory is used to compute the behavior of the failure scenario. The *explanation constraint* limits the combinations of the type of component and the type of revision on that component that must be considered to those that produce revised theories capable of explaining the anomalous observations. The *abstraction constraint* structures the space of revised theories by clustering theories under a common abstract hypothesis with the property that, if the abstract hypothesis is determined to be false on testing, then none of the theories (refined hypotheses) under it are valid. If the space of revised theories is such that large clusters can be formed and the abstract hypotheses can be independently tested, then the rejection of each abstract hypothesis obviates the generation and testing of the refined hypotheses.

Figure 3 shows the scheme (in bold type) adopted by COAST for generating revised theories for unexpected observations. Based on the explanation and abstraction constraints, the three types of abstract hypotheses for an unexpected observation are: (1) an active process causes the observed change; (2) an inactive process that could explain the observed change is active; and (3) a new process is active and causes the observed change. The scenario constraint defines the active and inactive processes that are considered under the scheme. Each of the abstract hypotheses corresponds to a large number of revised theories. For example, if an active process is causing the observation, then the corresponding revised theories include ones formed by adding a new influence or new relation to the process, extending the scope of an existing influence or relation, and inverting the effect of an existing influence or relation. Notice that the explanation constraint prevents the consideration of revisions to the conditions of the active process and limits the types of revisions on the effects of the active process to those that result in an expansion of the effects. Figure 3 also shows (in italics) the four abstract hypotheses and some of the refined hypotheses that COAST generates for the evaporation example introduced in the previous subsection. Table 3 shows the explanation for the observed decrease in the water temperature, which is based on the proposed addition of a new relation between the temperature of the evaporating liquid and its amount.

Table 3. Explanation for the anomaly based on a hypothesized relation.

```
(decrease (temperature water1))
  (decrease (amount-of water1))
    I-[(amount-of water1), (A (evaporation-rate
        (evaporation water1 vapor1)))]
      (active (evaporation water1 vapor1))
      (open? (container water1))
(Q+ (temperature water1) (amount-of water1))
  (active (evaporation water1 vapor1))
    (open? (container water1))
```

2.4 Designing Experiments

In spite of the imposed constraints, one may still generate a large number of revised theories, and many of these will be incorrect. Rather than passively waiting for disconfirming evidence, explanation-based theory revision seeks the additional information by designing experiments. Rajamoney and DeJong (1988) describe the procedure for testing the hypotheses in detail. Briefly, the procedure consists of three steps: (1) obtain predictions supported by each hypothesis using the inference engine, (2) design experiments using *discrimination*, which identifies incompatible predictions supported by the hypotheses and measures the quantity involved, and *transformation*, which constructs new scenarios in which discrimination can be applied; and (3) eliminate hypotheses whose predictions are not compatible with the outcome of the experiments.

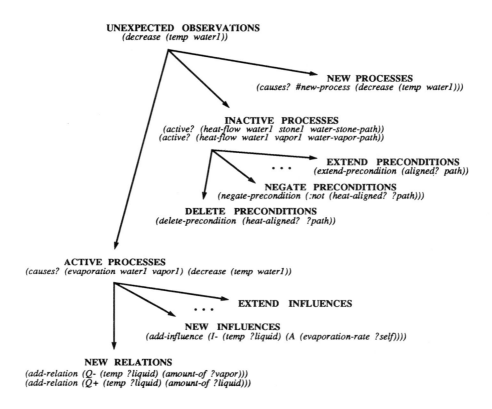

Figure 3. Generating hypotheses for unexpected observations.

Returning to the evaporation example, two of the abstract hypotheses generated by COAST are H_1: (active? (heat-flow water1 stone1 water-stone-path)) and H_2: (causes? (evaporation water1 vapor1) (decrease (temperature water1))). The system computes the predictions supported by the two hypotheses for the evaporation scenario. The former predicts that the temperature of the stone will increase due to the hypothesized heat flow, whereas H_2 predicts that the temperature will remain constant since it is not affected by the evaporation of the water, the only active process. COAST uses these conflicting predictions as the basis for a discrimination experiment that measures the temperature of the stone.

Alternatively, or if the previous experiment fails to produce conclusive results, the system can design additional experiments by transforming

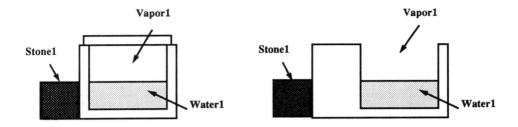

Figure 4. Scenarios obtained by applying transformation operators to the evaporation scenario.

the original evaporation scenario. COAST uses a set of knowledge-based operators to construct new scenarios. Two such transformation operators (paraphrased in English) are: "if an active process is hypothesized to cause an unexpected observation, then construct a new scenario in which one of the manipulable conditions of the active process is defeated" and "if an inactive process is hypothesized to be active, then construct a new scenario in which the rate of the process is enhanced by changing one of the parameters on which the rate depends."

An application of the first operator to the evaporation scenario produces a new scenario in which evaporation of the water is prevented by closing its container (Figure 4). For this new scenario H_2 predicts that the temperature of the water will remain constant since no evaporation occurs, whereas H_1 predicts that the temperature will decrease as before since closing the container has no effect on the hypothesized heat flow. Based on these two predictions, COAST designs a discrimination experiment to measure the temperature of the water in the transformed scenario.

An application of the second transformation operator produces a scenario in which the path of the heat flow through the container is extended (Figure 4). Since the rate of the heat flow is inversely proportional to the length of its path, H_1 predicts that the decrease in the temperature of the water in the transformed scenario will be slower than the corresponding decrease in the original scenario. However, since the change to the path does not affect evaporation, H_2 predicts that the decrease in the temperature of water will be the same for both the scenarios. COAST designs an experiment to compare the decrease in the temperature of the water in the two scenarios.

2.5   Reexplaining Exemplars

Explanation-based theory revision incorporates a second method, *exemplar-based theory rejection*, to test the validity of the proposed theories. Experiment design, by proposing interactions with the domain, actively searches for the information required to identify and eliminate incorrect hypotheses. In contrast, exemplar-based theory rejection tests the proposed theories for consistency with the data previously collected from the domain. The revisions proposed to the original theory may invalidate explanations for some of the previous observations. The revised theory is tenable only if alternative explanations can be constructed.

Exemplar-based theory rejection (Rajamoney, 1989b) collects *exemplars* that illustrate the use of the components of the theory to construct explanations for observations. An exemplar consists of four pieces of information: (1) the observation that is explained using the theory; (2) the layout of the scenario in which the observation is made; (3) the explanation for the observation constructed by the theory; and (4) the components of the theory that were used to construct the explanation. Exemplar-based theory rejection gradually builds an *exemplar space* (Figure 5), the collection of exemplars associated with the theory, by creating exemplars when the theory is used to successfully construct explanations and adding them to the exemplar space when they meet certain criteria.[3] During theory revision, exemplar-based theory rejection tests each proposed theory by determining which exemplars are affected by the proposed revisions, retrieving those exemplars, and constructing new explanations for their observations based on the revised theory. If a proposed theory cannot explain the observation of any one of the retrieved exemplars, then it is rejected.

For the evaporation example, three of the revisions that COAST proposes to explain the observed decrease in the water temperature involve modifications to the conditions of the heat flow process definition. They are:

- TR_1:   Deleting the *heat-aligned* precondition of the heat flow process.

- TR_2:   Negating the *heat-aligned* precondition of the heat flow process.

---

3. The interested reader is referred to Rajamoney (1989a) for further details on the criteria required to keep the size of the exemplar space manageable.

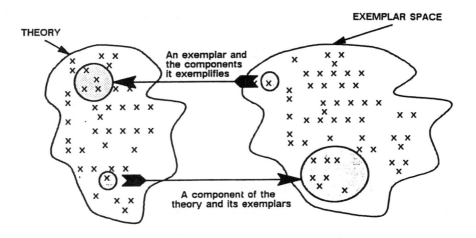

Figure 5. A theory and its exemplar space.

- TR₃: Extending the scope of the *heat-aligned* precondition of the heat flow process to a more general *aligned* predicate that is true if the objects are connected by a path.

Each of the revisions affects exemplars associated with the *heat-aligned* precondition of the heat flow process definition—specifically, exemplars involving scenarios in which the heat flow process was determined to be inactive because the *heat-aligned* precondition was not satisfied. COAST retrieves these exemplars to test the validity of the proposed theories.

Figure 6 shows one such exemplar scenario, in which an ice block is connected to a hot stone by a heat-insulated wooden rod. The temperatures of the ice block and the stone were observed to remain constant. According to the initial theory, since the wooden rod is heat-insulated, no heat flow between the ice block and the stone occurs, and therefore the temperatures of the two objects remain constant. However, each of the revised theories predicts that a heat flow from the ice block to the stone is active since the revised precondition is now satisfied in the exemplar scenario. Consequently, the temperature of the ice block is predicted to increase, and the temperature of the stone is predicted to decrease. Since these predictions contradict the previous observations, each of the revised theories is rejected.

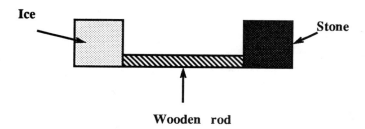

Figure 6. An exemplar scenario involving heat flow.

2.6   Rating Proposed Theories

The remaining theories are consistent with the anomalous observations, the experimental observations, and the exemplar observations. However, although the theories are of equivalent explanatory power, some may be preferred to others on dimensions like simplicity, elegance, and ease of use. Explanation-based theory revision defines three criteria to rate the remaining theories: *structural simplicity* evaluates the syntactic makeup of a theory; *explanation simplicity* evaluates the quality of the explanations constructed by the theory; and *predictive power* evaluates the predictions made by the theory.

For the evaporation example, three of the revised theories that COAST proposes are:

- T_1: A revised theory incorporating a new process that can explain the observed decrease in the temperature of the water.

- T_2: A revised theory in which the evaporation process is augmented by a new negative influence on the temperature of the evaporating liquid.

- T_3: A revised theory in which the evaporation process is augmented by a new relation specifying that the temperature of the evaporating liquid is directly, qualitatively proportional to the amount of the evaporating liquid.

COAST estimates the structural simplicity of a qualitative theory based on the number and the type of the components in the theory. According to this metric, a revised theory involving the modification of a component is preferred to one involving the addition of a component of the same type, and both of these are preferred to a revised theory involving the addition of a new process. Thus, on this dimension, T_2 and T_3 are superior to T_1.

The explanation simplicity criterion examines the explanations provided by each theory for the same collection of exemplar observations. The simplicity of an explanation is based on the number of links in the explanation. This metric prefers a revised theory involving a new influence to one involving a new equivalent relation since the former constructs explanations by direct effects that are simpler than the explanations by indirect effects constructed by the latter. Table 4 shows the explanations constructed by the T_2 and T_3 for the observed anomaly. Based on a comparison of the exemplar explanations, COAST prefers T_1 and T_2 to T_3.

Table 4. Explanations constructed by the theories T_2 and T_3.

(decrease (temperature water1))
  I-[(temperature water1), (A (evaporation-rate
      (evaporation water1 vapor1)))]
    (active (evaporation water1 vapor1))
    (open? (container water1))

(decrease (temperature water1))
  (decrease (amount-of water1))
    I-[(amount-of water1), (A (evaporation-rate
      (evaporation water1 vapor1)))]
     (active (evaporation water1 vapor1))
     (open? (container water1))
  (Q+ (temperature water1) (amount-of water1))
    (active (evaporation water1 vapor1))
    (open? (container water1))

Finally, the system estimates the predictive power of a theory based on the number of predicted changes for a common or standardized collection of exemplar scenarios. Revised theories that incorporate changes to relations make more predictions than those that incorporate changes to the influences. Consider the exemplar scenario shown in Figure 7, which involves a flow in addition to the evaporation. Since the amount of Solution1 decreases faster due to the combined effects of the evaporation and the flow processes, T_3 predicts that the decrease in the tempera-

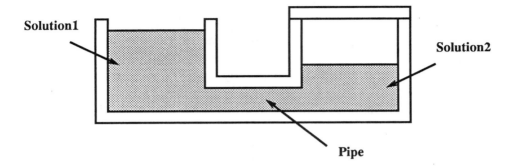

Figure 7. An exemplar scenario involving flow and evaporation.

ture will be correspondingly faster as compared with the decrease in
a scenario in which evaporation alone is active. However, T_2 predicts
that the decreases in the temperature are the same in the two scenarios since the temperature is influenced only by the evaporation process.
Consequently, T_3 is preferred since it makes additional predictions.

The overall rating of the theory depends on the goals of the system.
A system modeling the scientific discovery process might emphasize the
predictive power of the theory since incorrect surviving theories are more
readily identifiable, whereas a practical system might emphasize the
two simplicity metrics since simpler theories are easier to use. For the
evaporation example, COAST was run with the latter emphasis, and the
three metrics were combined simply by adding the normalized simplicity
metrics and subtracting the normalized predictive power metric from the
result. Based on this combination, COAST rates theory T_2 as the best
revised theory.

3.   A Detailed Example of COAST's Theory Revision

The gradual, incremental type of revision that COAST exhibits is best
demonstrated by an example involving a series of theory revision steps.
In this section, we trace the system's behavior when it is provided with
an initial theory of liquids. This theory consists of process definitions
for liquid flow, the condensation of vapor, the evaporation of liquids,
the absorption of liquids by solids, and the release of the absorbed liquids, and individual view definitions for solutions, liquids, and gases in
containers. COAST uses this theory to compute the behavior of such

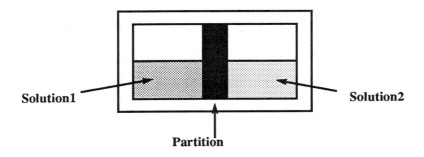

Figure 8. The osmosis scenario.

scenarios as water in two containers connected by a path, alcohol in an open container, and a sponge in contact with water.

3.1   Learning a New Process

Figure 8 shows one such scenario, in which COAST is asked to explain an observed decrease in the amount of Solution1. In the scenario, two solutions of different concentrations are placed in containers separated by a partition. Unknown to the system, the partition is *semipermeable* and thus allows *osmosis*, a flow of solvent from the solution of lower concentration to the solution of higher concentration through the semipermeable path. The initial theory predicts no changes in the given scenario, since all the potential processes are inactive; no flow occurs between the solutions because the paths connecting them (involving the partition and the wall of the containers) are not aligned for fluid flow; evaporation and condensation cannot occur because both containers are closed; and release and absorption cannot occur because neither of the solids involved (the wall of the containers and the partition) are absorbent. Consequently, the observed decrease in the amount of Solution1 is unexpected, and COAST invokes explanation-based theory revision.

The system proposes revisions to the theory to explain the anomalous observation, including abstract hypotheses that a flow, evaporation, or absorption process may be active despite the failed conditions or that a new process may be causing the observed drop in the amount. It also

Table 5.   A new process created by COAST and an explanation for the observed
decrease in the amount of Solution1 based on the revised theory.

Process8974 (?var8975 ?var8976 ?var8977)
    Individuals:
        ?var8975 (contained-fluid ?var8975) (contained-liquid ?var8975)
        ?var8976 (contained-fluid ?var8976) (contained-liquid ?var8976)
        ?var8977 (path ?var8977)
    Preconditions:
        (precondition8978 ?var8975 ?var8976 ?var8977)
    Quantity Conditions:
    Relations:
        (Q+ (process8974-rate ?self) (cross-sectional-area ?var8977))
    Influences:
        I+[(amount-of ?var8976), (A (process8974-rate ?self))]
        I-[(amount-of ?var8975), (A (process8974-rate ?self))]

(decrease (amount-of Solution1))
    I-[(amount-of Solution1),
        (A (process8974-rate Solution1 Solution2 Partition))]
        (Active (process8974 Solution1 Solution2 Partition))
        (precondition8978 Solution1 Solution2 Partition)

---

proposes refined hypotheses, such as that liquid flow does not require the
fluid-aligned precondition or that the *fluid-aligned* precondition should
be replaced by a more general *aligned* precondition. After rejecting
hypotheses that are inconsistent with the observations from the designed
experiments and the retrieved exemplars, COAST is ultimately left with
the new process hypothesis. This process consists of three participants
(two contained solutions and a fluid path connecting the two solutions),
a new precondition linking the participants (defined to be satisfied in
the given scenario), a qualitative proportionality between the rate of
the new process and the cross-sectional area of the path, a negative
influence on the amount of the source solution, and a positive influence
on the amount of the destination solution. Table 5 shows the definition
of the new process and the explanation of the anomaly constructed by
the revised theory.

Two important aspects of the revision must be noted. First, the
new process incorporates information obtained during the experimen-

tation step of the theory revision. For example, neither the increase
in the amount of Solution2 nor the qualitative proportionality between
the process rate and the cross-sectional area of the path were initially
specified; instead, they were discovered while conducting experiments.
Second, although the revised theory successfully explains the observed
changes, it is neither correct nor complete. Osmosis involves a flow of
solvent; it is active only when a difference in the concentrations of the
two solutions exists, and its rate depends on the concentrations of the
two solutions. Instead, COAST learns a skeletal, imperfect description of
osmosis that suffices to explain the observed behavior. This imperfect
theory is revised as more failures are encountered.

3.2   Correcting Influences

COAST is next asked to explain an observed increase in the concentration of Solution1 in the same scenario.[4] The revised theory cannot explain this observation since the only active process (process8974)
specifies that the solution as a whole flows through the path, so the
concentration of Solution1 should not be affected. As before, COAST
invokes explanation-based theory revision to deal with this failure. After proposing and eliminating many hypotheses (including the addition
of new relations and influences), the system is finally left with a revised
theory in which the scope of process8974's influence on the amount of the
source solution is restricted to the amount of its solvent. Table 6 shows
the revised theory and the explanation for the anomaly constructed using this theory. In a similar fashion, COAST corrects the other influence
of process8974 to explain an observed decrease in the concentration of
Solution2.

Again, two important points must be noted about the revision. First,
unlike the earlier revision, which dealt with an incomplete theory, this
revision deals with an incorrect theory. Second, exemplar-based theory rejection ensures that the revised theory can explain the previously
observed decrease in the amount of the Solution1. However, the ex-

---

4. The user gives the observations to COAST one at a time, after each round of
   revision. If multiple anomalous observations are provided to COAST, it would
   still process them one at a time. The current implementation of COAST cannot
   exploit relationships among multiple anomalous observations to limit the number
   of hypotheses generated.

Table 6. Correcting an influence of the new process and the explanation for the observed increase in the concentration of Solution1 based on the revised theory.

Process8974 (?var8975 ?var8976 ?var8977)
   Individuals:
      ?var8975 (contained-fluid ?var8975) (contained-liquid ?var8975)
      ?var8976 (contained-fluid ?var8976) (contained-liquid ?var8976)
      ?var8977 (path ?var8977)
   Preconditions:
      (precondition8978 ?var8975 ?var8976 ?var8977)
   Quantity Conditions:
   Relations:
      (Q+ (process8974-rate ?self) (cross-sectional-area ?var8977))
   Influences:
      I+[(amount-of ?var8976), (A (process8974-rate ?self))]
      I-[(amount-of (solvent-of ?var8975)), (A (process8974-rate ?self))]

(increase (concentration Solution1))
   (decrease (amount-of (solvent-of Solution1)))
      I-[(amount-of (solvent-of Solution1)),
         (A (process8974-rate Solution1 Solution2 Partition))]
     (Active (process8974 Solution1 Solution2 Partition))
     (precondition8978 Solution1 Solution2 Partition)
  (Q- (concentration Solution1) (amount-of (solvent-of Solution1)))
     (Active (solution Solution1))
     (greater-than (A (amount-of (solute-of Solution1))) 0)
     (soluble? (solute-of Solution1) (solvent-of Solution1))

---

planation (shown in Table 7) is significantly different from the earlier explanation (Table 5) due to the revisions.

3.3   Learning a New Quantity Condition

The system is next asked by the user to predict the behavior of a scenario similar to that in Figure 8, except that the concentrations of the two solutions are now equal. Based on the revised theory, COAST predicts that process8974 is active, resulting in a decrease in the amount of Solution1. However, the amount is observed to remain constant, and the system invokes explanation-based theory revision to eliminate this

Table 7. Explanation for the decrease in the amount of Solution1.

(decrease (amount-of Solution1))
  (decrease (amount-of (solvent-of Solution1)))
    I-[(amount-of (solvent-of Solution1)),
        (A (process8974-rate Solution1 Solution2 Partition))]
      (Active (process8974 Solution1 Solution2 Partition))
      (precondition8978 Solution1 Solution2 Partition)
  (Q+ (amount-of Solution1) (amount-of (solvent-of Solution1)))
    (Active (solution Solution1))
      (greater-than (A (amount-of (solute-of Solution1))) 0)
      (soluble? (solute-of Solution1) (solvent-of Solution1))

---

failure. After proposing and eliminating many hypotheses (primarily involving the addition or modification of the conditions of process8974), COAST finally chooses a revised theory that incorporates a new quantity condition into process8974 (Table 8). This condition requires that the concentration of the destination solution be greater than the concentration of the source solution. The revised theory correctly predicts that the amount of Solution1 in the scenario does not change; process8974 should no longer be active because the new quantity condition is not satisfied in the scenario.

As a result of the four cycles of failures and theory revision, COAST has acquired and incrementally refined a new process definition that approximates osmosis. The process description is not yet fully accurate; for example, it does not capture the dependencies between the rate of the process and the difference in concentrations of the solutions or the length of the path. COAST must encounter further anomalous observations before it acquires a more complete qualitative description of osmosis.

4.   Related Work on Theory Revision

COAST is descended from ADEPT (Rajamoney, 1986; Rajamoney, DeJong, & Faltings, 1985), another system that carried out theory revision. ADEPT's representation was ad hoc in many respects; consequently, it was severely limited in the number of examples it could handle. In contrast, COAST uses qualitative process theory—a general, well-defined representation with associated inference machinery. In addition, the

Table 8. Learning a new condition for the process to explain why the amount
of Solution1 remains constant in the scenario of Figure 8.

Process8974 (?var8975 ?var8976 ?var8977)
    Individuals:
       ?var8975 (contained-fluid ?var8975) (contained-liquid ?var8975)
       ?var8976 (contained-fluid ?var8976) (contained-liquid ?var8976)
       ?var8977 (path ?var8977)
    Preconditions:
       (precondition8978 ?var8975 ?var8976 ?var8977)
    Quantity Conditions:
       (greater-than (A (concentration ?var8976)) (A (concentration ?var8975)))
    Relations:
       (Q+ (process8974-rate ?self) (cross-sectional-area ?var8977))
    Influences:
       I+[(amount-of (solvent-of ?var8976)), (A (process8974-rate ?self))]
       I-[(amount-of (solvent-of ?var8975)), (A (process8974-rate ?self))]

---

current system incorporates more general methods for generating revised theories and experimentation than those used in ADEPT and uses
methods not used in its precursor, such as exemplar-based theory rejection and methods for rating theories.

Rose and Langley's (1986, 1989) REVOLVER system and its predeces-
sor, the STAHLp system, also perform theory formation and revision.
The REVOLVER system accepts input premises representing relations
between types of substances and uses heuristics to infer componential
models for substances. If inconsistencies are encountered, it revises the
premises underlying the inconsistencies. The REVOLVER system and the
COAST system are closely related: both systems detect problems, propose revisions, and use different metrics to select a revision. REVOLVER
addresses the more difficult task of revising the input premises as well
as the inferred models—it identifies the premises leading to erroneous
inferences and, by revising them, indirectly also revises the inferred componential models. COAST, on the other hand, assumes that the input
observations and the scenario description are correct (though they may
be incompletely specified) and proposes direct revisions to the theory to
explain anomalies. However, this additional assumption allows COAST
to test the validity of the proposed revisions through experiment de-

sign and exemplar-based theory rejection. Compared to REVOLVER's equationlike representation for premises and theories, COAST's representation of theories is more complex and structured, and therefore it requires additional methods such as abstraction to control the exploration of the much larger hypothesis space.

Karp (1989) describes two systems, HYPGENE and GENSIM, that perform hypothesis formation and prediction of experimental outcome, respectively. These two programs were used to model aspects of the discovery of *attenuation*, a mechanism of gene regulation in bacteria. Like COAST, Karp's systems also use qualitative representations to model the domain theory; consequently, the operators developed by Karp to revise the domain theory are very similar to those developed for COAST. Unlike COAST, HYPGENE does not use abstraction to control hypothesis formation; instead, it collects *reference experiments* whose outcome was successfully predicted and focuses on the difference between them and the anomaly to guide the heuristic search for hypotheses.

Kulkarni's (1989) KEKADA system also performs scientific discovery, and was used to model aspects of the discovery of urea synthesis in chemistry. The primary focus of the research appears to be the identification of general heuristics for selecting problems to solve, proposing hypotheses to test, and designing experiments. The strategy for designing experiments is very different from that adopted by COAST, which uses the proposed hypotheses to make predictions and designs experiments to refute the incorrect hypotheses, whereas KEKADA's experiment proposers appear to be focusing on the discovery of surprising phenomena and appear to be much less theory-driven. A complete theory formation system must have both styles of experiment design, the former to confirm theory predictions and refute incorrect hypotheses and the latter to uncover new anomalies and to investigate them.

Falkenhainer (1989) describes a method called *verification-based analogical learning* that addresses theory formation and revision. This approach relies on analogical inference to revise existing theories, whereas explanation-based theory revision relies on information obtained from the existing theory, the failure, and the failure scenario to constrain the generation of revised theories. Unlike Falkenhainer's method, which uses simulation, explanation-based theory revision designs experiments and uses exemplars to test the validity of the revised theories. The two methods are complementary, and Falkenhainer and Rajamoney (1988) describe initial efforts to integrate them into a more comprehensive theory development system.

Shrager's (1987) IE system uses an approach to theory revision called *view application*, which uses preexisting abstract schemata to restructure the theory, often in complex ways. Consequently, view application takes larger steps through the space of revisions than explanation-based theory revision. Also, Shrager intends IE as a model of human learning in the domain of complex devices, whereas COAST focuses on developing scientific theories of physical phenomena.

O'Rorke, Morris, and Schulenburg (this volume) describe a system called AbE that models the chemical revolution—the shift from the phlogiston theory to the oxygen theory. Their method discards questionable portions of a theory and attempts to rebuild the theory from more basic laws. However, they do not describe how the questionable portions of a theory (in this case, all portions of the theory that refer to phlogiston) are identified, what distinguishes them from the more "basic" laws, and how competing theories, if any, are tested or evaluated. Although their approach is more suited for modeling paradigm shifts that involve revolutionary changes, it could conceivably be applied to ordinary theory revision as well. However, the approach could prove expensive if it frequently discards large portions of the theory, since the effort expended in constructing it would be lost. COAST has not been used to model the chemical revolution, but it has been used to model simpler revisions to the phlogiston theory (like AbE, COAST also uses Forbus' qualitative process theory for representation and reasoning) that could explain an anomaly (Rajamoney, 1989a). These revisions were proposed by the proponents of the phlogiston theory, and they preceded the chemical revolution.

5. Discussion

This chapter proposed explanation-based theory revision as a computational approach to scientific discovery. However, this method is still a considerable distance from accounting for all the subtleties and nuances of theory revision. For instance, the history of science shows that a number of competing scientific theories are entertained prior to the dominance of a single theory (Kuhn, 1970). Each theory accounts for some of the known observations and either ignores the remaining anomalous observations or attempts to finesse them by ad hoc elaborations.

Explanation-based theory revision cannot model this aspect of theory development, since it zealously forces a theory to assimilate the anomalous observations. Also, it presently does not model the coexistence of competing theories of different explanatory power. In addition, the framework must be extended to incorporate quantitative information and extended qualitative vocabularies like those described by Raiman (1986).

Despite the limitations, explanation-based theory revision represents an important step toward understanding scientific discovery and toward automating this process. Future work on the framework will address these limitations and will attempt to replicate theory revision episodes from the history of science.

Acknowledgements

This work was supported by the Office of Naval Research under grant N-00014-86-K-0309 and a Faculty Research Initiation grant from the University of Southern California. I thank Gerald DeJong for his support, advice, and numerous discussions on the topic. Thanks are also due to Ken Forbus, Bob Stepp, Dedre Gentner, Uday Reddy, Brian Falkenhainer; to the present and past members of the machine learning group at the Coordinated Science Laboratory of the University of Illinois at Urbana-Champaign for helpful advice and discussions; and to Pat Langley and Jeff Shrager for providing copious comments and suggestions on drafts of this paper.

References

DeJong, G., & Mooney, R. (1986). Explanation-based learning: An alternative view. *Machine Learning, 1*, 145–176.

Dietterich, T. G. (1986). Learning at the knowledge level. *Machine Learning, 1*, 287–316.

Falkenhainer, B. (1989). *Learning from physical analogies: A study in the analogy and the explanation process*. Doctoral dissertation, Department of Computer Science, University of Illinois at Urbana-Champaign, Urbana.

Falkenhainer, B., & Rajamoney, S. (1988). The interdependencies of theory formation, revision, and experimentation. *Proceedings of the Fifth International Conference on Machine Learning* (pp. 353–366). Ann Arbor, MI: Morgan Kaufmann.

Forbus, K. (1984). Qualitative process theory. *Artificial Intelligence*, *24*, 85–168.

Karp, P. (1989). *Hypothesis formation and qualitative reasoning in molecular biology*. Doctoral dissertation, Department of Computer Science, Stanford University.

Kuhn, T. (1970). *The structure of scientific revolutions*. Chicago: University of Chicago Press.

Kulkarni, D. (1989). *The processes of scientific research: The strategy of experimentation*. Doctoral dissertation, Department of Computer Science, Carnegie Mellon University, Pittsburgh.

Mitchell, T., Keller, R., & Kedar-Cabelli, S. (1986). Explanation-based generalization: A unifying view. *Machine Learning*, *1*, 47–80.

Raiman, O. (1986). Order of magnitude reasoning. *Proceedings of the Fifth National Conference on Artificial Intelligence* (pp. 100–104). Philadelphia: Morgan Kaufmann.

Rajamoney, S. (1986). *Automated design of experiments for refining theories*. Master's thesis, Department of Computer Science, University of Illinois at Urbana-Champaign.

Rajamoney, S. (1989a). *Explanation-based theory revision: An approach to the problems of incomplete and incorrect theories*. Doctoral dissertation, Department of Computer Science, University of Illinois at Urbana-Champaign.

Rajamoney, S. (1989b). Exemplar-based theory rejection: An approach to the experience consistency problem. *Proceedings of the Sixth International Workshop on Machine Learning* (pp. 284–289). Ithaca, NY: Morgan Kaufmann.

Rajamoney, S., DeJong, G., & Faltings, B. (1985). Towards a model of conceptual knowledge acquisition. *Proceedings of the Ninth International Joint Conference on Artificial Intelligence* (pp. 688–690). Los Angeles: Morgan Kaufmann.

Rajamoney, S., & DeJong, G. (1987). The classification, detection and handling of imperfect theory problems. *Proceedings of the Tenth International Joint Conference on Artificial Intelligence* (pp. 205–207). Milan, Italy: Morgan Kaufmann.

Rajamoney, S., & DeJong, G. (1988). Active explanation reduction: An approach to the multiple explanations problem. *Proceedings of the Fifth International Conference on Machine Learning* (pp. 242–255). Ann Arbor, MI: Morgan Kaufmann.

Rose, D., & Langley, P. (1986). Chemical discovery as belief revision. *Machine Learning*, *1*, 423–452.

Rose, D., & Langley, P. (1989). A hill-climbing approach to machine discovery. *Proceedings of the Fifth International Conference on Machine Learning* (pp. 367–373). Ann Arbor, MI: Morgan Kaufmann.

Shrager, J. (1987). Theory change via view application in instructionless learning. *Machine Learning*, *2*, 247–276.

CHAPTER 9

Experimentation in Machine Discovery

DEEPAK KULKARNI[1]
HERBERT A. SIMON

1. Introduction

Although experimentation plays an important role in many fields of science, only a few studies have focused on the role of experimentation in scientific discoveries (Friedland, 1979; Karp, 1989; Rajamoney, this volume). This chapter describes KEKADA, a system capable of carrying out a complex series of experiments.

The study of surprising phenomena is an important research task in many experimental domains. This task can be formulated as:

- *Given:* A surprising phenomenon;
- *Given:* A set of facts and hypotheses about the domain;
- *Do:* Carry out an experimentation program, acquire the results of the experiments, and use these results to refine the domain theory incrementally.

Unlike the programs developed by Karp and Rajamoney described elsewhere in this volume, KEKADA is capable of detecting surprising phenomena and using these surprises to guide its attempts to revise the domain theory. In the next section, we describe the representation used in the system. After this, we describe the program's control and its basic

---

1. The current address of the first author is Artificial Intelligence Research Branch, Mail Stop 244-17, NASA Ames Research Center, Moffett Field, CA 94035. The paper was written while he was visiting the University of California, Irvine.

Table 1. Representation of an experiment in Kekada.

Current goal: Determine relevant factors of a phenomenon.
Independent entity: Name = Ornithine, Concentration = Medium.
Apparatus variables: Method = Tissue-slice, pH = 8.
Measure: Rates of production of the outputs of the reaction.
Dependent entity: Expected-Name = Urea, Expected-Rate = < 2, 10 >.

---

processes. In the fourth section, we present an example of Kekada's behavior on a particular problem. Then in the fifth section, we analyze the abilities of the system, after which we draw some tentative conclusions.

2. Representation in Kekada

We will first describe Kekada's representation of its knowledge, including the experiments it suggests and the theories it refines. An experiment is an operation carried out under well-specified conditions to determine an unknown effect. For example, one may immerse liver tissue slices in a solution of ornithine, maintained at pH 8, and carry out certain tests to measure the results. A substance like ornithine, which the experimenter includes in the experimental setup, is called an *independent entity*. An experiment may also result in production of new *dependent entities*.

Each entity has an associated set of independent, dependent, or apparatus variables. For example, ornithine might have *concentration* as an independent variable and *rate of consumption* as a dependent variable. An apparatus variable plays an auxiliary role in an experiment and is not a direct cause of the results of the experiment. In the scenario that we described earlier, the experimenter uses the tissue slice method and maintains the pH at 8. In this case, both *pH* and *method* are apparatus variables. Kekada's representation of this particular experiment is shown in Table 1.

In addition to the different entities and variables, this experiment has as a goal to find relevant factors in a previously observed phenomenon. Before any experiment is carried out, Kekada associates expectations about the values of one or more dependent entities and associated vari-

Table 2. Representation of an hypothesis in KEKADA.

```
Process:  Ornithine Cycle consists of P1, P2, and P3.
```
Process Name: P1
```
     Independent Entities:  Ammonia, Carbon dioxide, Ornithine
     Dependent Entities:  Citrulline, Water
```

Process Name: P2
```
     Independent Entities:  Citrulline, Ammonia
     Dependent Entities:  Arginine, Water
```

Process Name: P3
```
     Independent Entities:  Arginine, Water
     Dependent Entities:  Ornithine, Urea
```

---

ables. For numeric attributes, lower and upper bounds on the value are specified. For symbolic attributes, a single nominal value is specified. In the experiment in Table 1, the system expects urea to be produced at a rate between 2 and 10.

Whereas KEKADA represents experiments in such specific terms, it represents hypotheses at various levels of abstraction. A hypothesis is described as a sequence of processes, where each process has the same basic representation as an experiment. For example, consider a process that is a sequence of three reactions: P1, P2, and P3. Suppose that in P1, the substances ammonia, carbon dioxide, and ornithine combine to produce citrulline and water. In P2, citrulline and ammonia combine to produce arginine and water. In P3, arginine and water combine to produce ornithine and urea. The representation of this hypothesis is shown in Table 2.

Hypotheses can be even more abstract than the one shown in Table 2. Thus, one hypothesis might specify that ornithine donates the amino group to urea in a reaction, and another hypothesis might specify that ornithine acts as a catalyst, as in:

```
Type:  donates-group, Donor:  Ornithine, Group:  Amino, Receiver:  Urea.
```
```
Type:  is-catalyst, Reactant:  Ornithine.
```

Each hypothesis in KEKADA has an associated confidence vector, which is represented as a 5-tuple <success, failure, failed-effort, implied-success, implied-failure>. The success slot stores the number of experiments that have verified a hypothesis, whereas the failure slot stores the number of experiments that have failed to support a hypothesis. The implied-success slot stores the number of experiments that are a positive but inconclusive indication of the validity of a hypothesis. The failed-effort slot stores the amount of effort spent to find positive instances of an existential hypothesis. Finally, the implied-failure slot stores the number of experiments that indicate, but not conclusively, that a hypothesis is false.

3.   Control Structure and Processes

In this section, we will examine KEKADA's control structure and basic processes. The control structure incorporates two high-level techniques: *heuristic search through two problem spaces and use of surprises to direct the search*. We first discuss these techniques and then describe various components in the system.

3.1   Dual Space Search and Surprises

The basic source of KEKADA's new knowledge is the environment. The system carries out experiments on the external world to gather new information and to modify confidences in existing beliefs. Thus it searches two spaces, one containing hypotheses and the other containing experiments and results (Simon & Lea, 1974). On the basis of the current state of the hypothesis space (existing hypotheses and their confidences), the system chooses an experiment to carry out. It then interprets the outcome of the experiment, modifying its hypotheses and their confidences. Figure 1 shows this organization in a graphic form.

Let us consider an example of how KEKADA uses experiments. If the system is studying the hypothesis that a specific substance, ornithine, is acting as a catalyst in a given reaction, it may decide to carry out an experiment to verify this belief. Thus it may measure the amount of urea produced in the presence of a small amount of ornithine. If a large amount of urea is produced in this experiment, this lends evidence to the hypothesis.

Experimentation

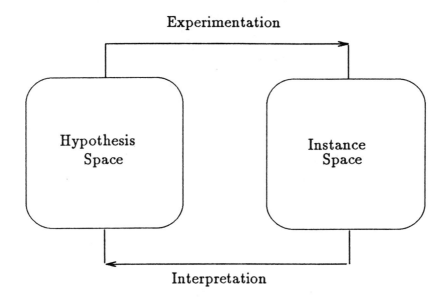

Interpretation

Figure 1. KEKADA's dual space search organization.

KEKADA focuses its attention on surprises to constrain its search. Surprises have played a central role in many important discoveries. For example, in the course of years of research that produced many important results, Priestley observed that "the first hints, at least of almost everything I have discovered, of much importance, have occurred to me in this manner (as unexpected phenomena)" (Conant, 1957). KEKADA attends to surprises, thereby searching the parts of its experiment space dense with useful phenomena. To this end, it associates expectations with each experiment. Thus, if it is carrying out an experiment on ornithine and ammonia in the liver tissue slices, its prior experiences may lead it to expect that the experiment would produce urea at some rate between 2 and 10 units. If urea is produced at the rate of 20 units, the outcome violates expectations. In this case, KEKADA lacked a whole body of knowledge and thus had wrong expectations about the results of the experiment. Focusing attention on this surprise puts a powerful constraint on the system's search for new knowledge.

3.2   Overall Control Structure

The overall control structure of KEKADA,[2] shown in Table 3, allows it
to carry out this dual space search, to detect surprises, and to focus
on them. The system has a set of strategies, each with three associated parts: *hypothesis generators, experiment proposers,* and *experiment
evaluators.* Each of these consists of a set of rules in the form of conditions and actions. The input to KEKADA is a surprising phenomenon:
the expectations and the results of an experiment. First, the hypothesis generators whose conditions are satisfied produce hypotheses. Then
the system chooses one of these hypotheses using a preference scheme
described in Kulkarni (1988). The strategy whose hypothesis generator
suggested this hypothesis also has components for proposing experiments and evaluating them. Next the experiment proposers suggest a
number of experiments, and a set of *expectation setters* generates expectations for them. At this stage, the user provides the system with
the results of these experiments. If these results violate the expectations for the experiment, KEKADA detects this as a surprise and makes
a recursive call to itself. If no surprise is detected, then the experiment
evaluators interpret the results of the experiment. The system repeats
this cycle, choosing an hypothesis and carrying out experiments until it
encounters a recursive call to itself or all the current hypotheses have
been ruled out.

The system can also be viewed as carrying out a mixture of breadth-
first and depth-first search through its hypothesis and experiment spaces.
Under normal circumstances, KEKADA iterates through each hypothesis
and through each of the experiments generated to test these hypotheses;[3]
one can view this behavior as a form of breadth-first search. However,
when the result of an experiment is surprising, the system abandons this
search and focuses attention on the newly found surprise. In essence,
KEKADA forgets its previous goals and attends entirely to the new phenomenon, using it to generate a new set of hypotheses, and using these

---

2. KEKADA is implemented in the production system language OPS5 (Brownston,
   Farrell, Kant, & Martin, 1985). A production system consists of two main components: a set of condition-action rules, or *productions*, and a dynamic working
   memory.
3. Kulkarni (1988) describes the conditions under which the system employs a different strategy.

Table 3. The control structure of KEKADA.

```
Inputs:
    R: the results of the experiment
    EXP: the expectations for the experiment
Procedure KEKADA (R, EXP)
  Generate a set of hypotheses HL using the hypothesis
      generators whose conditions are satisfied.
  Repeat
    Choose hypothesis HS from the set HL.
    Let S be the strategy whose hypothesis generator suggested
        the hypothesis HS.
    Generate a set of experiments EL using the experiment
        proposers associated with the strategy S.
    For each experiment E in the set EL,
        Generate expectations EXP' for the experiment E.
        Carry out the experiment E to get results (R').
        If results R' are not within the expectation EXP',
            Then KEKADA (R', EXP');
            Else interpret the results of the experiment E
                using the experiment evaluators associated
                with the hypothesis HS.
```

---

to guide further experimentation. One can view this control scheme as a form of depth-first search without backtracking, with the system being easily "distracted" by unexpected results. As we will see later, this counterintuitive strategy works quite well in the scientific domains we have examined.

3.3 KEKADA's Strategies

To constrain the search in the problem spaces, the system employs a small set of strategies to formulate a small number of good hypotheses and to suggest a few informative experiments. In addition to KEKADA's heuristic of focusing on surprises, these strategies are the system's basic source of power. We will now describe these strategies.

One strategy that KEKADA employs is to attempt to magnify a surprising phenomenon. For instance, if the system observes that switching on the electric current in a coil produces electric current in an adjacent coil, it would try changing the apparatus to increase the electric current in the second coil. If the surprising phenomenon has one or more apparatus variables associated with it, then the hypothesis generators suggest that the phenomenon may be magnified upon changing the value of the apparatus variables. When this hypothesis is chosen, the experiment proposers suggest a number of experiments in which one apparatus variable has a different value from that in the surprising phenomenon, and all the other variables have the same values.

If the experiment results in magnifying the surprising phenomenon,[4] then all the future experiments characterizing the surprising phenomenon are carried out with the new set of values of the apparatus variable. This increases the chances of making crucial observations on further experimentation with the surprising phenomenon. For instance, suppose that, in one of its attempts to magnify the electric current, the system increases the length of the first coil to its maximum value, and this manipulation results in a significantly larger current in the second coil. KEKADA would then carry out all the further experiments with this longer coil, enabling it to make some crucial experimental observations.

A second strategy is a specific implementation of the divide and conquer method. If the surprising phenomenon is known to contain subprocesses, the hypothesis generators create the hypothesis that one of the subprocesses is behaving in an unexpected fashion. When the hypothesis is chosen, the experiment proposers suggest a number of experiments on this subprocess in the same manner. If the results reveal that the subprocess is behaving in an unexpected manner, the experiment evaluators make a recursive call to KEKADA. For instance, suppose the system finds that alanine is not producing urea as expected, and it knows that this process has two subprocesses. In this case, the hypothesis generators would suggest the hypothesis that one of two subprocess is behaving in an unexpected manner and would invoke KEKADA to resolve the surprise.

A third strategy involves trying to assess the scope of the surprising phenomenon using domain-specific taxonomies. If one of the indepen-

---

4. In these experiments, the non-numeric variables can have all possible values, but the numeric variables can have only the maximum and the minimum values.

dent variables involved in the surprising phenomenon belongs to a general class, then the hypothesis generators would suggest a hypothesis that this phenomenon is exhibited generally by members of this class. When this hypothesis is chosen, the experiment proposers would select members of the class, based on cost and availability, and test experimentally whether the surprising phenomenon is also exhibited by them. When the number of members exhibiting this phenomenon exceeds the threshold value of three, the experiment evaluators interpret that the hypothesis is correct, generalize the description of the surprising phenomenon, and make a recursive call to KEKADA. In contrast, when the number of members that do not exhibit this surprising phenomenon exceeds three, the system inactivates this hypothesis. For example, when KEKADA studies the surprising phenomenon in which ornithine produces ammonia in kidney tissue, its hypothesis generators suggest a number of scope hypotheses. Upon carrying out various experiments, the system infers that the effect is exhibited generally by amino acids.

A fourth strategy involves finding the relevant factors in a surprising phenomenon. If the phenomenon has two or more independent entities, then the hypothesis generator would suggest a hypothesis that all the independent entities are not necessary to produce this effect. When this hypothesis is chosen, experiment proposers would suggest a number of experiments, each of which has all the independent entities in the phenomenon except for one. If one of these experiments exhibits the surprising effect, the experiment evaluators infer that the effect can be exhibited with fewer independent entities than those in the originally observed surprising phenomenon and then make a recursive call to KEKADA. For instance, suppose the system observes that, when electric current is switched on in an experimental setup, electric current appears in another coil. The system would then examine which factors are necessary to cause this effect. Now suppose it finds that switching electric current in a coil alone is sufficient to cause the effect. The system would then focus on this effect.

A fifth strategy involves looking for phenomena that are similar to the surprising phenomena in some way and then trying to find some relation between them. In particular, if two anomalous effects include the same variable, then the hypothesis generators suggest two hypotheses. The first hypothesis is that a larger class of values of this variable would exhibit anomalous effects. The experiment proposers and experiment evaluators associated with such a scope hypothesis were described earlier. The second hypothesis is that the two effects are part of a more complex mechanism of the form A → B → C. The experiment proposer for this hypothesis is specific to the domain of metabolic biochemistry in the early 1900s. It suggests that one should measure the rates of formation of C from A and from B. If the second rate is slower than the first, then the experiment evaluators increment the implied-failure attribute of the confidence vector for the hypothesis. In contrast, if the first rate is slower than the second rate, then the experiment evaluators increment the implied-success attribute of the confidence vector for the hypothesis.

KEKADA employs these strategies, along with a few others, to create a small number of hypotheses. Each hypothesis has a priority level associated with it, and the system prefers hypotheses with higher priority levels. Furthermore, it chooses between hypotheses with the same priority level using the confidence vectors. Kulkarni (1988) describes both the strategies and the preference scheme in detail.

4.   Rediscovery of Glutamine Synthesis

Now that we have described the representation, the control structure, and the processes in KEKADA, we will examine the behavior of the system on a particular problem from the history of science. In 1933, the biochemist Hans Krebs worked on the problem of understanding the nature of amino acid metabolisms. He established that the deamination of amino acids (i.e., the removal of the amino group from the amino acids) occurs by an oxidative reaction in the kidney, and not in the liver, as had been previously assumed. He also produced data on the deamination rates of various amino acids. Furthermore, he showed that glutamic acid combines with ammonia to produce glutamine, a substance that was not previously known to play any role in metabolism. The discovery of the glutamine reaction opened a whole set of new questions in metabolic biochemistry.

Here we examine KEKADA's behavior when it is given a problem similar to that faced by Hans Krebs. The problem can be stated as:

- *Given:* A surprising reaction in which ornithine produces ammonia in the presence of kidney tissue slices;

- *Given:* Three previously postulated reaction pathways about the deamination reaction (described later);

- *Given:* Background knowledge about the chemistry of various substances;
- *Do:* Revise the existing domain theory for amino acid metabolisms.

KEKADA's behavior on this research problem can be divided into three stages: characterization of the ornithine-in-kidney effect, study of the deamination reaction, and discovery of the glutamine reaction.

In the first stage, the system tries to characterize the surprising phenomenon in which ornithine produces ammonia in the kidney. Experiments reveal that other amino acids are also able to produce the effect, so KEKADA concludes that the phenomenon is a specific instance of the more general deamination reaction. During the second stage, the system carries out a variety of experiments that reveal details of the deamination reaction. In the process, the system comes across an unusual reaction in which the presence of arsenite increases the production of ammonia from the glutamic acid. In the final stage, KEKADA conjectures that glutamic acid combines with ammonia. The system verifies experimentally that this reaction occurs and that it produces glutamine as output. Below we describe this discovery process in greater detail.

4.1   Characterization of the Ornithine-in-Kidney Effect

In response to the surprise that ornithine produces ammonia in kidney, the hypothesis generators suggest a number of alternative explanations. One of the generators suggests assessing the scope of the phenomenon whenever the reactants in the observed phenomenon belong to a class of substances. In this case, ornithine belongs to the class of amino acids, amines, and carboxylic acids. Therefore, this generator suggests a hypothesis that the effect may be common to one of these classes. If the phenomenon has at least one apparatus variable associated with it, the *magnification generator* suggests attempting to magnify the effect. As ornithine has two amino groups and ammonia also has an amino group, another hypothesis generator suggests that ornithine is donating one of its amino groups to ammonia.

For reasons discussed in Kulkarni (1988), the preference scheme in KEKADA prefers the magnification hypothesis over other hypotheses. Thus the system attempts to magnify the observed effect. However, attempts to magnify the phenomenon by varying the apparatus variables fail, so next KEKADA decides to assess the scope of the surprising phenomenon. This leads to experiments on a number of amino acids, which reveal that other amino acids can also produce ammonia in the kidney. When four different amino acids are found to produce ammonia in the kidney, experiment evaluators generalize the original observation and make a recursive call to KEKADA.

4.2   Study of the Deamination Reaction

Now that the system has determined that the surprising phenomenon is exhibited by amino acids in general, the hypothesis generators suggest a number of new hypotheses. As the observed phenomenon shares a number of entities with deamination, one hypothesis generator suggests that the observed phenomenon may be deamination and thus is one of the three previously hypothesized deamination reactions:

- Hydrolytic: $R\text{-}CH\text{-}COOH + H_2O = R\text{-}CHOH\text{-}COOH + NH_3$
- Reductive: $R\text{-}CH(NH_2)\text{-}COOH + H_2 = R\text{-}CH_2\text{-}COOH + NH_3$
- Oxidative: $2\ R\text{-}CH\ (NH_2)\text{-}COOH + O_2 = 2\ R\text{-}CO\text{-}COOH + 2\ NH_3$

The original observation of the effect was specific to ornithine. In this case, KEKADA established the generality of the effect, which let it use general knowledge about amino acids to create additional hypotheses. The system first decides to verify the oxidative hypothesis, but it must choose an amino acid to carry out an experiment. KEKADA decides to carry out the experiment on alanine because this substance is both reactive and cheap. Here we see another advantage of the strategy of assessing the generality of a surprise. If the phenomenon turns out to be general, the experimenter has more choice in choosing a variable in the phenomenon. To verify the oxidative reaction, KEKADA carries out experiments on alanine and oxygen together. The results are consistent with the chemistry of the oxidative reaction.

The system next decides to gather more data on the deamination of amino acids. After carrying out experiments on a number of other amino acids, it carries out an experiment on glutamic acid. KEKADA expects the glutamic acid to deaminate in a similar way to other amino acids, as expectation setters in the system use the information of a class to set expectations about specific substances. However, the system finds that the rate of production of keto acid from glutamic acid is lower than expected. KEKADA notes this as a surprise and makes glutamic acid reaction the focus of attention.

As the system focuses on this surprise, the hypothesis generators associated with various strategies suggest a number of hypotheses. Using its preference scheme, KEKADA first chooses the magnification hypothesis. It then attempts unsuccessfully to magnify the phenomenon by changing the tissue and the aerobic conditions.

Next, the system selects the hypothesis that addition of an inhibitor substance would selectively block a side reaction consuming the keto acid. Expectation setters use the chemistry-specific knowledge about the action of inhibitors to make predictions about the results of the experiment. KEKADA does not predict whether the rate of production of keto acid will increase, but it does expect that the rate of production of ammonia will not be affected. When the reaction is carried out with arsenite, a particular inhibitor, the rate of production of both keto acid and ammonia increases. This violates the expectations, leading the system to focus its attention on this newly found surprising phenomenon.

4.3   Discovery of the Glutamine Reaction

At this point, the hypothesis generators associated with the various strategies in KEKADA suggest a number of different hypotheses. Among others, these include the following:

- The phenomenon may be common to some larger class, such as the class of carboxylic acids, amino acids, or amines.
- Ammonia may be reacting with a reactant in a side reaction.

Based on its preference scheme, KEKADA decides to test the hypothesis that the phenomenon may be common to the class of carboxyllic acids. However, after getting negative results for aspartic acid and other carboxylic acids, the system reduces its confidence in this hypothesis. At this point, the system decides to consider the hypothesis that ammonia is reacting with one of the other reactants in a side reaction. When it carries out a reaction with glutamic acid and ammonia, glutamine is produced. Thus the system discovers the important glutamine reaction.

KEKADA's operation on the ornithine-in-kidney problem produced a number of interesting results. It established that deamination of amino acids occurs by an oxidative reaction in the kidney, and not in the liver, as had been previously assumed. It produced data on deamination rates of various amino acids. Furthermore, it showed that glutamic acid combines with ammonia to produce glutamine, a substance that was not previously known to play any role in metabolisms. The basic source of KEKADA's power was its ability to create a small number of good hypotheses and experiments and to focus on surprises.

4.4 Summary of the Discovery Process

Table 4 shows the state of KEKADA's knowledge at five steps, labeled S1, S2, S3, S4, and S5, in the operation just described. Each step is associated with the knowledge KEKADA has at that point in the run. Initially (S1), the system lacks the knowledge about amino acid metabolisms that it would acquire at the end of the run (S5). At S1, it does not know the exact nature of the deamination reaction, that it occurs in the kidney, or that it involves a glutamine reaction.

At this point, the system attends to the surprise that ornithine produces ammonia in the kidney. This leads it to carry out experiments mainly on various amino acids in the kidney tissue slices. In doing so, it searches a small part of the problem space that is rich in informative experiments. Thus, the surprise about the ornithine reaction in the kidney constrains KEKADA's search. The system then discovers that amino acids deaminate in the kidney and focuses on this general effect (S2); at this point, it has generalized the ornithine effect that it knew at S1.

At a later stage, when the system detects the surprising glutamic acid effect, it attends to this effect instead of to the previously found surprises (S3). This leads it to carry out a number of experiments on glutamic acid. Again, the surprise about glutamic reaction constrains the system's search to a smaller problem space that contains the glutamine reaction. Finally, the system discovers this reaction (S5). In summary, KEKADA employs a greedy strategy of focusing on a surprise as soon as it is detected, and this constrains its search.

5. Discussion

In this section, we evaluate KEKADA's performance and compare the system with other programs. In addition, we identify the limitations of the system and the ways it could be improved.

Table 4. KEKADA's knowledge at various points in the discovery of glutamine synthesis.

S1: The system focuses attention on the ornithine-in-kidney effect.

```
ornithine → ammonia in kidney
```

S2: The system focuses attention on the amino-acids-in-kidney effect.

```
amino-acid → ammonia in kidney
```

S3: The system focuses attention on the glutamic acid effect.

```
amino-acid + oxygen → keto-acid + ammonia
glutamic acid → alpha-keto-glutarate (<< 1) +
   ammonia (< 1)
```

S4: The system focuses attention on the arsenite effect.

```
amino-acid + oxygen → keto-acid + ammonia
glutamic acid → alpha-keto-glutarate (<< 1) +
   ammonia (< 1)
glutamic acid + arsenite →
   alpha-keto-glutarate + ammonia
```

S5: The system has discovered the glutamine reaction.

```
amino-acid + oxygen → keto-acid + ammonia
glutamic acid → alpha-keto-glutarate (<< 1) +
   ammonia (< 1)
glutamic acid + arsenite →
   alpha-keto-glutarate + ammonia
glutamic acid + ammonia → glutamine
```

5.1 Evaluation of KEKADA's Performance

A scientist is judged by the results of his or her research, and KEKADA can similarly be judged by the discoveries it has made. We have tested the system on a number of different problems. In the previous section, we discussed the general problem of amino acid metabolisms and how KEKADA responds to this problem. We also tested the system on the problem of urea synthesis in the body, which had been an open problem until 1932. In that year, the biochemist Hans Krebs showed that urea is synthesized in a cyclic mechanism, which KEKADA was able to rediscover. Another historical problem involved elucidating the structure of common alcohol. In 1853, Williamson showed that this substance has an ethyl group attached to an hydroxyl group, and KEKADA was able to rediscover this structure. In yet another run, the system was able to rediscover magneto-electric induction.

These successes demonstrate the generality of the overall system, but it is also important that a system's components themselves be general. KEKADA has 43 heuristics, of which 28 are domain independent. The remaining 15 are specific to a domain, such as biochemistry, but none is specific to a particular problem, such as urea synthesis. The system used 31 of its 43 heuristics in solving more than one task in the runs.

The generality of the system is also implied by its psychological plausibility. KEKADA's behavior matches very well against the behavior of scientists; Kulkarni and Simon (1988) have shown that the system is a good model of the heuristics Hans Krebs used in his discovery. The generality of KEKADA's heuristics and their applicability to a variety of problems suggest that the system would be effective in solving a wide class of tasks.

5.2   Relation to Other Research

It is useful to compare KEKADA with related work on experimentation and theory revision in machine discovery. For instance, Friedland (1979) studied the problem of producing a plan for an experiment and developed MOLGEN, a program that uses skeletal plan refinement to this end. In contrast, KEKADA carries out a complex series of experiments, changing its goals significantly along the way, but its specification of an experiment is abstract. Thus, a specification, such as "carry out an experiment with ornithine and ammonia with certain concentrations in liver using the tissue slice method," leaves out many details about how this experiment is to be carried out. The MOLGEN system can refine an abstract description into a detailed plan for execution of the experiment.

Two outgrowths of the original MOLGEN project were the programs called GENSIM and HYPGENE (Karp, 1989, this volume). These systems simulate a number of discoveries in molecular genetics. Given an experiment and a theory, GENSIM predicts the results of the experiment. If there is a mismatch between these predictions and the observations, HYPGENE infers modifications in the theory and the initial conditions of the experiment. To do this, it reasons backward from the differences between the predictions and the observations. Another novel feature of these systems is a powerful representation that is used to reason qualitatively about complex functional relations between variables. Both systems require prior knowledge to resolve an anomaly, and neither can carry out an extensive experimentation program.

Another program that explains anomalous experiments is ADEPT (Rajamoney, this volume). This program uses qualitative processes to represent theories and produces an explanation of an anomalous observation in terms of these processes. It then carries out experiments to confirm it. Like HYPGENE, the system has a focused goal of explaining a particular phenomenon. In contrast, KEKADA has a more general goal of revising its theories by carrying out informative experiments. It uses the surprising effect as an opportunity to acquire more knowledge. Even in situations where a surprise cannot be explained using existing knowledge, KEKADA can be effective in revising the theories. Furthermore, it can detect new surprises and focus on these.

Other researchers (Klahr & Dunbar, 1988; Shrager & Klahr, 1986) have studied experimentation and theory revision in human subjects' understanding of complex devices. Their SDDS model accounts for fine-grained data in the form of think-aloud protocols; in contrast, KEKADA is based on coarse-grained data on historical discoveries. However, there are also commonalities between SDDS (Klahr & Dunbar, 1988) and KEKADA, such as their use of a dual-space framework.

5.3 Limitations and Future Work

Although we found that KEKADA's behavior matched well against that of scientists, our comparison also revealed some limitations of the system. We found that scientists appeared to employ a number of additional heuristics, some that were not useful on these particular problems but that would be useful on many other problems. As a result, they are able to solve significantly more problems than the system. One direction for future work is to develop a scientist's assistant by incorporating additional discovery mechanisms, along with a large body of knowledge about a given domain.

This would require representations that support both qualitative and quantitative reasoning, along with an architecture that can support a number of different mechanisms. A possible extension would be to adopt ideas from Nordhausen and Langley's IDS (this volume), a system that integrates taxonomy formation, discovery of qualitative laws, and discovery of numeric laws. Their approach should support KEKADA's scope and factor-analysis strategies with a few changes, and it also provides an extended representation for experiments. Future work should attempt to integrate insights from the KEKADA and IDS systems.

6.   Conclusion

In this chapter, we described KEKADA, a system that is capable of carrying out a complex series of experiments on problems from the history of science. The system incorporates a set of experimentation strategies that were extracted from the traces of the scientists' behaviors. KEKADA focuses on surprises to constrain its search, and uses its strategies to generate hypotheses and to carry out experiments. Some strategies are domain independent, whereas others incorporate knowledge of a specific domain. The domain-independent strategies include magnification, determining scope, divide and conquer, factor analysis, and relating different anomalous phenomena. KEKADA represents an experiment as a set of independent and dependent entities, with apparatus variables and a goal. It represents a theory either as a sequence of processes or as abstract hypotheses.

In this chapter, we described KEKADA's response to a particular problem in biochemistry. On this and other problems, the system is capable of carrying out a complex series of experiments to refine domain theories. Analysis of the system and its behavior on a number of different problems has established its generality, but it has also revealed the reasons why the system, in its present form, would not be a good experimental scientist. Nevertheless, we believe our work advances the state of research on scientific discovery by proposing a set of computational strategies that can be applied to a wide variety of domains.

Acknowledgements

Editorial comments by Pat Langley have greatly improved the presentation of this chapter. Our work has benefited from discussions with

Allen Newell, David Klahr, Tom Mitchell, Raul Valdes-Perez, and Peter Jansen. Thanks are also due to Kate McKusick for her comments on a draft of this chapter. The research was supported by Contract F33615-87-C-1499 from the Defense Advanced Research Projects Agency.

References

Brownston, L., Farrell, R., Kant, E., & Martin, N. (1985). *Programming expert systems in OPS5: An introduction to rule-based programming*. Reading, MA: Addison-Wesley.

Conant, J. (Ed.) (1957). Harvard case histories in experimental science. Cambridge, MA: Harvard University Press.

Friedland, P. (1979). Knowledge based experiment design in molecular genetics. *Proceedings of the Sixth International Joint Conference on Artificial Intelligence* (pp. 285–287). Tokyo, Japan: Morgan Kaufmann.

Karp, P. (1989). *Hypothesis formation and qualitative reasoning in molecular biology* (Technical Report KSL-89-52). Stanford, CA: Stanford University, Department of Computer Science.

Klahr, D., & Dunbar, K. (1988). Dual space search during scientific reasoning. *Cognitive Science*, *12*, 1–48.

Kulkarni, D., & Simon, H. A. (1988). The processes of scientific discovery: The strategy of experimentation. *Cognitive Science*, *12*, 139–175.

Kulkarni, D. (1988). *The processes of scientific research: The strategy of experimentation*. Doctoral dissertation, Department of Computer Science, Carnegie Mellon University, Pittsburgh, PA.

Shrager, J., & Klahr, D. (1986). Instructionless learning about a complex device: The paradigm and observations. *International Journal of Man-Machine Studies*, *25*, 153–189.

Simon, H. A., & Lea, G. (1974). Problem solving and rule induction: A unified view. In L. Gregg (Ed.), *Knowledge and cognition*. Hillsdale, NJ: Lawrence Erlbaum.

CHAPTER 10

Hypothesis Formation As Design

PETER D. KARP

1. Introduction

In the 1960s and 1970s, biologists discovered a new mechanism of gene regulation in bacteria called *attenuation*. Charles Yanofsky (1981) and his colleagues discovered attenuation in a gene regulation system called the *tryptophan (trp) operon*. This chapter describes a computational investigation of scientific reasoning that is based on an analysis of the biological research on the trp operon.

Karp (1989) performed a detailed historical study of the discovery of attenuation in which he reconstructed the different intermediate states of knowledge that the biologists possessed as their understanding of the trp operon evolved. He analyzed the differences among these states of knowledge to elucidate examples of how and why the biologists modified their theories, such as by postulating the existence of a new chemical reaction.

This chapter describes two computer programs. The GENSIM program provides a framework for representing theories of molecular biology and has been used to represent a theory of bacterial gene regulation. GENSIM can use these theories to predict the outcomes of biological experiments. The HYPGENE program formulates hypotheses that improve the predictive power of GENSIM theories, given experimental data. Both programs have been tested on examples from the history of attenuation.

I argue that it is productive to treat the task of hypothesis formation as a *design* problem because AI methods developed for design and planning are well suited to the task of hypothesis formation. In order to treat hypothesis formation as a design problem, I view a hypothesis as an *artifact* to be synthesized. Its synthesis is performed subject to *design constraints,* such as the constraint that predictions generated in the context of the hypothesis should match experimental observations. HYPGENE is a designer of hypotheses. It uses *design operators* to modify a theory to satisfy design constraints. I derived these operators from the study of the history of attenuation and by considering the space of allowable syntactic changes to the GENSIM representation language. This approach to hypothesis formation is theory-driven because it assumes that one of HYPGENE's inputs is a good — but not perfect — theory for predicting experimental outcomes.

I have used the history of attenuation as a testbed for developing computational models of scientific reasoning. Scientists perform a number of different reasoning tasks, such as experiment design and hypothesis formation. Figure 1 provides a rough outline of these tasks and how they are related. The work described in this chapter addresses only a subset of the tasks in Figure 1—the representation of theories and experiments in molecular biology, the use of a theory to predict experimental outcomes, and the formulation of hypotheses that improve the predictive power of a theory. I do not consider other important tasks, such as designing experiments, selecting experimental goals, or maintaining libraries of significant experiments.

This chapter begins with a description of the GENSIM program. Section 2 outlines GENSIM's framework for representing theories and experiments in molecular biology and describes the manner in which GENSIM uses a theory to predict experimental outcomes. Section 3 states the hypothesis-formation problem, which occurs when GENSIM's prediction of an experimental outcome does not match the observed outcome of the experiment. A sample hypothesis-formation problem is presented, followed by a trace of HYPGENE's derivation of a solution. The problem is specified as a *design goal,* and HYPGENE utilizes *design operators* to satisfy that goal. The remainder of the chapter presents the methods that the program uses to generate hypotheses, including how it represents and manages its design goals, what design operators HYPGENE employs, and how the application of these operators is controlled.[1]

---

1. It is important to note that I seek to model the *competence* of biologists, but not their *performance.* That is, I wish to identify reasoning mechanisms that are sufficient to solve hypothesis-formation problems from the history of attenuation; these reasoning mechanisms may or may not be valid psychologically.

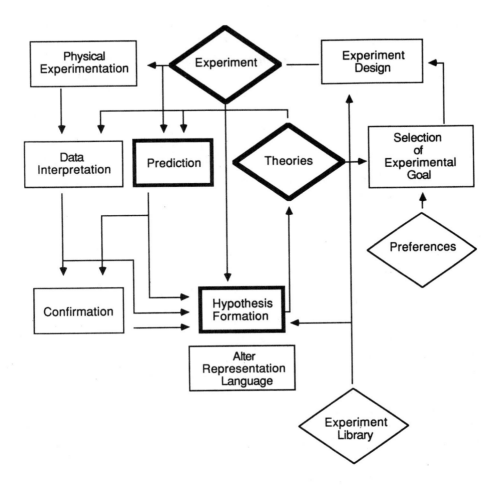

Figure 1. An overview of scientific reasoning tasks and their relationships. This chapter addresses only the tasks that are outlined in bold.

2. GENSIM

GENSIM embodies a *qualitative chemistry*—an ontology for chemical objects and chemical reactions that provides a framework for simulating chemical processes. A user describes an experiment in the GENSIM framework by specifying what objects are present at the start of the experiment. Each object represents a *population* of molecules. Chemical reactions are probabilistic events that *fork* populations of reactant molecules into two parts: a subpopulation that participates in the reaction and a subpopulation that does not react. Although in the general case an entire population of molecules could participate in a reaction, GENSIM assumes that its simulations occur over a sufficiently small interval of time that reactant populations are never fully consumed by a reaction. This implies that GENSIM predictions are monotonic computations — they add new objects to those present at the start of an experiment. All objects present at the start of an experiment will also be present at its end because objects are never deleted during a simulation.

GENSIM uses a general biological theory to describe specific experiments. One component of the theory is a *class knowledge base* (CKB) containing a taxonomic hierarchy of over 300 classes of biological objects, including strains of bacteria, genes, enzymes, and amino acids. Many of these objects have complex internal structures that are represented by GENSIM. Each class is represented as a frame in the KEE language (Kehler & Clemenson, 1984). An actual experiment is described by creating frames in a separate knowledge base (the *simulation knowledge base,* SKB) to represent the objects present at the start of the experiment. Each object in the simulation knowledge base is a member of a class in the class knowledge base.

The second component of GENSIM's general theory is a process knowledge base, known as either PKB or T. GENSIM processes comprise a theory of chemical reactions and are similar to those in Forbus' (1984) qualitative process theory. A sample process is shown in Table 1; the predicates used in processes are explained in Table 2. Each process is represented as a frame that lists:

- The classes of chemical objects that participate in its reaction (termed its *parameter objects*)
- Conditions that must be true of the parameter objects for the reaction to occur (the process *preconditions*)
- The *effects* of the reaction, which usually involve the creation of new chemical objects

GENSIM does not reason about quantities such as concentrations and reaction rates. The program predicts *what* objects exist at the end of an experiment but not their concentrations. Predictions are computed by a *process interpreter* that applies processes to the objects present

Table 1. Process **Trp-Repressor.Binds.Operator** describes a reaction between the trp-repressor protein and the trp-operator region of DNA. Parts of the process have been omitted for brevity.

Parameters: ($A, type Trp.Repressor) ($B, type Trp.Operator)

Preconditions:

> *Check that $B contains an active site that interacts*
>
> *with objects of $A's type. Call the site $BSite.*

(EXISTS $Bsite

(AND (IS.PART $Bsite $B)

      (OBJECT.EXISTS $Bsite Active.Sites)

      (EXISTS $site.interaction.class

        (AND (MEMBER $site.interaction.class

               (Potential.Interacting.Objects $Bsite)

          (OBJECT.EXISTS $A $site.interaction.class)))))

> *Check that $Bsite is not occupied.*

(NOT (EXISTS $object

      (AND (MEMBER $object (Object.Interacting.With.Site $Bsite))

          (OBJECT.EXISTS $object (Potential.Interacting.Objects

$Bsite)))))

> *Check that $Bsite does not contain a mutation that*
>
> *disables the current reaction.*

(NOT (EXISTS $mutation

        (AND (IS.PART $mutation $Bsite)

           (OBJECT.EXISTS $mutation Mutations)

           (MEMBER $Current.Process (Processes.Disabled $mutation)))))

Effects:

> *Create a new object that contains $A and $B as parts.*

(BIND $Complex (CREATE.COMPLEX RepOp.Complexes ($A $B)))

> *Record that $A is interacting with $Bsite.*

(PUT.VALUE $Bsite Object.Interacting.With.Site $A)

> *Record that the promoters controlled by $B are*
>
> *no longer able to bind RNA Polymerase.*

(PUT.VALUE (Promoters.Controlled $B) Receptive.To.Polymerase NO)

Table 2. The meanings of the predicates and functions used by GENSIM and
HYPENE.

PREDICATE OR FUNCTION	MEANING
(OBJECT.EXISTS x y)	Object **x** exists as a member of class **y**
(IS.PART x y)	Object **x** is physically contained by object **y**
(MEMB x y)	**x** is a member of list **y**
(GET.VALUES x s)	Return the value of slot **s** of object **x**

in the experiment to detect reactions and simulate their effects. The
interpreter is much like a production system.

A GENSIM prediction is shown in Figure 2. The initial conditions
of this biological experiment include a subset of the objects in the trp
operon: the molecule tryptophan (trp), a protein called *trp-aporepressor,*
and a region of DNA called the *trp operator.* The trp-aporepressor protein contains two binding sites; in the first reaction predicted by GEN-
SIM, tryptophan binds to one of these binding sites. The reaction yields
a modified version of the protein called *trp-repressor.* When bound to
tryptophan, the protein is in an active form that allows a second reaction
to occur—the trp operator binds to the other binding site within trp
repressor to form a *repressor-operator complex.* This complex contains
three formerly separate molecules; its part-whole structure is modeled
by GENSIM as shown in Figure 3. The second reaction is described by
the process in Table 1. This process has fairly complicated preconditions that verify that the repressor protein actually contains a binding
site, that the binding site is not already occupied by another molecule,
and that the binding site does not contain a mutation that prevents this
reaction from occurring.[2]

3. A Design System For Hypothesis Generation

I treat hypothesis generation as a synthetic process that is heavily goal
directed—as a process of design. A hypothesis is an artifact that scientists design to improve their predictions of experimental outcomes. The
artifact is synthesized from the representation-language primitives used

---

2. Mutations are structural changes to proteins or DNA that can alter their behaviors in many different ways.

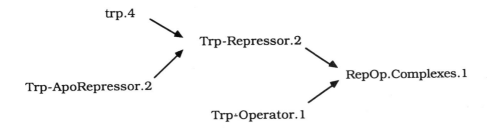

a

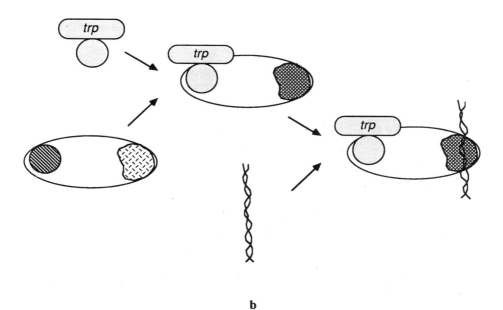

b

Figure 2. A sample GENSIM prediction involving two reactions. Part (a) shows
the names of the reacting objects; part (b) shows pictoral representations of the objects. Here a simple molecule called trp binds to
a binding site within the trp-aporepressor protein. The protein is
then in a form that allows its other binding site to bind to a region
of DNA called the trp operator.

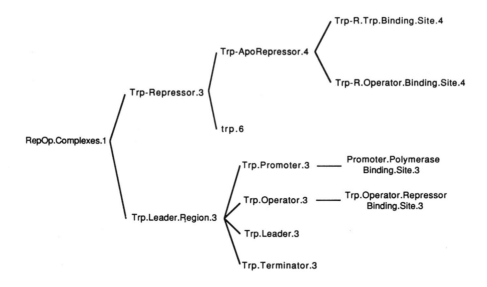

Figure 3. The internal structure of the RepOp.Complexes.1 object. When a
line connects two objects, it indicates that the object to the right
is a part of the object to the left. This repressor–operator complex
contains a region of DNA and a repressor protein, both of which
contain their own internal structures.

to describe theories and initial conditions of experiments. In general,
a designer seeks to synthesize an artifact given goals and constraints
that describe the form and function of the artifact. In the design of
a hypothesis, these goals and constraints include a description of the
experimental anomaly that the hypothesis must rectify.

I have constructed a computer program called HYPGENE[3] that formulates hypotheses through a process of design. HYPGENE's output is
a set of hypotheses. Its input includes a set of *design goals* (discussed in
Section 3.3) that describe an error in a GENSIM prediction. As in many
AI programs, the computation performed is a search procedure. HYP-
GENE's search incrementally synthesizes hypotheses by applying *design
operators*. These operators satisfy design goals by extending partial
hypotheses. The design operators are discussed in Section 3.4. The
remainder of Section 3 describes different aspects of the search process
itself, including the use of an assumption-based truth maintenance sys-

---

3. For *hypothesis generator* or *hypotheses for genetics*, but not *hype generator*.

tem (de Kleer, 1986) to support a best-first search, the representation used for HYPGENE's goal stack, the method the program uses to decide how to expand a search node, and the criteria it uses to recognize when goals have been satisfied.

The techniques used to implement the designer are similar to those developed in the course of previous AI research in planning (Fikes & Nilsson, 1971). HYPGENE is called a *designer* rather than a *planner* because the output of a planner is usually a temporal sequence of actions. The hypotheses generated by HYPGENE have no temporal component and thus more are similar to the class of artifacts synthesized by the process of design. The distinction between planning and design is fuzzy, however, because the techniques the two employ are similar.

3.1 Specification of HYPGENE's Behavior

Let us consider HYPGENE's behavior more precisely. The program takes five inputs (Figure 4). The first is a description of the initial conditions of an anomalous experiment, termed I_A. Although GENSIM represents objects in an experiment using frames, it is convenient to think of I_A as a list of predicate-calculus assertions that describe the initial objects.[4] The second input is GENSIM's prediction of the outcome of this experiment—termed P_A—plus dependency information that records how GENSIM computed P_A from I_A. The dependency trace indicates the processes that each object caused to fire and the new objects produced by these process firings. The third input, called $Error_A$, describes the error in GENSIM's prediction. $Error_A$ is the initial goal for HYPGENE; it specifies the incorrect assertions contained by P_A and the correct assertions that P_A lacks. $Error_A$ is expressed as an arbitrary predicate-calculus formula that may include quantification. The fourth input is T, the theory that GENSIM used to compute its prediction. The fifth input is the class knowledge base (CKB), which describes known classes of biological objects. Thus, the input is a tuple $\{I_A, P_A, Error_A, T, CKB\}$.

By definition of the process of prediction, the theory and initial experimental conditions taken together entail a predicted outcome: $I_A \cup T \models P_A$. An anomalous experiment is one whose outcome is not correctly predicted by theory: $P_A \neq O_A$.

---

4. Nilsson (1980) describes how to convert a frame representation to predicate-calculus form.

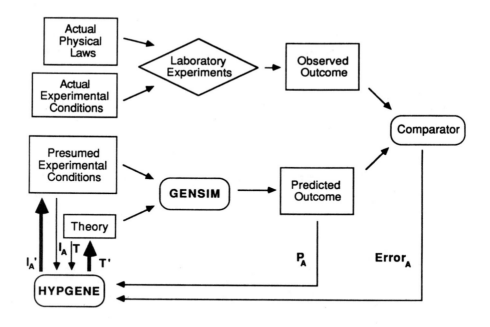

Figure 4. The relationship between the GENSIM and HYPGENE programs. GENSIM predicts the outcome of an experiment. Its prediction is compared with the observed outcome of the experiment. HYPGENE's input is a tuple that describes the experiment and its outcomes; it output is a hypothesis that aligns prediction with observation.

The prediction error can be expressed as

$$Error_A = (P_A \cup O_A) - (P_A \cap O_A)$$

or equivalently as

$$Error_A = (P_A - O_A) \cup (O_A - P_A).$$

Figure 5 diagrams these two breakdowns of $Error_A$.

HYPGENE's output is a set of hypotheses, where each hypothesis is a tuple $\{I_A', T'\}$. A hypothesis is a modified version of the initial conditions I_A (that is, a change in what objects the experimenters *thought* were present at the start of the experiment) and/or the theory T, such that when GENSIM uses T' to predict the outcome of I_A', the new prediction P_A' matches the observed outcome of the experiment: $I_A' \cup T' \models P_A'$ and $P_A' = O_A'$.

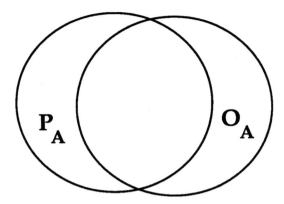

Figure 5. The error in the prediction is the sum of the two crescents and can be expressed as the area of both disks minus the area common to both. One crescent represents the assertions that the prediction should have but lacks; the other represents assertions that the prediction includes but should lack.

Some readers may find it odd to think of modifying the initial conditions of an experiment to align its predicted and observed outcomes. While studying the history of attenuation, however, we observed that biologists often modified I_A because their knowledge of I_A was uncertain. One reason for this uncertainty is the incredible complexity of the objects taking part in these experiments. One element of I_A is a description of bacterial DNA, which contains millions of bases and whose structure has only been partially determined by biologists. Another source of uncertainty is that experimental conditions are often tailored using laboratory techniques whose effects cannot be predicted with complete certainty, such as gene-splicing techniques.

3.2   A Sample Hypothesis-Formation Problem

This section contains an edited trace of HYPGENE's solution of a hypothesis-formation problem. The best way to understand HYPGENE's methods is to skim this example now, then to read the later sections that describe HYPGENE's methods, and finally to read through this example a second time with the general description of the methods in mind.

The biological experiment involved is that depicted in Figure 2. This experiment involves two reactions involved in the regulation of the genes in the trp operon (the actual experiment contains more reactions, but they are omitted to simplify this discussion). GENSIM predicts that, when trp is incubated with the trp-aporepressor protein, plus a region of DNA that includes the trp operator, two reactions take place. Trp activates the aporepressor (as shown in Figure 2), and then the repressor binds to the trp operator. Although the prediction in Figure 2 is often correct, one can perform experiments with mutant bacteria in which no repressor–operator complexes are observed. Such experiments are described by Hiraga (1969). In this example, we instruct HYPGENE to formulate hypotheses to explain the missing repressor–operator complexes. The initial goal is shown below. The expressions under "Outstanding constraints" show HYPGENE's goal stack at different points in time; these are predicate-calculus formulas that the program attempts to satisfy. The labels "universal variables" and "existential variables" indicate which variables in the outstanding constraints are universally quantified and which variables are existentially quantified (variable names begin with "$").

HYPGENE starts with the goal that no repressor–operator complexes should exist:

```
Universal variables: ($object)   Existential variables: ( )
Outstanding constraints:
  (NOT (OBJECT.EXISTS $object RepOp.Complexes))
```

In what follows, we step through the problem-solving actions that HYPGENE takes and observe how its goal stack changes in response to these actions. This line of reasoning leads to one of the 15 solutions HYPGENE finds to this problem; the other solutions are summarized in Section 4.1.

HYPGENE first determines if its universally quantified goal is satisfied and finds that the goal is violated by the object RepOp.Complexes.1. HYPGENE therefore creates a subgoal to remove the assertion that this object exists:

```
Outstanding constraints:
    [(Remove.Assertion (OBJECT.EXISTS RepOp.Complexes.1
                                      RepOp.Complexes]
```

The operator called "Remove Assertion" consults the GENSIM simulation dependency-trace and finds that RepOp.Complexes.1 was created by a process activation called Trp-Repressor.Binds.Operator.PACT.1.

To delete this object, HYPGENE determines that its next goal should be to prevent the process that created RepOp.Complexes.1 from firing.

```
Outstanding constraints:
    ((Prevent.Process.From.Asserting
        (Trp-Repressor.Binds.Operator.PACT.1 XCINITEXPT)))
```

The system uses several methods to prevent a process from firing; one method is to violate one of the preconditions of the process. HYPGENE *refines* the Prevent.Process.From.Asserting goal to use this method:

```
Outstanding constraints:
    (Violate.Process.Condition.To.Prevent.Assertion
        (Trp-Repressor.Binds.Operator.PACT.1 XCINITEXPT))
```

HYPGENE next executes the operator "Violate Process Condition To Prevent Assertion" on the process activation record Trp-Repressor.-Binds.Operator.PACT.1. It determines that the following formula expresses a condition that, if satisfied, will violate the preconditions of the Trp-Repressor.Binds.Operator process. It is not important to understand every clause within this goal; simply note that the expression is a disjunction and thus expresses a number of alternative ways of satisfying the goal. HYPGENE computed the disjunction by negating the preconditions of the process Trp-Repressor.Binds.Operator and converting the resulting expression to disjunctive-normal form.

```
Outstanding constraints:
    Universal variables:  ($Asite $site.interaction.class0329)
    Existential variables: ($mutation0327 $object0328)
    [(OR (NOT (OBJECT.EXISTS  Trp.Operator.1
                                $site.interaction.class0329))
        [NOT (MEMBER  $site.interaction.class0329
                    (GET.VALUES $Asite
                                Potential.Interacting.Objects]
        (NOT (OBJECT.EXISTS  $Asite  Active.Sites))
        (NOT (IS.PART.R  $Asite  Trp-ApoRepressor.3))
        [AND (MEMBER  $object0328
                    (GET.VALUES  $Asite
                        Object.Interacting.With.Site))
            (OBJECT.EXISTS  $object0328
                            (GET.VALUE $Asite
                                Potential.Interacting.Objects]
        (AND (IS.PART  $mutation0327  $Asite)
            (OBJECT.EXISTS  $mutation0327  Mutations)
            (MEMBER  Trp-Repressor.Binds.Operator
                    (GET.VALUES $mutation0327 Processes.Disabled]
```

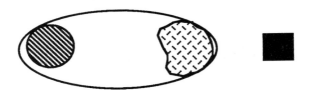

Figure 6. HYPGENE creates a separate mutation object, distinct from the trp-aprorepressor protein.

HYPGENE attempts to satisfy this disjunction by instantiating some of its universally-quantified variables and then considering each disjunct separately since each disjunct represents a different way of satisfying the overall expression. In what follows, HYPGENE considers the last disjunct, which involves a mutation in the site within the trp-repressor protein that binds to the trp operator:

```
Outstanding constraints:
    Existential variables: ($mutation0327)
    [(IS.PART  $mutation0327  Trp-R.Operator.Binding.Site.3)
     (OBJECT.EXISTS  $mutation0327  Mutations)
     (MEMBER  Trp-Repressor.Binds.Operator
             (GET.VALUES $mutation0327  Processes.Disabled]
```

To satisfy the preceding existentially-quantified condition, HYPGENE first creates a new mutation object called Mutations.17 and binds the variable $mutation0327 to Mutations.17. This object is depicted in Figure 6.

```
Actions:
    [(ASSERT (OBJECT.EXISTS Mutations.17 Mutations]

Outstanding constraints:
    [[Add.Assertion (MEMBER Trp-Repressor.Binds.Operator
                            (GET.VALUES Mutations.17
                                          Processes.Disabled]
     (Add.Assertion (IS.PART Mutations.17
                            Trp-R.Operator.Binding.Site.3]
```

The next condition that HYPGENE considers specifies that the Processes.Disabled slot of the Mutations.17 object should have a certain value. Since the Mutations.17 object was added to the initial conditions of the experiment, HYPGENE modifies I_A to achieve this goal, as depicted in Figure 7.

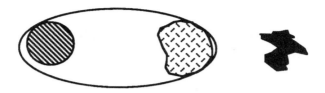

Figure 7. HYPGENE alters the specificity of the mutation, shown here as a change in shape. Within GENSIM, this change in specificity is represented by listing what processes this mutation disables.

```
Outstanding constraints:
  [Modify.Initial.Conditions.To.Add.Assertion
     (MEMBER Trp-Repressor.Binds.Operator
        (GET.VALUES Mutations.17 Processes.Disabled]
```

The operator "Modify Initial Conditions To Add Assertion" modifies I_A by adding the required value to the slot of `Mutations.17`:

```
Actions:
  [(ASSERT (MEMBER Trp-Repressor.Binds.Operator
              (GET.VALUES Mutations.17 Processes.Disabled]
```

The remaining goal attempts to modify an object (`Trp-R.Operator.Binding.Site.3`) that was created by a process; since this object was not present in I_A, HYPGENE cannot use the operator "Modify Initial Conditions To Add Assertion" to modify `Trp-R.Operator.Binding.Site.3`.

```
Outstanding constraints:
  (Modify.Process.Input.To.Add.Assertion
     (IS.PART Mutations.17 Trp-R.Operator.Binding.Site.3))
```

The process `Trp-ApoRepressor.Binds.Trp` created this binding site by copying it from the object `Trp-R.Operator.Binding.Site.2`, so HYPGENE posts a goal to modify the latter object, on the assumption that the modification will be copied by the process:

```
Outstanding constraints:
  (Add.Assertion (IS.PART Mutations.17
                      Trp-R.Operator.Binding.Site.2))
```

This goal is refined to a goal to modify I_A.

```
Outstanding constraints:
  (Modify.Initial.Conditions.To.Add.Assertion
     (IS.PART Mutations.17 Trp-R.Operator.Binding.Site.2))
```

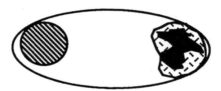

Figure 8. HYPGENE inserts the mutation object into a binding site within the trp-aprorepressor protein.

HYPGENE next makes the `Mutations.17` object a component of the binding site, as depicted in Figure 8. When the `Trp-R.Operator-Binding.Site.2` object is modified, HYPGENE finds all processes that the original `Trp-R.Operator.Binding.Site.2` had caused to fire and deletes all objects created by these processes (including `RepOp.Complexes.1`). This step is required because the modified object may yield different reactions with different products. HYPGENE recomputes what processes will fire in the presence of the modified `Trp-R.Operator.Binding.Site.2`. The modified repressor protein can still bind trp but cannot bind to the trp operator because of the mutation in `Trp-R.Operator.Binding.Site.2`. Thus, the original goal is satisfied because, in the new version of the experiment, only one of the two earlier reactions can occur, so no repressor–operator complex is created.

```
Actions:
  [(ASSERT (IS.PART Mutations.17
                    Trp-R.Operator.Binding.Site.2]
  (Unfiring (Trp-ApoRepressor.Binds.Trp.PACT.1 XCINITEXPT))
  (Unfiring (Trp-Repressor.Binds.Operator.PACT.1 XCINITEXPT))
  (Process Trp-ApoRepressor.Binds.Trp created Trp-Repressor.22)
```

We next discuss the methods HYPGENE employs to design hypotheses by considering how the program represents its goals, what operators it uses to satisfy its goals, how the program applies operators to goals, and how it detects when its goals are satisfied. Most of the discussion focuses on the design operators because they most strongly distinguish HYPGENE from other agenda-based reasoning systems.

3.3   The Design Goals

Design goals specify any of three types of differences between the prediction (P_A) and the observation (O_A): (1) that objects with specified properties should be added to or removed from P_A, (2) that properties of existing objects in P_A should be modified, or (3) that the concentration of objects in P_A should be modified (of course, concentration is simply a distinguished property of an object). Missing objects are specified with existentially-quantified formulas; universal quantification is used to specify that no objects with certain properties should exist.

Two typical hypothesis-generation tasks are:

1. The observed rate of transcription is higher than that predicted in the current experiment. Generate hypotheses to explain this observation.
2. Measurements show that no operator–repressor complexes are observed in an experiment, although their existence is predicted. Generate hypotheses to explain this observation.

The design goals for these hypothesis generation problems are:

1. `(INCREASE-QUANTITY XCinit.Complex.1)`
2. `(NOT (OBJECT.EXISTS RepOpComplex.1 RepOpComplexes))`

The first goal asks the designer to modify I_A or T in ways that will increase the predicted amount of transcription-initiation complex. The second instructs the designer to modify I_A or T such that GENSIM no longer predicts that the repressor–operator complex `RepOp.Complex.1` is formed.

3.4 Design Operators

HYPGENE satisfies goals by employing two types of *design operators*—operators that modify the initial state of the experiment (I_A), and operators that modify the theory T embodied by the process knowledge base. The designer is able to examine its outstanding (as yet unsatisfied) goals and to choose operators whose execution will satisfy one or more goals. To execute an operator, the designer makes the changes to I_A and T that the operator specifies. Operators often specify conditions under which they may be applied. When the program applies an operator to satisfy a goal, it adds the conditions under which the operator may be applied to the goal stack. A *valid hypothesis* is a set of operator applications that satisfies all goals on the stack. The system produces valid hypotheses as its output.

HYPGENE employs four different types of operators:

- Initial condition operators, which modify I_A

- Process-modification operators, which modify T
- Quantity-hypothesis operators, which address goals related to the quantities of objects present in P_A
- Class KB operators, which modify the classes in the CKB

Figure 9 shows HYPGENE's initial condition modification operators and process modification operators. Although the operators are implemented as LISP functions, they are organized conceptually in a class–subclass hierarchy that is used to control their execution (discussed in Section 3.5.3). Only the leaves of this tree are operators that can be applied during problem solving.

3.4.1   INITIAL CONDITION DESIGN OPERATORS

An assertion A could be present at the end of the simulation (and thus be a member of P_A) for only one of two reasons:

- Because A was present in I_A (the presence of A in P_A follows from the monotonicity of GENSIM simulations);
- Because a process fired (executed) that asserted A.

Thus, HYPGENE has an operator that simply adds an assertion to I_A and has a class of operators that modify the experiment such that a process fires and adds an assertion to P_A. Preventing an assertion from existing in P_A is essentially the converse of adding an assertion to P_A; A can be removed from I_A, or the process that asserted A can be prevented from firing. However, removing A from P_A is a bit more complex than adding A because the assertion could have been *both* present in I_A *and* asserted by one or more processes; all justifications of A must be eliminated if A is to be absent from P_A. Because GENSIM's simulations are monotonic, it is impossible for the firing of a process to remove an assertion from P_A, so there is no operator that attempts to remove an assertion from P_A by firing a process. We now describe in detail each operator for modifying initial conditions.

3.4.2   MODIFY INITIAL CONDITIONS DIRECTLY

The operators "Modify Initial Conditions To Add Assertion" and "Modify Initial Conditions To Prevent Assertion" have no conditions for their application. They satisfy a goal directly, either by adding an assertion

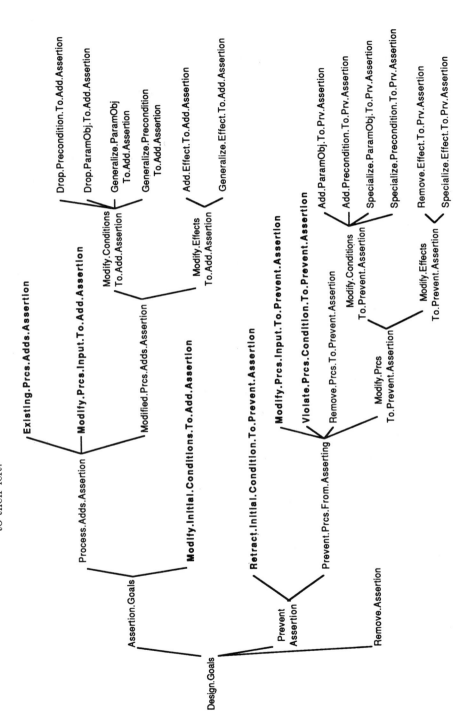

Figure 9. The hierarchy of initial-condition design operators and process-design operators. Operators that have been implemented are in bold. Leaves in the tree are executable operators that satisfy goals to their left.

to the initial conditions I_A or by removing it from I_A. For example, the first operator would satisfy the goal

```
(NOT (OBJECT.EXISTS trp.1 trp))
```

by deleting the object `trp.1` from the initial conditions of the experiment.

Since goals are expressed in predicate calculus but the experiment is represented in frames, HYPGENE must be able to translate between the two in order to make a predicate true in the simulation knowledge base (to *invert* the predicate). To do so, each predicate used in the predicate-calculus expressions that specify goals has two LISP functions associated with it—one to alter the frame representation of I_A to *satisfy* an atomic formula containing the predicate and one to alter the representation of I_A to *violate* the predicate (for negated goals). The predicates used by HYPGENE and their satisfying and violating inversions are as follows:

1. `(OBJECT.EXISTS object class)`
   - To satisfy: Add `object` to `class`
   - To violate: Delete `object`

2. `(IS.PART component container)`
   - To satisfy: Make `component` part of `container`
   - To violate: Remove `component` from `container`

3. `(MEMBER atom list)`
   - To satisfy: Make `atom` a member of `list` (usually the list is a slot value, so this operation adds a value to the slot)
   - To violate: Delete `atom` from `list`

3.4.3   ENABLE A PROCESS TO ADD AN ASSERTION

The operator "Existing Process Adds Assertion" attempts to add an assertion A to P_A by searching the process knowledge base for a process whose effects assert A. For example, if the goal is `(EXISTS $x (OBJECT.EXISTS $x RepOp.Complexes))` then this operator begins by finding all processes in the process knowledge base that create an object of type `RepOp.Complexes`.

Once the operator has found a process that can satisfy a goal, it must ensure that the process will fire within this experiment. To do so, it posts a new design goal that specifies the conditions that must be true for this process to fire. These conditions include the preconditions of the process and the existence of objects that are instances of the parameter object classes of the process.

For STRIPS-like planners, the task of finding a process with effects that satisfy a given goal is relatively simple. These planners encode the effects of their operators using an add list and a delete list. These lists specify assertions that are added to and deleted from working memory when an operator is executed. To find an operator that satisfies a particular goal, STRIPS-like planners match the goal against the add list of each operator. This task is conceptually simple because the goal and the add list are represented in the same declarative language.

This is not the case for HYPGENE; although some processes do assert simple fixed lists of propositions, the effects of other processes call LISP functions that may perform complicated tasks, such as copying objects with complex internal structures. These functions may contain recursion and cannot be represented using fixed add and delete lists.

HYPGENE employs an incomplete solution to this problem. A preprocessing phase computes an add list for most processes (no delete list is needed because of simulation monotonicity) by simulating the execution of processes on typical parameter objects. For each process R, the preprocessor creates an object from each parameter object class of R (in an otherwise empty experiment knowledge base). Then the preprocessor executes R with the created objects as process parameter objects. The add list of R is created from an analysis of the objects created by R.

The weakness of this method is that it assumes that the effects of R are always the same because the add list it computes for R is based on a single set of prototypical parameter objects. More precisely, this method will work when either of the following conditions are satisfied by a process R:

- The process executes on a finite set of parameter objects, where objects of the same type that have different properties are considered to be different;
- The process executes on a potentially infinite set of parameter objects, but the effects of the process are not dependent on the properties of these objects.

These conditions hold for most but not all of the processes in the process knowledge base for the trp system.

3.4.4   DISABLE A PROCESS TO REMOVE AN ASSERTION

The operator "Violate Process Condition To Prevent Assertion" is used to remove an assertion from P_A that was asserted by a process R. It does so by preventing R from firing. A process fires when its parameter objects are present in the simulation and its preconditions are satisfied. To prevent R from firing, the designer must make one of these conditions false. To do so, this operator posts a new goal to negate the predicate-calculus formula that specifies the conditions under which R fires.

3.4.5   MODIFY PROCESS INPUTS TO ALTER PROCESS EFFECTS

The operators "Modify Process Input To Add Assertion" and "Modify Process Input To Prevent Assertion" are employed when a design goal requires the modification or deletion of an object created by a process. Such an object can be modified by changing the properties of the parameter objects of the process (assuming that the properties of the objects created by the process are a function of the properties of the process parameter objects). This type of operation has been studied in planners by Pednault (1988) and Chapman (1987). To infer what modifications (if any) will be sufficient, HYPGENE must reason about the effects of the process. As stated earlier, this is very difficult in general because of the complexity of some process effects. These operators therefore employ a heuristic that has fairly wide applicability in this domain. Many processes copy their parameter objects and then modify the copies in some way (the processes fork the reactant objects). If a parameter object B is copied to object B', the properties of B' are largely derived from the properties of B, and modifying B will produce a similar modification to B'. These operators work by replacing all occurrences of B' in the design goal stack by B. Since *all* properties of B' were not derived from B — and in fact, B' may not have been copied from any parameter object — this approach sometimes fails. HYPGENE uses this heuristic in the example in Section 3.2. To satisfy the goal of modifying a part of the repressor within the repressor–operator complex in Figure 2, HYPGENE modified the same binding site within the trp-aporepressor protein.

3.4.6 QUANTITY-HYPOTHESIS DESIGN OPERATORS

The GENSIM model of the trp system has no explicit representation of dependencies among object state variables, such as concentrations, and does not generate quantitative predictions for the concentrations of the objects that are created during a simulation. Despite this, HYPGENE can generate hypotheses about what factors would raise or lower the concentrations of different objects. Normally, such goals would result from a difference between prediction and observation, but, since quantitative predictions are not made by GENSIM, the user must estimate how the quantitative prediction made by a theory would fail when he creates a quantitative goal for HYPGENE. These goals direct the program to formulate hypotheses about how to produce more or fewer of specified objects.

Figure 10 shows the set of operators that is used to generate quantitative hypotheses. Figure 11 shows a hypothetical object-dependency graph computed by GENSIM that will be used as an example in this discussion. Imagine that we wish to generate hypotheses about how to increase the amount of G present in the experiment in Figure 11. In general, there are three ways to increase the amount of an object:

1. Increase the amount present at the start of the experiment (accomplished by the operator "Modify Initial Conditions To Increase Quantity")

2. Increase the amount produced by processes during the experiment ("Increase Production Rate To Increase Quantity")

3. Decrease the amount consumed by processes during the experiment ("Decrease Consumption To Increase Quantity")

Case 1 is self-explanatory. For case 2, there are several ways to increase the amount of G produced by a process (as shown in Figure 11). The first is to alter the existing processes that produce G. Implicit in the semantics of processes for chemical reactions is the property that increasing the concentrations of any of $\{A, B, C, D\}$ ("Increase Influencing Variable To Increase Quantity") will increase the rate at which G is produced (with a restriction discussed below). In addition, we can think of each process as having an *intrinsic rate*, which corresponds to the rate constant for its chemical reaction. Increasing the intrinsic rate of either P_1 or P_2 will increase the production of G ("Speed Pro-

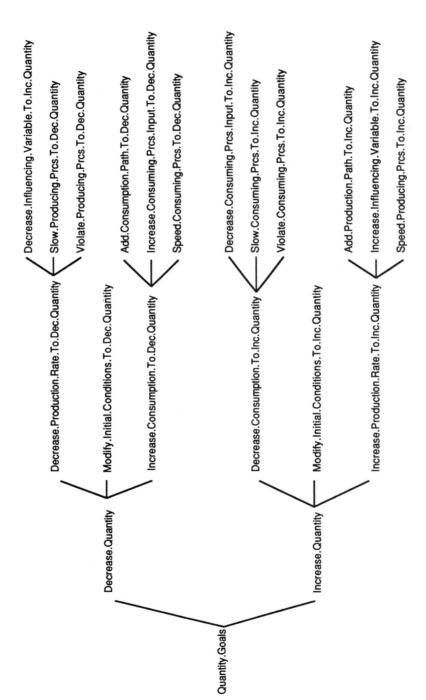

Figure 10. The hierarchy of quantity-hypothesis design operators. The leaves in this tree are executable operators that can achieve the goals to their left.

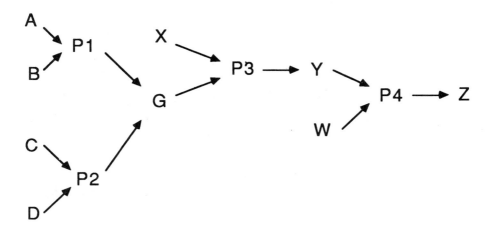

Figure 11. A sample reaction network. Here the process P_1 specifies that A and B react to form G.

ducing Process To Increase Quantity").[5] The second way of increasing the amount of G produced by a process is to produce G through an additional reaction that is not currently occurring, and one that does not consume any of $\{A, B, C, D\}$ ("Add Production Path To Increase Quantity") — in which case, the net effect on G would be unclear.

Finally, in case 3 we increase G by decreasing the amount of G consumed by processes. Case 3 is somewhat analogous to case 2: we can modify the consumption of G by existing processes by decreasing the intrinsic rates of processes that consume G (P_3), or by decreasing the concentration of objects that G combines with (X). In addition, the amount of G will be increased if P_3 no longer fires at all, which would occur if X ceased to exist or if a precondition of P_3 were violated. This case actually shares a restriction with case 2 — the fact that X is a parameter object for P_3 does not necessarily imply that the chemical reaction represented by P_3 consumes the object. For example, enzymes participate in many reactions but are not consumed by the reactions they catalyze. For HYPGENE to determine whether a given object is

---

5. What altering the rate of a process means in physical terms depends on the process in question. If the process described is an enzymatic reaction, we could alter its rate by changing the physical properties of the enzyme, for example, by introducing mutations in the enzyme. HYPGENE does not have this type of domain-specific knowledge.

consumed by a process, it must either be told this information explicitly or deduce it by comparing the chemical composition of the parameter objects of a process with the objects created by the process. In addition, this analysis cannot be local to a single process but must be a global analysis that is applied to all objects that result from reactions involving G (in this example, Y and Z are the relevant objects). The processes that are activated by G must consume G, and none of the later processes may produce G, if we are to deduce that this network is a net consumer of G. HYPGENE does not currently perform this type of chemical analysis, but it would be fairly straightforward to implement.

In the preceding discussion, we considered how to generate hypotheses to account for increased quantities of G; accounting for a decrease in G is similar. By analogy to case 1, we can decrease the amount of G present in the initial conditions (if G was present in I_A). By analogy to case 2, we can decrease the production of G by decreasing any of $\{A, B, C, D\}$, or by decreasing the intrinsic rates of P_1 or P_2, or by preventing P_1 or P_2 from firing by violating a condition on which they depend. By analogy to case 3, we can increase the consumption of G by increasing X, by increasing the intrinsic rate of P_3, or by firing a new process that consumes G. All objects produced by firing such a consuming process must be analyzed to determine whether they do yield a net consumption of G. They may neither produce any precursor of G (such as A) nor consume anything that reacts with G (for example, W must not be equivalent to X).

3.4.7 POSSIBLE OPERATORS FOR CLASS AND PROCESS DESIGN

Karp (1989) describes operators that would eliminate a difference between O_A and P_A by modifying processes in the process knowledge base, and by modifying object class descriptions in the class knowledge base. The process-design operators modify the effects, preconditions, and parameter objects of a process. They do so to achieve the usual goals—removing assertions from and adding assertions to P_A. The operators for the class knowledge base perform such functions as creating new classes, removing classes, and creating new slots in existing classes. These operators have not been implemented within HYPGENE.

3.5 Implementation Details

Here we consider several aspects of the implementation of HYPGENE, including its search procedure, its detection of solution hypotheses, and its verification of solutions.

3.5.1 ATMS SUPPORT FOR BEST-FIRST SEARCH

HYPGENE conducts a best-first search of the design space defined by its operators. This design space is the set of possible combinations of operator applications—the set of all partial hypotheses. The nodes in the design space are called *design states*. Each state D contains several pieces of information, including:

1. A description of the outstanding goals at D

2. Bookkeeping information, such as the name of the state from which D was derived

3. Complete descriptions of I_A', T', and P_A'. (Each design state represents a different hypothesis that postulates different modifications to I_A and T and hence alters the prediction P_A.)

KEEworld's assumption-based truth maintenance system (ATMS) is used to represent design states compactly (de Kleer, 1986; Intellicorp, 1986). Each state is associated with a KEE world, which by default inherits information of types 1 and 3 from its parent state, but to which arbitrary changes to the goal lists and to $\{I_A, P_A, T\}$ can be made. The ATMS is used to compactly represent *both* the experiment state and the goal stack in each design state.

3.5.2 THE AGENDA

HYPGENE's search is controlled by an agenda that records the unexpanded design states for the current problem—the boundary of the best-first search. Every state on the agenda has a rating; the state with the highest rating is expanded next (or *pursued*). A simple evaluation function rates a design state based on the syntactic complexity of its unsatisfied goals and the number of objects modified in the hypotheses that produced D (a measure of the cost of computing D).

3.5.3   EXPANSION OF SEARCH NODES

HYPGENE expands a design state in one of several different ways, depending on the types of goal elements in the `Constraints.Outstanding` slot of the state. The elements on the goal stack can be in one of three forms:

1. An atomic formula or negated atomic formula containing unbound variables;

2. A well-formed formula in disjunctive normal form containing unbound variables;

3. An expression of the form (`design-operator parameter`), where `design-operator` is the name of a design operator from Figure 9 or Figure 10 and `parameter` is a parameter of that design operator.

Different design operators have different parameters; often the parameter is simply an atomic formula. A sample expression of type 3 is

```
(Modify.Initial.Conditions.To.Retract.Assertion
       (IS.PART Foo.1 Bar.1))
```

This statement is a goal to retract the assertion that `Foo.1` is part of `Bar.1` by modifying I_A. The goal also names an operator that can be executed to achieve the goal.

In the simplest case, all goals are of the third type: (`design-operator ground-condition`). In this case, HYPGENE pursues the topmost goal, G, by considering whether the design operator named in G is a leaf or a nonleaf node in the hierarchy in Figure 9 or Figure 10. For a leaf node, the named operator is executed on its parameter. For a nonleaf node, the program *refines* the design state by creating a new child state for each descendant of the design operators shown in Figures 10 and 11. A refined version of G is then pushed on the goal stack of each new design state (the name of each descendant operator is substituted for the operator in each refined G).

The other cases are more complicated because some element of the `Constraints.Outstanding` contains unbound variables when the element is of type 1 or 2. Karp (1989) describes the methods that are used to satisfy goals of this form.

3.5.4 Hypothesis Simulation

HYPGENE must compute the modified prediction P_A' associated with the complete or partial hypothesis $\{I_A', T'\}$ in each design state. To facilitate this computation, each design operator summarizes the changes it has made to I_A by indicating what objects it has modified (those objects it created, deleted, or altered). The modifications to these objects can cause new processes to fire, can prevent processes that fired previously from firing, and can alter the effects of processes that fired previously and that still fire under the new I_A.

To compute P_A', HYPGENE retracts the firing of any process that acted on a modified object (as recorded by GENSIM's dependency structures) and then calls GENSIM to execute those processes activated by the modified objects.

3.5.5 Goal Satisfaction

To fully understand how HYPGENE detects when a hypothesis satisfies its design goals requires detailed knowledge of how the goal stack is represented, which is beyond the scope of this chapter (see Karp, 1989). Briefly, after each operator application, the program checks all goal elements on its stack against P_A' and removes satisfied elements from the stack. When the stack is empty, we have found a solution. Note that this approach allows us to detect when operators executed to satisfy one goal element serendipitously satisfy another element.

Karp (1989) describes a number of other issues relevant to HYPGENE's design of hypotheses, including the detection of goal loops, what to do when previously satisfied goals are violated, and details of how the goal stack is implemented.

3.6 Controlling HYPGENE's Search

Because HYPGENE searches a very large hypothesis space, we must control its search such that more plausible hypotheses are generated before less plausible ones. This section describes several types of knowledge that can be used to guide the application of design operators. A more complete discussion of this topic is contained in Karp (1989).

HYPGENE always chooses to pursue the simplest partial hypothesis on its agenda using the heuristic notion of simplicity discussed in Section 3.5.2. Various types of domain knowledge could be used to prune and rank hypotheses. Biologists have knowledge about what types of chemicals are likely to be present in certain experiments, about the structural configurations in which different objects are likely to be found, and about general properties that chemical reactions must satisfy (such as conservation of mass). This type of knowledge is not currently used by HYPGENE.

While studying the history of attenuation, we noted that biologists tend to employ initial-condition modification operators before process-modification operators. This preference reflects the intuition that there is often greater uncertainty in the initial conditions of an experiment than in an established theory. We also observed that biologists seem to prefer to create new processes by instantiating general classes of existing processes rather than by designing them from scratch. HYPGENE could use this knowledge of operator precedence to prioritize the design states on its agenda.

Finally, *reference experiments* provide a powerful source of information for controlling HYPGENE's search. A reference experiment E_R is one that is similar to E_A, but for which the predicted outcome of the experiment matches the observed outcome:

$$I_R \cup T \models P_R$$

$$P_R = O_R$$

E_R can be used in two ways. First, we assume that E_R is similar enough to E_A that any unknowns present in E_A are also present in E_R. Thus, the modifications to I_A and T that produced the hypotheses $\{I_A', T'\}$ should also be applied to I_R. However, since the outcome of I_R was predicted correctly to begin with (by definition), any hypothesis should not change the predicted outcome of this experiment. Thus, any hypothesis for which it is not true that $I_R' \cup T' \models O_R$ should be rejected. For example, imagine that HYPGENE modified I_A by postulating that it contained an additional object U. We would also postulate that U was present in I_R and determine if the presence of U alters the reactions that occur in the prediction P_R.

The second use for I_R requires a new reasoning mechanism. Instead of generating hypotheses by reasoning backward from $Error_A$, it is possible (and sometimes more efficient) to generate hypotheses by reasoning *forward* from the difference $(I_A - I_R)$. Since the difference between the initial conditions of the two experiments is presumably responsible for the erroneous prediction P_A, we focus on this difference by reasoning forward from it in an attempt to derive $Error_A$ from the difference. For example, imagine that the difference in initial conditions is that an object U is present in I_A but not in I_R and that the error in P_A is that P_A fails to predict the presence of an object V. This suggests (among other hypotheses) that T is missing a process that states that U reacts (perhaps with other objects) to form V.

4. Discussion

4.1 Experimental Results

Section 3.2 contained a trace of the reasoning HYPGENE used to find a single solution to the transcription-initiation hypothesis-formation problem. The program finds a total of 15 solutions to this problem, after searching approximately 300 design states in two hours of execution time on a Xerox 1132 (Dorado). The solutions fall into the five classes, shown in Figure 12. These solutions are correct, and they are complete in the sense that they include all hypotheses that biologists generate for this problem. However, some of these solution classes are redundant from a biologist's viewpoint. For example, one class of solutions postulates that the specificity of a protein binding site has changed, whereas another class postulates that a mutation exists within a binding site. In fact, changes in binding site specificity are usually caused by mutations, but HYPGENE does not have this causal knowledge.

HYPGENE has been tested on several other experiments, most notably an experiment by Jackson and Yanofsky (1973). The program found four of the five solutions that the biologists formulated for this problem and generated approximately ten additional solutions that can be pruned using information from a reference experiment. Karp (1989) describes these examples in more detail. In general, HYPGENE finds most if not all the solutions to the problems it has been tested on and has found a few incorrect solutions whose analysis yielded interesting potential refinements to my hypothesis-formation methods.

Class 1: Postulate that certain objects were in fact not
present in the initial conditions of the experiment.

Class 2: Postulate that binding sites were absent
from specific objects.

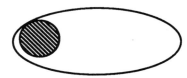

Class 3: Postulate that specific binding sites were
occupied, thus prohibiting them from participating
in reactions.

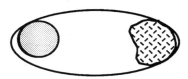

Class 4: Postulate that specific binding sites contained mutations
with specificities that interfere with the reactivity of the binding
sites.

Class 5: Postulate that the specificities of binding site were
altered in a manner that prohibits them from participating
in reactions.

Figure 12. HYPGENE finds 15 solutions for the transcription-initiation
hypothesis-formation problem. These solutions can be grouped into
the five classes shown.

4.2 Limitations

The HYPGENE program is a designer of scientific hypotheses. HYPGENE focuses on the difference between prediction and observation—$Error_A$—to design changes to the initial conditions of the experiment that will make a GENSIM prediction compatible with observation. Here we evaluate the strengths and weaknesses of this approach, the assumptions it is based on, and its generality.

An important class of assumptions made by HYPGENE concerns its inputs. They must exist, and they must have certain properties. The existence condition is not trivial because many scientific reasoning problems seek one or more of the entities that the program takes as input, such as a detailed description of the initial conditions of the experiment. The supplied descriptions of I_A and T must allow GENSIM to compute a predicted outcome for the experiment, for comparison with the experiment's observed outcome. The prediction and observation must be represented in a form that allows their difference to be computed. Some scientific reasoning problems meet this criteria, and some do not.

Another assumption concerns the invertibility of domain predicates. In order to invert goals, such as (IS.PART X Y), HYPGENE must possess a LISP function that can invert the predicate; in this example, the function would make X a component of Y. Some domain predicates do not have such an inverse.

The methods described here have been developed and tested in the domain of molecular biology, but these techniques appear to be sufficiently general to apply in other domains as well. HYPGENE has design operators for modifying I_A, T, and the class knowledge base, and for designing quantity hypotheses. The first three of these are very general, in that they mention no domain concepts; they are concerned with making syntactic modifications to the simulation knowledge base, to processes, and to the class knowledge base. The quantity-hypothesis operators also do not contain domain concepts, but they do contain assumptions about the quantitative behavior of GENSIM's chemical processes that are not true of processes in some other domains. HYPGENE's planner is very general because it contains no domain-specific knowledge but manipulates arbitrary predicate-calculus assertions.

An important limition of HYPGENE comes from its model of time and its ontology for chemical reactions. In the real world, processes are active longer than the short intervals that GENSIM and HYPGENE assume, in which case the assumption that populations of objects are never fully consumed by chemical reactions does not hold. These techniques will require revision in domains concerned with longer periods of time.

HYPGENE has a limited ability to regress design goals through process effects to determine what contents of the simulation knowledge base will cause a process to achieve a given goal. Some processes call recursive LISP procedures, which are notoriously difficult to reason about. This limitation has not been a major problem for the test cases we have run because the program contains heuristics to solve this problem in a limited way.

HYPGENE's performance is very closely tied to what it knows. For example, the design operator "Existing Process Adds Assertion" attempts to satisfy a design goal by finding a process whose effects satisfy the goal. The success or failure of this operator is highly dependent on the contents of the process knowledge base. This dependency is a double-edged sword, in the sense that if the process knowledge base does not describe reactions relevant to the problem at hand, then the program will make very slow progress in solving the problem; on the other edge, HYPGENE's performance in an area of molecular biology can easily be improved by giving it knowledge of important processes and objects. We expect no more and no less of molecular biologists.

4.3   Comparison with Previous Work

This section compares our approach to hypothesis generation with approaches that other AI researchers have used to solve similar problems. Although most of the work discussed in this section loosely falls under the topic of scientific theory formation, these projects have addressed several different problems.

4.3.1   DIETTERICH'S PRE

Dietterich's (1984) PRE system addresses the problem of *theory-driven data interpretation*. Given descriptions of various UNIX file-system commands, such as more and ls (analogous to HYPGENE's T), plus the output obtained when a UNIX command C was executed (analogous to the observed outcome of an experiment, O_A), PRE must infer the state of the UNIX file system (I_A) as it existed before the execution of C.

PRE uses the language developed for the Programmer's Apprentice system (Rich & Shrobe, 1976) to represent the functionality of UNIX commands. In that language, programs are structured like logic circuits; they contain primitive components (corresponding to logic gates) whose input and output ports are wired together to transmit the outputs of one component to the inputs of others. A primitive element is a LISP procedure such as CONS. New primitives can be constructed by treating a circuit as a black box with an assigned name and input–output ports.

A given data-interpretation problem consists of the program circuit describing a UNIX command plus observed values for the output ports of the circuit. PRE infers the inputs to the command (the state of the UNIX file system) by reasoning backward through the logic circuit for the command from the values on its output wires. The system uses constraint-propagation techniques to infer the inputs to primitive elements within the circuit from the outputs of the circuit element until the inputs of the command itself are derived.

The following similarities and differences exist between our approaches. First, PRE computes I_A from $\{O_A, T\}$, whereas HYPGENE computes $\{I_A', T'\}$ from $\{I_A, P_A, Error_A, T, CKB\}$. Thus, PRE is concerned with interpreting a correct prediction, whereas HYPGENE seeks to debug a faulty prediction. Although our system could compute $\{I_A', T'\}$ from $\{P_A, T\}$ alone, I_A' is usually sufficiently similar to I_A that it is easier for HYPGENE to compute I_A' by modifying I_A than to compute it from scratch.

PRE reasons backward from O_A because it is not able to compute a prediction—it has at most a partial description of I_A. In contrast, HYPGENE reasons backward from the *difference* between P_A and O_A. PRE propagates the known output of a circuit gate back through that gate to deduce what input(s) to the gate could have caused the output. HYPGENE instead uses its operators to work backward from an unpredicted or incorrectly predicted assertion to postulate experimental conditions or theory modifications that would undo the error. Whereas PRE examines a fixed set of circuit components, HYPGENE's operators give it the power to modify the inputs to existing components, to wire known types of components into the biochemical-reaction "circuit" with appropriate inputs (by postulating that additional processes fire), or to create new types of components (by postulating the existence of new objects) and to wire them into the reaction network. Finally, PRE's constraint-propagation approach is a deductive framework that cannot be extended easily to the task of modifying PRE's theory T. A modification to T would alter the set of constraints in the statement of the problem; thus it is not a use of constraint propagation per se. HYPGENE's framework does include operators that modify T.

4.3.2   RAJAMONEY'S COAST

Rajamoney (1988) has addressed the problem of theory revision in the domain of naive physics. More specifically, his system COAST formulates hypotheses to revise a theory when its predictions do not match observation, designs new experiments to discriminate between hypotheses, evaluates competing hypotheses based on past experiments, and evaluates hypotheses based on aesthetic criteria.

Rajamoney's approach to hypothesis formation focuses exclusively on revising his domain theory T (which is expressed in qualitative process theory) when the predicted rate of change of some quantity (such as the rate of evaporation of a fluid) does not match the observed rate at which the quantity changes. Theory revisions are performed by operators that are strikingly similar to HYPGENE's process modification operators. For example, if the domain theory predicts that a quantity decreases when it is observed to stay constant, his system would locate the process responsible for predicting the decrease and would postulate that the process is inactive (corresponding to HYPGENE's operator "Modify Condition To Prevent Assertion"). Specific ways of rendering the process inactive include creating new process preconditions and narrowing the scope of existing preconditions, which correspond to HYPGENE's operators "Add Precondition To Prevent Assertion" and "Specialize Precondition To Prevent Assertion." Rajamoney does not define operators for modifying the Individuals field of a process (the equivalent of GENSIM's `Parameter.Object.Classes`), so it would appear that his system is not syntactically complete.

One important difference between COAST and HYPGENE is that the former is able to generate abstract hypotheses, test them for correctness, and then further refine those abstract hypotheses that are not rejected. The advantage of this approach is that, if an abstract hypothesis is rejected, all specializations of that hypothesis have been ruled out in one fell swoop. Although HYPGENE does not currently include this capability, such reasoning should be easy to incorporate into its framework. HYPGENE already generates abstract hypotheses, such as "Process Adds Assertion," but it does not test them before refining them.

Rajamoney does not define operators that modify either the initial conditions of the experiment or the class knowledge base. COAST is not able to formulate hypotheses to account for the presence or absence of objects. In addition, although Rajamoney (1988) says very little about the implementation of his system, it does not appear to include a sophisticated planner. This would prevent COAST from reasoning about processes with complex preconditions or about complex networks of processes. Most of his sample hypothesis-formation problems involve very simple predictions in which only a single process has executed.

Rajamoney's use of previously stored experiments to filter hypotheses is very similar to the manner in which HYPGENE uses reference experiments to filter hypotheses. However, he does not discuss the possibility of reasoning forward from the difference between I_A and I_R. He does address the problem of designing new experiments to test hypotheses, whereas HYPGENE does not.

Rajamoney also addresses the issue of discriminating among competing theories using "aesthetic criteria," after the above empirical criteria have been exhausted. He proposes aesthetic evaluation functions that are based on such properties as syntactic features of the processes contained in a theory, the syntactic complexity of explanations generated by the theory, and the number of predictions made by the theory. I have not addressed this issue in this chapter. Unfortunately, Rajamoney's criteria are defined too imprecisely to evaluate in the context of the discovery of attenuation.

4.3.3 SIMMONS' GORDIUS

Simmons (1988) addresses geologic *interpretation* problems, which involve a geologic region that has changed over time due to the effects of geologic processes. Given descriptions of the region before and after the change, his system proposes a sequence of geologic processes that could have transformed the starting state to the ending state (such as faulting and erosion). Simmons terms a proposed sequence of processes an *interpretation*. GORDIUS uses a rule-based system to generate an approximate interpretation, then tests this solution with a simulator.

If the interpretation is incorrect, a debugger attempts to eliminate its flaws.

Just as HYPGENE is invoked when GENSIM's prediction does not match an observation, the GORDIUS debugger is invoked when the final geologic state predicted by the GORDIUS simulator does not match the known final state of the region. The debugger considers each bug in the prediction (for example, the height of a region might be incorrect) and examines dependency structures produced by the simulator to determine what assumptions the bug depended on (such as the amount of erosion that occurred at some time). GORDIUS considers assumptions about the events (processes) that occur, the values of parameters (such as the height attained by an uplift process), the relative ordering of events, the persistence of attribute values, the persistence of objects, and the existence of objects of a given type. These assumption types are similar to the operators implemented for HYPGENE. For example, event-occurrence hypotheses would be generated by "Existing Process Adds Assertion" and by "Violate Process Condition To Prevent Assertion." Simmons does not propose assumptions that correspond to HYPGENE's process-modification operators, although he says that adding new classes of assumptions to GORDIUS is easy to do.

One disadvantage to the use of assumptions rather than operators is that the GORDIUS simulator explicitly inserts all possible types of assumptions at every possible point in the ATMS dependency structure. For example, a persistence assumption is associated with every object value at every point in time. In large simulations, GORDIUS generates many assumptions, at a significant cost. HYPGENE's operators let it represent these assumptions implicitly. For example, the "Violate Process Condition To Prevent Assertion" operator causes HYPGENE to attempt to retract an assertion A from P_A by violating some precondition of every process that asserted A. GORDIUS would have to record each satisfied precondition explicitly as an assumption.

4.3.4 THE EARLY CHEMISTS

Langley, Simon, Bradshaw, and Zytkow (1987) have explored the discovery process with four different AI systems, all of which have reproduced discoveries from early chemistry.[6] BACON finds quantitative constraint equations (laws) consistent with quantitative measurements of a physical system; GLAUBER finds qualitative chemical reaction laws consistent with a set of observed chemical reactions; and DALTON and STAHL postulate container–component relationships among the chemicals in a reaction. In addition, Rose and Langley (1986) have described STAHLp, an extension of the latter system.

These systems differ from HYPGENE in terms of both the scientific reasoning tasks they address and the methods they use to perform these tasks. The chemistry programs rely heavily on data-driven reasoning to solve problems. Their inputs are data from experiments, and their outputs are laws and theories that explain the data. Unlike HYPGENE, they do not begin with a theory and use it to predict experimental outcomes. Thus, their methods apply to a scientific field in its infancy, when no theory exists to predict experimental outcomes or to guide the hypothesis-formation process.

The chemistry systems can consider data from many experiments simultaneously, whereas HYPGENE considers at most two experiments at once. For example, STAHLp computes hypotheses that are consistent with a large number of simple experiments (an experiment might involve a single reaction), whereas HYPGENE computes hypotheses that are consistent with a small number of complex experiments. Their methods are similar, in that both manipulate reasoning dependency structures. HYPGENE reasons about dependencies between reactions in a single GENSIM prediction to infer how that prediction might be altered. In contrast, STAHLp uses techniques similar to constraint propagation to infer component structures from descriptions of several experiments. These experiments are sometimes inconsistent; when the constraint propagator detects an inconsistency, the system uses a dependency structure created by the constraint-propagation procedure to hypothesize changes to the experiments that will render them consistent. Thus, HYPGENE reasons about the correctness of a single experiment, whereas STAHLp reasons about the consistency of multiple experiments.

HYPGENE works *backward* from the error in a prediction to formulate hypotheses and is thus guided by the dependency structure created by

---

6. See also the work of Kulkarni (this volume) in the domain of biochemistry, which is more concerned with strategies for experimentation than with hypothesis formation.

GENSIM. The chemistry programs apply their operators for generating laws and component structures in a forward search that has a small amount of heuristic guidance. For example, when the DALTON program searches for the component structure of water, it blindly generates chemical formulas such as HO, H_2O, HO_2. This approach would presumably take a long time to find the formula of the compound glucose: $C_6H_{12}O_6$. HYPGENE uses the difference between the initial conditions of an anomalous experiment and a reference experiment to focus its search. Since the chemistry programs do not predict the outcomes of the experiments they consider, they have no notion of a correctly predicted versus an incorrectly predicted outcome and so cannot use this information.

The expressiveness of the language that the chemistry programs use to represent qualitative laws (chemical reactions) is extremely limited. For example, they cannot represent reactions that are predicated on any type of logical condition, such as the states of the reactants. All their processes can be expressed in a single line, such as

$$\text{copper} \longrightarrow \text{phlogiston} + \text{vitriolic\_copper},$$

whereas HYPGENE processes are sometimes a page in length. HYPGENE reasons about processes whose preconditions include complex predicate-calculus formulas; such reasoning is beyond the capabilities of the chemistry programs.

The operators used by the chemistry programs to generate qualitative laws and component models are fairly domain specific; for example, they explicitly refer to the predicate `component-of`. The HYPGENE operators do not mention domain-level concepts but manipulate processes and predicate-calculus assertions. Thus, they appear to be more general.

5. Summary and Conclusions

This chapter has advanced the idea that we can view hypothesis formation as a process of design. The value of this analogy is that it instructs us to use knowledge of design to solve the hypothesis-formation problem, under the assumption that design is a better understood problem. A program called HYPGENE solves hypothesis-formation problems using design methods. Design goals specify the difference between the predicted outcome of an experiment and its observed outcome. HYPGENE satisfies such goals by applying design operators. Each operator satisfies specific types of goals (such as adding an assertion to a prediction) by making certain types of changes to the initial conditions of the experiment or to the theory used to derive predictions. The program conducts a best-first search of the hypothesis design space and uses an ATMS to represent the states of this space compactly.

HYPGENE provides a flexible and efficient framework for designing hypotheses. The flexibility arises because its operators are able to modify both the initial conditions of an experiment and the theory embodied by the process knowledge base. The operators are syntactically complete, in that they can make all possible syntactic changes to process descriptions. Another reason is that HYPGENE's operators are largely domain independent because they refer only to arbitrary predicate-calculus assertions and process definitions; they mention no domain concepts. A final reason for the system's flexibility is its ability to manipulate arbitrary predicate-calculus assertions, which gives it the ability to reason about complex processes and design goals. The HYPGENE framework is efficient because its search focuses on errors in a prediction; the system works backward from prediction errors to create hypotheses that account for the errors. In addition, the program could easily employ other types of knowledge, such as general knowledge of chemistry, to guide its search. Finally, one could use results from a second *reference experiment* to filter solutions and to guide the search process. I have demonstrated the validity of these techniques by testing the HYPGENE program on hypothesis-formation problems from the domain of molecular biology that are dominated by a historical study of the discovery of a new mechanism of gene regulation in bacteria.

Acknowledgements

Robert Engelmore, Jeff Shrager, and Pat Langley provided helpful comments on this chapter. This work benefitted greatly from many discussions with Peter Friedland, Bruce Buchanan, and Edward Feigenbaum. I am indebted to Charles Yanofsky for his collaboration on this project. Lyn Dupre provided editorial assistance. This work was supported by NSF grant MCS83-10236, NIH grant RR-00785, and DARPA contract N00039-83-C-0136.

References

Chapman, D. (1987). Planning for conjunctive goals. *Artificial Intelligence*, *32*, 333–377.

De Kleer, J. (1986). An assumption-based TMS. *Artificial Intelligence*, *28*, 127–162.

Dietterich, T. G. (1984). *Constraint propagation techniques for theory-driven data interpretation*. Doctoral dissertation, Department of Computer Science. Stanford University, Stanford, CA.

Fikes, R., & Nilsson, N. (1971). STRIPS: A new approach to the application of theorem proving to problem solving. *Artificial Intelligence*, *2*, 189–208

Forbus, K. (1984). Qualitative process theory. *Artificial Intelligence*, *24*, 85–168.

Hiraga, S. (1969). Operator mutants of the tryptophan operon in *E. coli*. *Journal of Molecular Biology*, *39*, 159–179.

Intellicorp (1986). *KEEworlds reference manual*. Mountain View, CA: Intellicorp.

Jackson, E., & Yanofsky, C. (1973). The region between the operator and first structural gene of the tryptophan operon of *E. coli* may have a regulatory function. *Journal of Molecular Biology*, *76*, 89–101.

Karp, P. (1989). *Hypothesis formation and qualitiative reasoning in molecular biology*. Doctoral dissertation, Department of Computer Science, Stanford University, Stanford, CA.

Kehler, T., & Clemenson, G. (1984). An application development system for expert systems. *Systems Software*, *3*, 212–224.

Langley, P., Simon, H. A., Bradshaw, G. L., & Zytkow, J. M. (1987). *Scientific discovery: Computational explorations of the creative processes*. Cambridge, MA: MIT Press.

Nilsson, N. (1980) *Principles of artificial intelligence*. San Mateo, CA: Morgan Kaufmann.

Pednault, E. (1988). Extending conventional planning techniques to handle actions with context-dependent effects. *Proceedings of the Seventh National Conference on Artificial Intelligence* (pp. 55–59). St. Paul, MN: Morgan Kaufmann.

Rajamoney, S. (1988). *Explanation-based theory revision: An approach to the problems of incomplete and incorrect theories.* Doctoral dissertation, Department of Computer Science, University of Illinois, Urbana.

Rich, C., & Shrobe, H. (1976). *Initial report on a LISP programmer's apprentice* (Technical Report TR-354). Cambridge, MA: Massachusetts Institute of Technology, Laboratory for Artificial Intelligence.

Rose, D., & Langley, P. (1986). Chemical discovery as belief revision. *Machine Learning, 1*, 423–451.

Simmons, R. (1988). *Combining associational and causal reasoning to solve interpretation and planning problems.* Doctoral dissertation, Laboratory for Artificial Intelligence, Massachusetts Institute of Technology, Cambridge, MA.

Yanofsky, C. (1981). Attenuation in the control of expression of bacterial operons. *Nature, 289*, 751–758.

CHAPTER 11

Diagnosing and Fixing Faults in Theories

LINDLEY DARDEN

"Heuristic counterexamples ... spur the growth of knowledge."
(Lakatos, 1976, *Proofs and Refutations*, p. 86)

1. Introduction

Resolving anomalies in scientific theories is a diagnostic reasoning task. Just as a repair person diagnoses faults in a device or a physician diagnoses a disease in some part of the body, a scientist engaged in resolving an anomaly has to locate the fault in a component of the theory. Providing a new hypothesis to resolve the anomaly is like fixing a faulty component in a device or providing treatment for a disease. Representing the theory as a series of steps in a normal process provides a schematic flow diagram for the normal case. Such a diagram can be used in the reasoning to localize the fault indicated by an anomaly. An important distinction can be made between what I call *monster* and *model* anomalies. Monster anomalies are problems that indicate where typical steps fail in unusual cases. Model anomalies show that the diagram for normal cases is incorrect and in need of modification, that is, the theory needs to be changed. Whether an anomaly is a monster or a model may be difficult to determine until it is successfully resolved.

The scientific case discussed here is the development of the theory of the gene in Mendelian genetics in the early twentieth century. First, I will discuss T. H. Morgan's depiction of that theory in terms of both exemplary problem solutions and in a precise verbal statement in his 1926 book, *The Theory of the Gene*. Next I present a representation of the theory in terms of the steps in normal hereditary processes, which give the mechanism of transmission of hereditary characteristics from parents to offspring. Exemplary problem solutions can be abstracted to yield several schemas to represent typical crosses in artificial breeding. This set of abstract problem-solving schemas makes up the explanatory repertoire of the Mendelian geneticist. Crosses like the exemplary ones could be explained by Mendelism, but when anomalies were encountered, the geneticist had to decide how to diagnose the fault. An analysis of the typical steps in exemplary cases of hybrid crosses aids in localization of the fault, as we will see from two different anomalies faced by the theory of the gene prior to 1926. I will spend most of the time on the anomaly that Cuénot found in 1905 while breeding yellow mice, which turned out to be a monster anomaly. Then I will contrast the outcome of the monster case with the model anomalies that indicated the typical phenomenon of linkage of some genes in inheritance. The linkage cases added a new exemplar to the explanatory repertoire to explain the inheritance of linked genes.

2.   The Representation of Gene Theory

Morgan (1926) discussed typical hybrid crosses and their results and then presented an explicit statement of the theory of the gene. In other genetic literature, the theory was rarely stated separately in declarative sentences; instead it was used in the context of explaining data from crosses made by artificial breeding.

Mendel's cross with different varieties of garden peas was the common exemplar to illustrate crosses between varieties differing in the characters of one trait. Tall peas were crossed with short ones (see Figure 1), with the first filial generation (F_1) showing all tall peas. Self-fertilization of the hybrids in the F_1 yielded three tall to one short in the F_2 generation, or, more precisely, one pure tall, two hybrid tall, and one pure short. The theory explained the 3:1 ratios by postulating an unobservable difference between the germ cells in hybrids. The theory claimed that each germ cell carried a gene for either tall or short, but not both. This central claim was called the "purity of the germ cells," or the "segregation" of genes in the formation of germ cells. Segregation was the new discovery of Mendelism. A Punnett square was used to illustrate the combinations between eggs and pollen that occurred during self-fertilization in the F_1 hybrids. Morgan's (1926) illustration of this cross

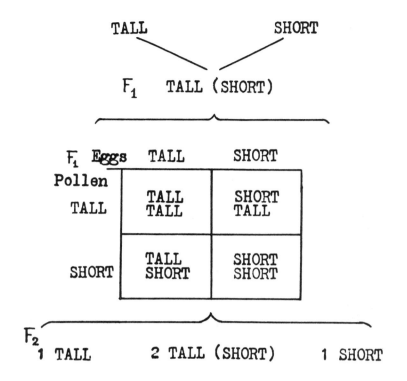

Figure 1. Cross between tall and short peas, from Morgan (1926, Figure 1, p. 2).

and its explanation is shown in Figure 1. Note how the initial conditions of the cross, the theoretical explanation, and the data to be explained are all represented in the diagram.

Very soon after Mendelism began in 1900, exceptions were found to the dominance of one character over the other in the F_1 generation. Morgan illustrated a case of "blending" in the F_1 generation with a cross between red and white four-o'clocks that produced pink flowers in the F_1. The F_2 generation showed one red to two pink to one white; Figure 2 depicts the cross, with red represented by black, white by white, and pink by gray. Morgan abstracted the details from the four o'clock case to produce an abstract diagram to show the transmission of genes from parents to offspring through three generations, as shown in Figure 3. The characters of the flowers (the "phenotype") were illustrated by the large circles, and bead-like particles representing genes were depicted by

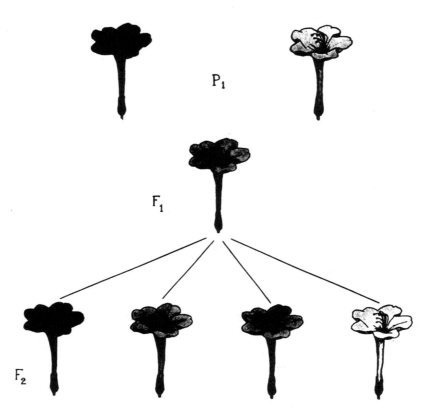

Figure 2. Cross between red and white four-o'clocks, from Morgan (1926, Figure 5, p. 6).

the small circles. Two of the same type were present in the parents; one of each type was represented in the genes of the F_1 hybrid. Small circles with a cross between them illustrated the role of genes in the formation of germ cells and fertilization in the hybrid. Finally, the circles depicted in the F_2 and F_3 generations graphically showed the genes associated with the 1:2:1 ratios of the phenotypic characters.

The abstract diagram in Figure 3 represented numerous implicit assumptions that were components of the theory. Genes were bead-like; they occurred in pairs in homozygous and heterozygous forms; they segregated in a pure form during the formation of germ cells in hybrids (with no contamination between red and white); they occurred in equal numbers; they combined randomly; and all the gene combinations were

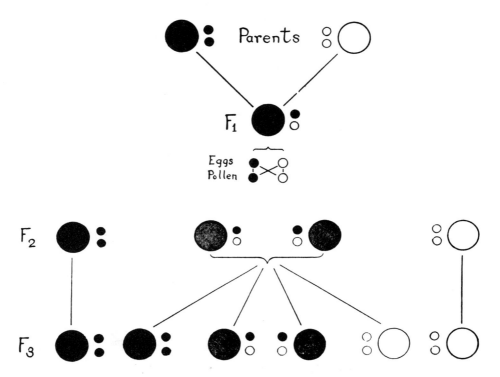

Figure 3. Abstraction from four-o'clock exemplar, from Morgan (1926, Figure 6, p. 7).

viable. A picture is worth a thousand words; I will not try to state them all.

Morgan continued his examples of typical genetic phenomena by showing the phenotypes and the Punnett square for the cross between peas that differed in two traits, yellow round and green wrinkled. Such a cross produced ratios of nine yellow round to three yellow wrinkled to three green round to one green wrinkled (see Figures 4 and 5). But numerous exceptions to the 9:3:3:1 ratios had been discovered, such as the linkage in the sweet pea between purple flowers and long pollen grains and the linkage of white flowers with round pollen grains, as shown in Figure 6. Linkage anomalies were explained by postulating that genes were parts of chromosomes, visible thread-like structures found in the nuclei of cells. Genes that were part of the same chromosomes were claimed to be linked in inheritance. Sometimes parts of the chromosomes switch, or "cross-over," so the linkage is not complete and small

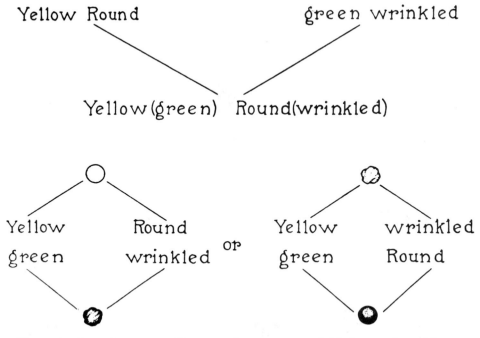

Figure 4. Cross between yellow round and green wrinkled peas, from Morgan (1926, Figure 7, p. 8).

cross-over classes result, such as the two cross-over classes illustrated by the purple flowers with round pollen and the white flowers with long pollen in Figure 6.

After presenting these and a few other examples of exemplary crosses and their explanation in terms of unobservable genes, Morgan stated the theory of the gene (which, it should be noted, contained no mention of chromosomes):

> We are now [after discussing the examples above] in a position to formulate the theory of the gene. The theory states that the characters of the individual are referable to paired elements (genes) in the germinal material that are held together in a definite number of linkage groups; it states that the members of each pair of genes separate when the germ-cells mature in accordance with Mendel's first law, and in consequence each germ-cell comes to contain one set only; it states that the members belonging to different linkage groups assort independently in accordance with Mendel's

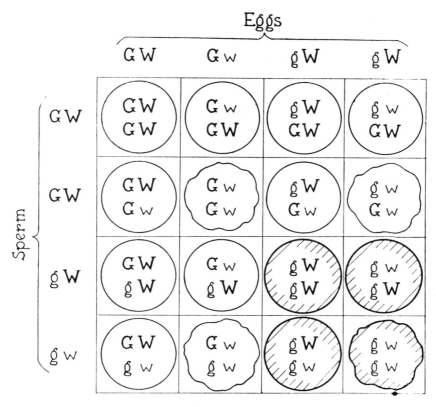

Figure 5. Punnett square for fertilization producing 9:3:3:1 ratios, from Morgan (1926, Figure 8, p. 9).

second law; it states that an orderly interchange—crossing-over—also takes place, at times, between the elements in corresponding linkage groups; and it states that the frequency of crossing-over furnishes evidence of the linear order of the elements in each linkage group and of the relative position of the elements with respect to each other. These principles, which, taken together, I have ventured to call the theory of the gene, enable us to handle problems of genetics on a strictly numerical basis, and allow us to predict, with a great deal of precision, what will occur in any given situation. In these respects the theory fulfills the requirements of a scientific theory in the fullest sense. (Morgan, 1926, p. 25, italics omitted)

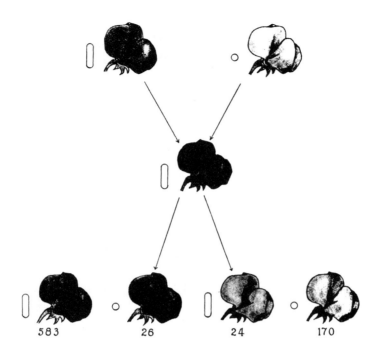

Figure 6. Linkage in sweet peas, from Morgan (1926, Figure 9, p. 11).

This statement of the theory provided an abstract characterization of the "typical" and "occasional" (crossing over) behavior of the unobservable genes during germ cell formation. The *abstract* verbal statement summarized all the cases that the separate diagrams illustrated in *concrete* examples. The theory was illustrated by *hybrid crosses* between parents with differing characters, but Morgan's abstract statement was more general. He postulated a general theory of heredity for *all* hereditary characters of all organisms. The theory thus consisted of general claims about the nature of unobservable details in the steps of hereditary processes between parents and offspring. Morgan stressed that all the claims made in the theory of the gene about the unobservable "elements" in the germ cells were based on data from hybridization experiments. Furthermore, the theory made predictions that had been tested by additional experiments. After presenting the theory as based on data from breeding experiments, he introduced discussion of the chromosomes that pass "through a series of manoeuvres that go far toward supplying a mechanism for the theory of heredity" (Morgan, 1926, p. 45).

The various diagrams of typical hybrid crosses depicting the role of genes illustrate what Kuhn (1970) called "exemplars," which are concrete examples that serve as model cases for applying the theory to specific phenomena. The kinds of exemplars constitute the explanatory repertoire of the theory, namely, the kinds of problem solutions that the theory can provide. Segregation, independent assortment, linkage, and crossing-over constituted the explanatory repertoire of the Mendelian geneticist as of 1926. As Kuhn indicated, students learn the theory, just as you the reader may have done, by seeing how it applies in the concrete examples. The next step is to see how a given problem-solving schema can be abstracted and applied to other similar cases. The verbal statement of the theory is less important than the abstractions from the concrete problems. The steps that incorporate both initial conditions, the theory, and the data to be explained, such as Figure 3 does abstractly, are the important kinds of representations for use in applying the theory to additional cases of that kind. When 1:2:1 ratios are encountered, then that schema should be instantiated to explain them.

My analysis of the theory of the gene here extends a view of theories that is emerging in the philosophy of science. A theory is represented as one or more abstract schemas that may be instantiated to explain specific instances in its domain. Kitcher (1981) argued that theories give unified explanations by supplying abstract schemas and instructions for instantiating the variables in a specific way. This view of scientific theories as a set of problem-solving schemas has been given an implementation in an AI system by Thagard (1988). Neither Kitcher nor Thagard discussed visual, diagrammatic representations of abstract schemas; however, my analysis of diagrams as representing abstract steps in typical processes is compatible with, and extends, their approaches. I will now use such a representation of the theory of the gene (in terms of abstract steps in typical hereditary processes) to show how anomalies for the theory can be localized and resolved.

3. Localizing and Fixing Monster Anomalies

This section will explore an analogy between the theory of the gene, represented as filling in steps in diagrams of typical hereditary processes, and structural or functional diagrams, depicting components of a device or stages in a process. This analogy will aid in an analysis of how to resolve problems for theories during their development. Three kinds of problems for a theory will be distinguished: (1) an *incompleteness* that indicates the need for further work to fill in holes in the diagram; (2) a *monster anomaly* that indicates that the normal process is failing in a particular instance, such as a disease indicates the body is not functioning normally in a given individual; and (3) a *model anomaly* that indicates that the theory is incorrect and the diagram needs to be changed or refined.

Diagrams are often useful for abstractly representing the structural components of a system or the functional stages in a process. Examples are electric circuit diagrams for representing flow of electricity in a device and diagrams of the circulatory or immune system in humans. When repair persons or doctors diagnose a fault or disease, one way of doing it is to try to localize the problem within a part of the normally functioning system, represented by a component of the diagram. Such diagrams can represent various levels of detail, just as theories can. One can imagine going up or down in levels of organization with varying amounts of detail at any one level within a diagram. "Black boxes" can occupy places in the diagram when no more detail about that step is known or when it is irrelevant for some purpose served by the diagram.

For a repair person diagnosing a fault in an electrical device, knowledge of the normal functioning may be important information. A schematic diagram of circuits shows which components are present and provides possible sites of malfunction. The modularity of the system is important. Separable components can be localized as sites of failure. The role the components play in normal functioning can aid in localization. The nature of the malfunction can be used to focus attention on some parts of the system rather than others as the most likely sites for failure. Similarly, physicians go through long training in the normal functioning of systems in the human body, such as the circulatory and immune systems. Symptoms of disease allow localization in one or more malfunctioning body parts.

Research in AI has investigated methods for diagnostic reasoning. The method that uses a model of normal functioning to aid in localization of a fault is called "model-based reasoning" (Davis & Hamscher, 1988) or reasoning based on "deep" knowledge (Chandrasekaran, Smith, & Sticklen, 1989). This methodology can be applied to anomaly resolution for theories when the theory can be represented as a model that depicts normal functioning.[1]

Viewing the problem of heredity as the need to fill in details in a diagram of steps between parents and offspring allows an analogy to be made to such schematic diagrams. A step between parents and offspring in a pedigree diagram may raise the question, what is passed from one to the other? This question is the scientific problem of heredity. Thus, scientific research problems can be found by considering steps in natural processes for which details have not yet been investigated. Hypotheses can be made as to how to fill in the blanks in the diagram. Not all the "black boxes" need be filled in at once. It was possible to represent hereditary characters as being passed from generation to generation in pedigree diagrams before it was known what, exactly, was passed from parent to offspring that caused, for example, blue eyes to be inherited. Thus, an *incompleteness* in a theory (as opposed to an incorrectness indicated by a model anomaly) can be represented by a black box or place holder in the diagram of typical steps. Constructing such a diagram and noticing blanks is a strategy for finding unsolved scientific problems.[2]

The theory of the gene can be viewed as filling in unobservable details in steps in typical hereditary processes. The abstract steps in a typical segregation process can be diagrammed. One way of constructing such a diagram is to drop out the details in a diagram of a specific cross, as Figure 3 is an abstraction from the cross of differently colored four o'clock flowers (compare Figure 2). Consider the aspects of a segregation process illustrated in Figure 3. The theory postulates the existence of bead-like genes, that they occur in pairs, that the parents had pairs of identical genes (were homozygous) but the F_1 hybrid had two genes of different types (was heterozygous), that the eggs and pollen contained one or the other but not both genes, that the genes in the egg and pollen were pure (like one of the parental types), that the different types of

---

1. Other methods for diagnostic reasoning exist besides appeal to functioning of a normal system, such as a fault dictionary that simply lists faults with probable diagnoses. These methods are less relevant for anomaly resolution in scientific theories than the model-based method. For more discussion of diagnostic reasoning methods, see Buchanan and Shortliffe (1984), Chandrasekaran and Mittal (1983), Davis and Hamscher (1988), Reggia, Nau, and Young (1983), Schaffner (1985), and Sembugamoorthy and Chandrasekaran (1986).
2. For more discussion of strategies for finding scientific problems, see Langley, Simon, Bradshaw, and Zytkow (1987).

germ cells combined randomly, and that all the combinations developed into organisms in the F_2.

The theory of the gene did not fill in all the details of the steps and structures that were used in its representation. For example, alternative allelomorphic genes were represented by different symbols, even though it was unknown how they differed. Such "promissory notes" in the early stages of theory formation are easily represented by some difference in the symbols (such as black and white beads) and signify *incompleteness*. Such problems of incompleteness may be solved at later stages of theory formation, perhaps by filling in details at even lower levels of organization.[3]

However, some problems are not problems of black-box incompleteness but result from a failure of the diagram to apply to a case of heredity. For example, for normal segregation processes, the theory predicts 3:1 ratios in the F_2 generation, and finding a 2:1 ratio is an anomaly. Thus, anomalies arise when an exemplar fails to apply in a case that it was expected to represent. The first step in anomaly resolution is to localize the fault in one or more of the modular steps in the abstract schema for depicting segregation. Each step of a normal process that occurs prior to the location in the diagram representing the anomalous data may be a candidate for being the problematic site. The nature of the anomaly itself is very important in localizing plausible problematic steps. For example, when the anomaly of *2Aa:1aa* ratios were found, then the focus was on the steps in segregation involving the production of the *AA* forms that were missing.

Cuénot (1905) found an exception to 3:1 ratios while breeding mice. Yellow mice bred with those having other colors showed that yellow was dominant. However, when the hybrids were bred, the percentage of yellow in the F_2 generation was smaller than expected. Cuénot found ratios between 2:1 and 3:1. When he bred the F_2 yellows, he was unable to obtain any homozygous yellows, that is, no pure dominants were found. Symbolizing the yellow as *A* and the other colors as *a*, then no *AA* was found, only *Aa* and *aa*. Others confirmed the anomalous data and improved its accuracy, showing that the ratio was close to *2Aa:1aa*.

Historically, three different hypotheses were proposed to account for the 2:1 anomaly. The hypotheses were localized in different steps in the

---

3. See Shapere (1977) for an example of black-box incompleteness in the Bohr model of the atom.

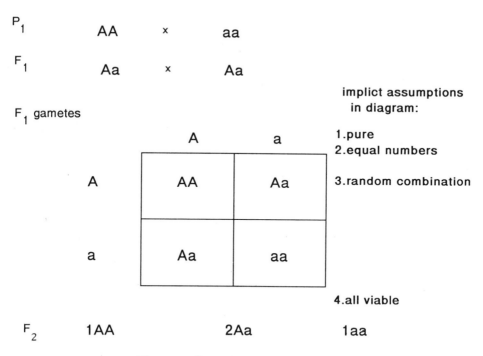

Figure 7. Steps in a typical 3:1 cross.

normal process of segregation, illustrated in Figure 7. Cuénot (1905) explained the anomaly by denying step 3, random fertilization. He proposed that selective fertilization occurred—germ cells with A selectively did not fertilize each other. Thus, the only combinations were Aa and aa, as depicted in Figure 8(b).

Morgan (1905, 1909) proposed an alternative hypothesis that was localized in a different step of segregation. He questioned step 1 in Figure 7, the pure segregation of parental genes during germ cell formation in hybrids. Purity of the germ cells was the startling new discovery of Mendelism. It was indeed puzzling that F_1 hybrids could give rise to pure breeding forms in the F_2 that showed no evidence of hybrid ancestry. The idea was an old one, that mongrels would show effects of the cross-breeding in their past. Morgan believed that Cuénot's case in which no pure dominants were found was a case that indicated that, in general, pure breeding forms did not reemerge after a hybrid cross. In the yellow mouse case, he claimed, the impurity just showed up sooner

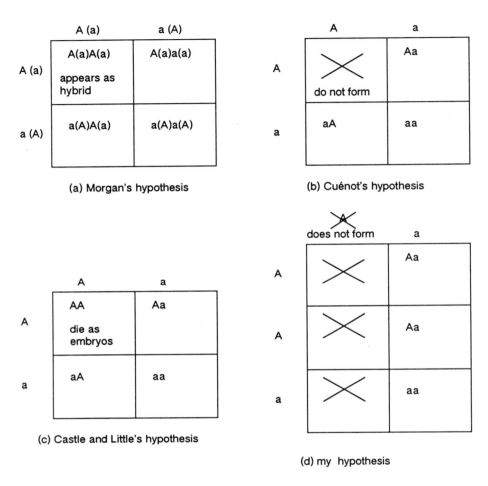

Figure 8. Alternative hypotheses to explain 2:1 ratios.

than in other cases. More specifically, Morgan proposed that the germ cells of hybrids contained both parental genes (alleles), one in an "active" form, the other "latent." Thus, instead of pure *A* and *a* germ cells, he symbolized the germ cells of hybrids as *A(a)* and *a(A)*.

Usually, Morgan claimed, the latent genes did not manifest themselves for several generations, and thus Mendelians had been misled into believing that pure breeding forms emerged after hybridizations. In the Cuénot case, the genotype *A(a)A(a)* (Morgan's representation for what others designated *AA*) for some reason showed its hybrid character earlier, and thus Cuénot did not find any seemingly pure *AA* forms, as shown in Figure 8(a). Morgan predicted that the seemingly pure recessive, *aa*, would also show the effects of hybridization. However, when he did additional breeding experiments with mice, he did not find such effects, and he abandoned his impurity hypothesis. Shortly thereafter, based on additional evidence, Morgan became a strong defender of Mendelism and the assumption of purity of germ cells.

Yet another hypothesis was proposed by Castle and Little (1910) that was localized in a later step. They suggested that the embryos formed by the mating of two germ cells with genes for yellow were inviable. Given this hypothesis, one would expect 2:1 ratios: *AA* died, leaving *2Aa:1aa*, as shown in Figure 8(c). As evidence for their hypothesis, they appealed to the smaller numbers of young produced in these crosses and pointed out that decreased viability had been found in other cases. Castle and Little's hypothesis was accepted and confirmed beyond a doubt with dissection of dead embryos (Kirkham, 1919).

Each hypothesis was localized in a different step in the normal segregation process. Morgan proposed that the germ cells of the hybrid were not pure and thus challenged the step of pure segregation in general (step 1 in Figure 7). Cuénot suggested that pure germ cells formed but, in this anomalous case, did not combine randomly in fertilization. His alternative was selective fertilization, which was localized in the step subsequent both to segregation and to the formation of types of germ cells in equal numbers (step 3 in Figure 7). The alternative that was eventually confirmed was that of Castle and Little, who proposed that the homozygous dominant form (*AA*) was not viable. In the lethal hypothesis, all steps in segregation functioned normally until after fertilization (they thus denied step 4 in Figure 7). The malfunction occurred sometime during embryological development (the exact stage was debated for decades, see Eaton & Green, 1962). Thus, in the Castle and Little hypothesis, all the normal steps of pure segregation and random fertilization were preserved, and a new, previously implicit assumption was delineated—equal viability of all zygotes.

Although I stated that viability was an implicit assumption in the diagram in Figure 7, equal viability was not stated as an explicit assumption of Mendelism in the early genetic literature. This contrasts with the explicit discussion of the purity of germ cells and occasional mention of the assumptions of equal numbers and random fertilization. Finding implicit assumptions, such as the equal viability assumption that is subverted in the case of lethals, involves realizing that important steps are implicitly represented in a diagram. For example, Morgan's abstract segregation figure (Figure 3) does not show the step of normal embryological development that occurs after fertilization but before the adult phenotypic characters form. Filling in more details in the diagram about unrepresented steps in the process is a method for uncovering implicit assumptions during anomaly resolution. Anomalies may often drive the process of uncovering previously implicit assumptions.

No historical evidence exists to show that geneticists actually used such a diagrammatic representation in generating the alternative hypotheses. However, this analysis shows that such a stepwise representation can be a useful guide in hypothesis generation. Explicitly laying out the normal steps in segregation shows their modularity and aids in the localization of potential sites of the fault. The three historical hypotheses were localized in different steps of segregation. One step, as I have analyzed them, was not used in formulating a historical hypothesis, so far as I know—the step of the formation of germ cells in equal numbers (step 2 in Figure 7). I constructed my own hypothesis at that location by proposing that, in one sex, no gametes of A type formed, whereas in the other sex a double dose of gametes with A formed, as shown in Figure 8(d). Such an unequal number of types of gametes would result in the anomalous 2:1 ratios, with no AA combination. My ability to form this other hypothesis, which was not proposed historically, shows the value of an explicit delineation of separable steps in a process with a systematic consideration of alternatives to each step.

Explicit delineation of the steps leading to 3:1 ratios allowed me to localize the potential sites for modification in the face of the 2:1 anomaly. However, once the steps are localized, additional strategies are needed for developing the alternative hypotheses to account for the anomaly. First, the nature of the anomaly provided a constraint; not just any change to the segregation components would produce two parts heterozygotes to one part pure recessive. If, for example, the anomaly had been one pure dominant to one pure recessive with no heterozygotes, then the alternative hypotheses would have been different. The postulated process had to operate to produce two heterozygotes to one pure recessive.

But the constraints provided by the anomaly itself are not sufficient for developing the alternatives. Additional strategies for hypothesis generation are needed. In each case, an alternative was developed by proposing the opposite of the normal theoretical component. The four alternative hypotheses can be analyzed as built on the following opposites: purity-impurity (Morgan's); random combination-selective fertilization (Cuénot's); equal viability-unequal viability (Castle and Little's); and equal numbers-unequal numbers (mine). But the opposite of the normal step, such as selectivity, is just the beginning point, just a seminal idea, just a type of hypothesis. The idea of the opposite has to be fleshed out to become a plausible explanatory hypothesis to account for the anomaly.[4]

Denials of a general claim can take a range of forms. At the most extreme, a general claim can be completely denied, as Morgan did. He claimed that no pure germ cells were ever formed and that "all pure" became "all impure." Less extremely, a general claim can be specialized. "All combine randomly" was specialized by Cuénot to "some combine selectively" (while some others still do combine randomly). Similarly, I claimed that "all form in equal numbers" was to be specialized for this case to "some do not form in equal numbers." Even less challenging to a central tenet of a theory is the claim that just in some monstrous cases is equal viability violated. Thus, alternative hypotheses range in severity from (1) complete denial of a theoretical component to (2) merely limiting the generality of a component to (3) explaining away the quirky case without changing any theoretical component.

There was disagreement as to whether the 2:1 anomaly was a model anomaly that required a change in theory or just a quirky, monstrous case. Morgan viewed it as an indication that no step of pure segregation ever occurred after hybrid crosses. Thus, no exemplar of pure segregation (and perhaps no Mendelian theory at all) would have been used in explaining hybrid crosses had Morgan's hypothesis been confirmed. Had Cuénot's hypothesis been correct, then another exemplar would have become part of Mendelism—cases showing selective fertilization. Thus, in addition to the exemplar of yellow and green peas producing 3:1 ratios, the set would have included yellow and nonyellow mice producing 2:1 ratios, with the genes represented by A and A failing to combine with each other.

---

4. See Darden (1987) and Darden and Cain (1989) for further discussion of abstract theory types and their instantiation to form explanatory hypotheses.

In contrast, Castle and Little's hypothesis of homozygous dominant lethals viewed the anomaly as a monstrous event, occurring after all the typical steps had occurred. No components of the theory needed to be altered, but sometimes the normal development failed. Homozygous lethals were like a disease, a known malfunction class. When additional cases of 2:1 ratios were found, a possible explanation was to suggest that they were yet another instance of homozygous lethals. Thus, the field of Mendelism expanded as a result of this case, not by changing the theory but by expanding the set of known malfunctions—the typical theoretical steps were shown to apply, but some other quirky process prevented the normal result from occurring.

The anomaly was thus resolved without altering the claim that the segregation steps were typical of normal processes; the theory did not have to be modified in the face of Cuénot's monster anomaly. Tests confirmed the inviability hypotheses, and additional cases of lethal gene combinations were found. Thus, the 2:1 anomaly was resolved, not by changing any claim in the theory but by explaining the anomaly away as a monster.

Monster anomalies are like a malfunction in a device or a failure in a part of the body; the normal process is captured by the schematic, and the monster anomaly is a failure in a part. For quirky, unusual cases, the failure is unlike other known failures. In such a case, diagnosing the failure may be a difficult problem for the repair person or physician who has never before seen that kind of failure or disease. For malfunction classes, such as lethal gene combinations, the monstrous anomaly is like a known kind of failure in a device or like a disease. Seeing a problem as an example of a well-known malfunction class is a routine puzzle-solving task. Finding a new malfunction class is more interesting and adds to the knowledge about how things can go wrong. Scientists often receive credit for discovering the first instance of a malfunction class. Bauer is credited with discovering the first case of lethals (Dunn, 1965, p. 102). Castle and Little (1910) credited him with having found a similar anomaly, in their arguments that their hypothesis was not *ad hoc* but was a kind of anomaly for Mendelism found in other cases. Similarly, medical researchers are credited with the discovery of new diseases. Technicians are less likely to be rewarded for finding a malfunction that causes a device to be recalled, but engineers may be grateful to learn where their design is routinely failing.

I have taken the term *monster anomaly* from Lakatos' (1976) term *monster-barring*. He discusses the strategy of monster-barring as a way of preserving a generalization in the face of a purported exception; if the exception could be barred as a monster (shown not to be a threat to the generalization after all), then it could be barred from causing a theoretical change. Lakatos was concerned with distinguishing between legitimate exception-barring instances and illegitimate barring of instances that really did require a theory change. My usage of *monster* is a similar one. The "local counterexamples" he discussed would be included in my category of model anomalies; however, because he was not discussing scientific theories as represented by exemplary problem schemas, the term *model* would have been less appropriate for the mathematical, "local" counterexamples that he discussed. I have no examples of anomalies that served to falsify the entire theory of the gene; these would correspond to what Lakatos called "global counterexamples."

4. Localizing and Fixing Model Anomalies

Model anomalies, in contrast to monsters, require a change in the claims about what is normal (or general) in hereditary processes. To continue the analogy with diagrams, model anomalies show that the proposed schematic diagram of normal processes is incorrect. Presumably such a change is not required for schematic diagrams of artifacts, if the draftsman drew it properly. In contrast, in the history of medicine, coming to understand diseases may lead to new (or corrected) information in anatomy or physiology. For example, information about abnormal behavior resulting from brain damage (e.g., dyslexia) may aid in determining the normal functional architecture of the brain (Bub & Bub, 1988).

The model anomaly requires a change in the theory, a change in what is claimed to be normal. Once the theory is changed to accommodate it, the model anomaly is no longer anomalous; such model anomalies, requiring theory change, become exemplars ("models") for typical processes that had not been known to be typical prior to their discovery. Anomalies to 9:3:3:1 ratios turned out to be model anomalies. Prior to their discovery, segregation and independent assortment had not been clearly delineated as two different processes. Mendel (1865/1966) did not formulate what Morgan later called Mendel's two laws (Monaghan & Corcos, 1984). Along with the early Mendelians in the period before about 1910, Mendel believed that crosses involving two different traits simply showed that segregation operated in those crosses as well. When 3:1 ratios are considered for crosses with two different traits, the ratios are 9:3:3:1 ratios. There was no need to distinguish two separate laws until anomalies were found that showed 3:1 ratios for each trait considered separately but showed exceptions to the 9:3:3:1 ratios when the two traits were considered together.

Bateson, Saunders, and Punnett (1906) are credited with the first discovery of the incomplete linkage of Mendelian factors, although they used different terminology to label the phenomenon. In crosses with sweet peas that differed as to flower color and pollen shape, they found 1528 purple long, 106 purple round, 117 red long, and 381 red round.[5] They called this phenomenon "coupling." Although Bateson did not do so, the ratio of the numbers can be calculated to yield about 13.7 to 1 to 1 to 3.4. Such a ratio was obviously an exception to the expected 9:3:3:1 ratios, but the ratios of purple to red showed approximately 3:1 ratios, as did those of long to round. Segregation was not being violated, but independent assortment was.

Once again, the steps in the formation of the hybrid germ cells were implicated as the site in need of modification. The assumption that the genes segregated purely was left intact, as was the step of random combination. But the equal number assumption for all the combinations of the four different genes was denied. In normal cases, the germ cells were claimed to form in equal numbers, that is, $1AB + 1Ab + 1aB + 1ab$, to produce 9:3:3:1 ratios. Instead of equal numbers of gametes with the various combinations of genes, Bateson and his colleagues proposed unequal gametic ratios: $7AB + 1Ab + 1aB + 7ab$. Random combinations of such germ-cell ratios would yield ratios of about 11.8 to 1 to 1 to 3.2. They claimed that these predicted ratios were sufficiently similar to the anomalous ratios found in sweet peas to be a good fit for the hypothesis.

The numerical values of the anomalous phenotypic character ratios allowed calculation of germ-cell ratios. Once again, the specific nature of the anomaly aided in the construction of the hypothesis to account for it. In this case of a numerical anomaly, the anomalous character ratios enabled the geneticists to fix the parameters of the purported anomalous germ-cell ratios. For other anomalous character ratios, Bateson was able

---

5. I am unsure where Morgan obtained the numbers he reported for Bateson's cross, but see Figure 6.

to calculate other gametic proportions, such as 3:1:1:3. Once again, the strategy of "propose opposite" can be seen in the denial of equal numbers of types of germ cells. The proportions of unequal numbers could be calculated based on the nature of the anomaly.

I considered whether another step could be denied, rather than the equal number step. But a denial of random fertilization would require very complicated processes of selective fertilization to produce the anomalous ratios. I soon abandoned trying to formulate a workable hypothesis that denied randomness. Thus, it is not surprising that no such hypothesis, localized in the randomness step, appears in the historical record. It would be fascinating to have evidence that some geneticist also tried to formulate that type of hypothesis but failed. Unfortunately, unfruitful paths of reasoning that produce no hypotheses are rarely preserved in the written scientific record.

Bateson did not stop with merely proposing that unequal numbers of types of germ cells formed. He and his colleagues also formulated hypotheses to try to account for the production of unequal numbers. They proposed one hypothesis, abandoned it, and then proposed another. Their first hypothesis was that genes coupled prior to the formation of germ cells instead of randomly assorting. In other words, *A* and *B* coupled with each other, thereby producing more germ cells with the *AB* and *ab* combination and fewer with the *Ab* and *aB* than would be expected on random assortment. However, they soon formulated what they considered to be a better explanation, not at the level of the genes themselves but at the level of the germ cells. They claimed that all combinations formed in equal numbers but that some germ cells "reduplicated," whereas others did not. So the germ cells with *AB* and *ab* reduplicated to produce more of those types (Bateson & Punnett, 1911).

Bateson and Punnett's hypothesis that germ cells selectively reduplicated was not based on cytological data about cell divisions; no series of selective germ cell divisions were known in the usual process of germ-cell formation. They suggested that perhaps the germ cells formed early in embryological development and that cytologists had not yet found that stage when the duplications occurred.[6]

As a result of his work with fruit flies, T. H. Morgan converted from a critic of Mendelism to a supporter. In 1910, he began to investigate

6. For further discussion of the reduplication hypothesis, see Darden (1980).

coupled characters in fruit flies. As a result of his own discoveries, he proposed a different explanation for unequal numbers of germ-cell types. His hypothesis was thus localized in the same theoretical step, the production of unequal numbers of types, but his explanation was at yet another level of organization. Instead of hypothesizing coupling genes or reduplicating germ cells, he proposed that genes were parts of chromosomes. Those genes that were part of the same chromosome were likely to produce characters showing coupling in inheritance. But the characters were not completely coupled. He made use of the cytological observation that chromosomes intertwined during the process of germ-cell formation, claiming that occasionally pieces of paired chromosomes switched places or crossed over. The cross-overs produced the germ cells that occurred in small numbers. Thus, Bateson's sweet pea case was explained by claiming that genes for purple flowers and long pollen grains were part of the same chromosome. Similarly, the other coupled characters, red flowers and round pollen grains, were produced by genes on the other chromosome.[7] Occasionally, Morgan claimed, cross-overs between chromosomes resulted in the combination of purple with round, as well as the other infrequent combination.

Thus, all three hypotheses to account for linked characters were localized in the same step in the hereditary process—the formation of equal numbers of germ cells. All three hypotheses accounted for the anomalous ratios by proposing the opposite of equal numbers. Each alternative was of the "deny equal numbers" type. But three different hypotheses, located at three different levels of organization, were proposed to explain the production of unequal numbers—at the level of coupling genes, at the level of linkage groups on chromosomes, and at the level of reduplicating germ cells. A strategy for hypothesis formation can be formulated from these examples. The strategy is to introduce more details into the diagram to represent physical processes at various levels of organization. Finding such additional detail may require use of findings from another field, such as cytology, the study of cells and their parts. Alternative hypotheses can then be generated by postulating causes at different levels of organization. As it turned out, Morgan's successful linkage hypothesis involved postulation of a new level of organization, the linkage group—a higher level of organization than separate genes themselves but a lower level than all the genes in a germ cell.

---

7. Chromosomes occur in homologous pairs; two alleles for a character are parts of the two paired homologues.

Adding more detail to a diagram of normal steps and moving to new levels of organization may require substantial additional knowledge not originally supplied by the theory itself. Morgan's appeal to known cytological phenomena produced the hypothesis of linkage along chromosomes that was eventually confirmed. His appeal to chromosomes and their intertwining appearance produced the more plausible hypothesis of linkage and crossing-over than did Bateson's appeal to unobserved reduplication of selected germ cells. Thus, when using the strategy of forming hypotheses by using interrelations to another field, appeal to known information in the other field is better than predicting as yet unknown phenomena.[8]

Linkage emerged as a very general phenomenon. The linkage anomalies thus turned out to be model anomalies that required substantial changes to the theory to account for normal crosses involving two traits. Exemplary crosses illustrating linked traits (exceptions to 9:3:3:1) were added to the normal set of exemplars as "models." Morgan's statement of the theory of the gene in 1926 reflected the changes to the theory to include linkage and cross-overs, as we discussed in Section 2. By 1926, the theory of the gene and the claim that genes were parts of chromosomes had much evidence in their favor. Numerous anomalies had been successfully resolved by the theory.

5.   Conclusion

One can analyze a theory as representing typical, modular steps, which can be failing in a particular case (a monster) or failing in numerous cases (a model). This method of theory analysis is analogous to diagnostic reasoning, and one can study a theory's development by documenting the ways that a scientist localized and fixed faults in parts of a theory.

A series of stages characterize the reasoning in anomaly resolution. The first stage represents the theoretical steps diagrammatically and uses the diagram as an aid in diagnosing the location of anomalies. The nature of the anomaly aids in localization; only some steps affect a given outcome. The next stage generates alternative hypothesis at the potentially problematic steps. In generating such alternatives, strategies for producing new hypotheses can be invoked, such as specializing an old

---

8. See Darden and Maull (1977), Darden (1980), and Darden and Rada (1988) for further discussion of the strategy of using interrelations between two bodies of knowledge in hypothesis formation.

theoretical component or adding another one by proposing the opposite. Generation could proceed in different ways. One could systematically generate numerous alternatives at the potentially problematic steps, or some guidance might be introduced to focus generation (e.g., in hypotheses that would require the least theory modification). In other words, one might assume that the problem is due to a monster anomaly and only generate hypotheses that would explain it away as a quirky case. Alternatively, one might systematically generate numerous alternatives and design tests to choose among them. After hypothesis generation, the next task is to assess the adequacy of the various alternatives. If the anomaly turns out to be a model anomaly, then the new hypothesis becomes a new component of the general theory. If a new theoretical component is added, it should function as a step in a typical process to produce a now expected (as opposed to anomalous) end result.

The diagrammatic representation of typical steps supplied by the theory lets one depict the theory at various levels of detail and lets one represent incomplete portions by boxes yet to be filled. Such an explicit representation of steps aids (and is sometimes driven by) anomaly resolution. Anomalies may be shown to be either monsters not requiring a change in the "typical" steps or model anomalies that require a change in one or more steps of the theory and provide new exemplars. Some anomalies (or incompletely filled boxes) may point to the need for new theories at other levels of organization in order to resolve them.

The physical nature of the gene, how genes reproduced, and how they functioned to produce characters were unsolved problems for the theory of the gene, blanks not filled in for the steps of reproduction between parent and offspring. Biologists in the early twentieth century attempted to resolve them without much success by appealing to other theories at other levels of organization. Molecular biology eventually filled in these blanks at a lower level of organization.

Because anomaly resolution and diagnostic reasoning have much in common, the study of methods for diagnosis and trouble shooting in AI can be used to analyze reasoning in theory change. One such method represents functional steps in a device, indicating at each step its inputs and outputs (Chandrasekaran & Mittal, 1983; Chandrasekaran, Smith, & Sticklen, 1989). Underlying this functional representation may be information about the structure of the components and their behavior. Then diagnosis to localize a fault proceeds by determining where a step failed to produce an expected output. Queries about inputs and outputs at each step are generated, based on the details of the functional representation. This pattern of reasoning can also be applied to the gene case. For example, typical steps of segregation can be represented, and queries can aid in localizing an anomaly for segregation in one of the steps. Additional empirical investigation may be needed to localize the step of failure, as was the case for the discovery of lethal gene combinations in mice. The queries, based on the functional representation, may serve as research problems to guide the new empirical investigations.

This method is useful for the problem of localizing a failing theoretical component. However, little has been done on how to generate the new hypotheses at the site of failure. Historical cases, such as the development of the theory of the gene, may aid in uncovering strategies to use in generating new hypotheses to resolve an anomaly, such as proposing the opposite of an old assumption. Thus, a fruitful task for computational philosophy of science is to use and expand the AI work on diagnostic reasoning in order to make anomaly resolution tractable computationally.

Acknowledgements

My thanks to John Josephson, Dale Moberg, B. Chandrasekaran, and others at the Laboratory of Artificial Intelligence Research at Ohio State University for helpful comments on these ideas. My visit to their laboratory helped me with developing the analogy to diagnostic reasoning and schematic diagrams. Jeffrey Bub provided me with information about cognitive neuropsychology. I also thank Pamela Henson, Joel Hagen, Robert Witkin, Jim Platt, Pat Langley, and Jeff Shrager for their helpful editorial comments on an earlier draft. An extended version of this chapter will appear in a forthcoming book by L. Darden on strategies for theory change by Oxford University Press.

References

Bateson, W., Saunders, E. R., & Punnett, R. C. (1906). Experimental studies in the physiology of heredity (Reports to the Evolution Committee of the Royal Society, Report III). London: Harrison and Sons. Reprinted in R. C. Punnett (Ed.) (1928). *Scientific papers of William Bateson* (Vol. 2). Cambridge: Cambridge University Press.

Bateson, W., & Punnett, R. C. (1911). On the interrelations of genetic factors. *Proceedings of the Royal Society, B*, 84. Reprinted in R. C. Punnett (Ed.), *Scientific papers of William Bateson* (1928) (Vol. 2, pp. 216–220). Cambridge: Cambridge University Press.

Bub, J., & Bub, D. (1988). On the methodology of single-case studies in cognitive neuropsychology. *Cognitive Neuropsychology, 5*, 565–582.

Buchanan, B., & Shortliffe, E. (Eds.) (1984). *Rule-based expert systems: The MYCIN experiments of the Stanford heuristic programming project.* Reading, MA: Addison-Wesley.

Castle, W. E., & Little, C. C. (1910). On a modified Mendelian ratio among yellow mice. *Science, 32*, 868–870.

Chandrasekaran, B., & Mittal, S. (1983). Deep versus compiled knowledge approaches to diagnostic problem-solving. *International Journal of Man-Machine Studies, 19*, 425–436.

Chandrasekaran, B., Smith, J. W., & Sticklen, J. (1989). "Deep" models and their relation to diagnosis. *Artificial Intelligence in Medicine.* In K. S. Zadeh (Ed.), (Vol. 1), No. 1, (pp. 29–40).

Cuénot, L. (1905). Les races pures et leurs combinaisons chez les souris. *Archives de Zoologie Experimentale et Generale*, 4 Serie, *T.111*, 123–132.

Darden, L. (1980). Theory construction in genetics. In T. Nickles (Ed.) *Scientific discovery: Case studies.* Dordrecht: Reidel.

Darden, L. (1987). Viewing the history of science as compiled hindsight. *AI Magazine, 8* (2), 33–41.

Darden, L., & Cain, J. A. (1989). Selection type theories. *Philosophy of Science, 56*, 106–129.

Darden, L., & Maull, N. (1977). Interfield theories. *Philosophy of Science, 44*, 43–64.

Darden, L., & Rada, R. (1988). Hypothesis formation using part-whole interrelations. In D. Helman (Ed.), *Analogical reasoning*. Dordrecht: Reidel.

Davis, R., & Hamscher, W. C. (1988). Model-based reasoning: Troubleshooting. In H. E. Shrobe (Ed.), *Exploring artificial intelligence*. San Mateo: Morgan Kaufmann.

Dunn, L. C. (1965). *A short history of genetics*. New York: McGraw-Hill.

Eaton, G. J., & Green, M. M. (1962). Implantation and lethality of the yellow mouse. *Genetica, 33*, 106–112.

Kirkham, W. B. (1919). The fate of homozygous yellow mice. *Journal of Experimental Zoology, 28*, 125–135.

Kitcher, P. (1981). Explanatory unification. *Philosophy of Science, 48*, 507–531.

Kuhn, T. (1970). *The structure of scientific revolutions* (2nd Ed.). Chicago: University of Chicago Press.

Lakatos, I. (1976). *Proofs and refutations: The logic of mathematical discovery*. In J. Worrall & E. Zahar (Eds.), Cambridge: Cambridge University Press.

Langley, P., Simon, H. A., Bradshaw, G. L., & Zytkow, J. M. (1987). *Scientific discovery: Computational explorations of the creative processes*. Cambridge, MA: MIT Press.

Mendel, G. (1865/1966). Experiments on plant hybrids. In C. Stern & E. Sherwood (Eds.), *The origin of genetics: A Mendel source book*. San Francisco: W. H. Freeman.

Monaghan, F., & Corcos, A. (1984). On the origin of the Mendelian laws. *Journal of Heredity, 75*, 67–69.

Morgan, T. H. (1905). The assumed purity of the germ cells in Mendelian results. *Science, 22*, 877–879.

Morgan, T. H. (1909). Recent experiments on the inheritance of coat colors in mice. *American Naturalist, 43*, 494–510.

Morgan, T. H. (1926). *The theory of the gene*. New Haven, CT: Yale University Press.

Reggia, J., Nau, D., & Young, P. (1983). Diagnostic expert systems based on a set covering model. *International Journal of Man-Machine Studies, 19*, 437–460.

Schaffner, K. (Ed.) (1985). *Logic of discovery and diagnosis in medicine.* Berkeley: University of California Press.

Sembugamoorthy, V., & Chandrasekaran, B. (1986). Functional representation of devices and compilation of diagnostic problem-solving systems. In J. Kolodner & C. Reisbeck (Eds.), *Experience, memory, and reasoning.* Hillsdale, NJ: Lawrence Erlbaum.

Shapere, D. (1977). Scientific theories and their domains. In F. Suppe (Ed.), *The structure of scientific theories.* (2nd ed.). Urbana: University of Illinois Press.

Stern, C., & Sherwood, E. (Eds.) (1966). *The origin of genetics: A Mendel source book.* San Francisco: W.H. Freeman.

Thagard, P. (1988). *Computational philosophy of science.* Cambridge, MA: MIT Press.

APPENDIX A

An Implementation Note

Dale Moberg
John Josephson

1. The Normal Scientific Development of the Gene Theory

Kuhn's (1970) notion of "normal science" applies to phases of a research program during which the fit of theory with observation improves. Normal science can begin when there are exemplar problem-solving patterns that serve as the theoretical core of a research tradition. This chapter has presented a detailed account of the scientific growth of gene theory and has identified both exemplar problem-solving patterns and an increasingly better fit of theory with observation. Darden's analysis and description of this case provides us with the inspiration for a computational account of scientific discovery within normal science.

We see theories as constructs that help us in the performance of various cognitive tasks. Revising scientific knowledge is analogous to redesigning tools or other devices, in that scientists use anomalies in diagnosing faults in a theory and then propose fixes for those faults. In this appendix, we describe work that explores the perspective of "theory as device" and its applications to Darden's account of the development of the gene theory.

2. Anomaly-Driven Diagnosis and Redesign

The basic anomaly-driven redesign task satisfies the description:

> Given a theory in the form of a basic scheme for problem solutions and an observational anomaly for that theory, construct a modified theory that no longer has the given anomaly and that retains previous explanatory successes.

This task is not uniquely specified, in that many modified theories may be constructed that retain successes and eliminate the anomaly. Some of the different directions for theory modification are associated with different strategies. Ultimately we hope to investigate alternative programs for theory redesign with respect to a sequence of anomalies and selections of strategies. However, our present implementational efforts are directed toward finding a good representation for a theory's content and localizing potential redesign sites. If such localization is possible, the redesign task is greatly simplified.

3.   Functional Representation of a Theory's Causal Processes

Our approach to representation focuses on the characteristic causal processes studied in the theory of the gene. We have used our representational tools to automatically generate a diagnostic classification hierarchy from a representation of the gene theory in terms of the causal process of inheritance in sexually reproducing organisms.

The functional representation (FR) language was created to represent the functioning of devices. One goal of the language was to support problem-solving activities in troubleshooting and predicting changes in function when there is a change in the components, structure, or constituent behaviors of a device (Sembugamoorthy & Chandrasekaran, 1986). The language has been used to represent both concrete devices (such as bodily systems in medical applications and manufactured devices and manufacturing processes for engineering applications) and abstract "devices" (plans and programs) (Chandrasekaran, Josephson, & Keuneke, 1986; Chandrasekaran, Josephson, Keuneke, & Herman, 1989).

To some extent, the advantages of a particular form of representation are revealed in the support that is provided for characteristic problem-solving activities that use that representation. The functional representation facilitates aspects of diagnostic reasoning by providing automatic generation of diagnostic classification hierarchies and for qualitative simulation to test the success of redesign. Therefore, the intriguing analogy of a theory with a device suggests that FR will be a promising tool for expressing the content of a theory's underlying causal processes.

The case that we have used to investigate the advantages of the FR language is that of the anomaly of 2 to 1 ratios of inherited traits. The "flat" functional representation of the gene theory content (for a single gene locus) is shown in Figure 9.

The separate stages of the causal processes that lead from parents to a full representative generation of offspring are distinguished and la-

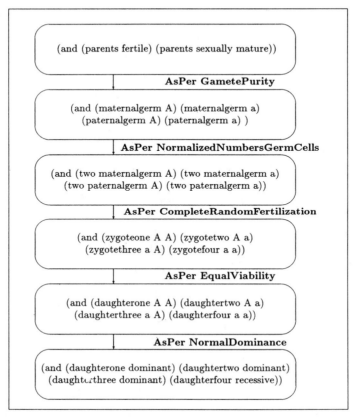

Figure 9. Functional representation of Mendelian inheritance in gene theory.

beled in the figure. The labels can be arbitrary symbolic expressions but are normally Boolean expressions. The state labels must support operations that enable the problem-solving system to assess whether the state has been entered. The transitions between states are normally causal links, and in functional representations are indicated using **By**, **Using function of**, and **AsPer** (Keuneke, 1989; Sembugamoorthy & Chandrasekaran, 1986). An **AsPer** link does not appeal to subfunctions or subbehaviors in providing a causal description of the transition between states but instead points to a chunk of "knowledge" that applies to the causal transition being represented. For example, the phrase **GametePurity** refers to one chunk of our conjectured knowledge of meiosis that can be summarized by the rule that genes separate into dominant or recessive alleles and not into intermediate blends with partial weighting of traits. Each other `AsPer` link label points to an additional chunk of knowledge according to which its state transition occurs. When there is an anomaly, one or more of these knowledge chunks requires modification.

This representation can automatically be converted into a diagnostic classification hierarchy (Bylander et al., 1987; Bylander & Mittal, 1986). This hierarchy is a directed acyclic graph produced by a standard recursive descent parsing of the functional representation. The state labels serve as the basis for constructing "knowledge groups" whose values, along with the control messages (normally, an "establish and refine" message), govern the fault localization process. According to the establish and refine control strategy, only those nodes whose knowledge group satisfies an "establish" threshold and that have received a "refine" message will be pursued; other nodes are rejected, and a saving in search effort is then obtained by not pursuing their subnodes.

In our present use of these tools, the knowledge group generates queries that are asked of the human user of the program.[9] The generation of useful and appropriate queries is then the most significant contribution of the diagnostic hierarchy to the discovery process. The user's answers permit the system to localize a potential site of fault in the theory and thus to provide focus for the computationally costly process of theory redesign (Goel & Chandrasekaran, 1989).

The operation of the diagnostic hierarchy can be grasped by tracing through the process of generating the "correct" answer for the anomaly of 2 to 1 ratios. First, the need to engage in diagnostic reasoning is established by asking whether fertile sexually mature organisms have mated and whether 3 to 1 ratios have been observed. Replying that the initial causal state (fertile and mature organisms mating) had been entered but that the final state of 3 to 1 ratios of a trait in the offspring population was not observed establishes the existance of an anomaly. This drives the problem solver to refine its diagnostic hypothesis about what parts of the theory might plausibly be mistaken. At present, we traverse the classification hierarchy from the final state toward the initial state, inquiring at each point whether the intermediate state had been entered. In this case, localization occurs when the system is in-

---

9. Automatic use of the diagnostic system would involve a database that can be suitably queried to arrive at a value allowing the knowledge group node to establish or reject.

formed that four zygotes were produced but that no homozygous (AA) offspring was produced. This information establishes that the causal behavior referring to "equal zygote viability" (`AsPer EqualViability`) is at fault.

4. Extensions to the Current Implementation

The above case shows how the causal process specified by the theory of the gene can be represented using the FR language and how diagnostic reasoning about theory faults is supported by the availability of a compiler for a diagnostic classification hierarchy. The diagnostic queries are generated through a control regimen of establish and refine, and thus carry out a search for knowledge needed to localize a fault. This search may reveal the need for additional experimental work. However, a full implementation of normal scientific discovery involving anomaly-directed redesign involves both proposing new hypotheses and testing to see that old success is retained while the anomalous result is removed. Such implementation will require considerable further effort and analysis.

Our representation of a theory's central causal processes does not yet take advantage of several features of the functional representation language that would provide considerable power. We have not yet used a full analysis of components, structure, and behavior or the provision for simulation. A functional representation normally involves a hierarchy whose vertical organization groups details of state sequence, subcomponents, and knowledge. These groups then organize the rich and complex knowledge of causal process sequences at different levels of detail. These in turn are the parts that can be selected as loci of failure and thus become targets of potential redesign.

What is the basis for such organization? One possible organizational principle applicable to gene theory is that stages of inheritance that involve theoretical and micro features or entities can be separated from stages that involve observable and macro features or entities. This modularity reflects the distinction of some philosophers of science between the theory's core assumptions and the auxiliary assumptions (those needed to relate theoretical states to observable ones). An alternative organization might follow more domain-specific distinctions within the process, such as meiosis, fertilization, and gestation. Decisions on organization and abstraction should be made to facilitate problem solving.

For example, if the strategies are formulated in Lakatos' (1978) vocabulary of the "hard core" and "protective belt" so as to implement the conservative strategy of always directing modifications away from the core, the representation can provide appropriate organization to implement this distinction. If the strategies are instead formulated in terms of parts of a causal story, then the organization can be changed to focus redesign efforts on those parts of the causal story thought most problematic. Consequently, philosophical analyses of alternative strategies will become increasingly important for fuller and deeper applications of functional representation to normal scientific discovery.

Acknowledgements

This work has been supported by Grant 87-0090 from the Air Force Office of Scientific Research, by Grant #1 R01 HL38776-01 from the National Heart, Lung, and Blood Institute, and by the Defense Advanced Research Projects Agency under RADC Contract F306034-85-C-0010. We wish to thank Lindley Darden, B. Chandrasekaran, Jeff Shrager, and Pat Langley for their discussions and comments; we are grateful to Dean Allemang for his help with the functional representation compiler.

References

Bylander, T., Chandrasekaran, B., & Josephson, J. (1987). The generic task toolset. *Proceedings of Second International Conference on Human-Computer Interaction*. Honolulu, Hawaii.

Bylander, T., & Mittal, S. (1986). CSRL: A language for classificatory problem solving and uncertainty handling. *AI Magazine, 7*, 66–77.

Chandrasekaran, B., Jospehson, J., & Keuneke, A. (1986). Functional representations as a basis for generating explanations. *International Conference on Systems, Man, and Cybernetics*.

Chandrasekaran, B., Josephson, J., Keuneke, A., & Herman, D. (1989). Building routine planning systems and explaining their behavior. *International Journal of Man-Machine Studies, 30*, 377–398.

Goel, A., & Chandrasekaran, B. (1989). Functional representation of designs and redesign problem solving. *Proceedings of the Eleventh International Joint Conference on Artificial Intelligence*. Detroit, MI: Morgan Kaufmann.

Keuneke, A. (1989). *Machine understanding of devices: explaining diagnostic conclusions.* Doctoral dissertation, Department of Computer Science, Ohio State University, Columbus.

Kuhn, T. (1970). *The structure of scientific revolutions,* (2nd ed.). Chicago: University of Chicago Press.

Lakatos, I. (1978). *Methodology of scientific research programmes.* Cambridge: Cambridge University Press.

Sembugamoorthy, V., & Chandrasekaran, B. (1986). Functional representation of devices and compilation of diagnostic problem solving systems. In J. Kolodner & C. Riesbeck (Eds.), *Experience, memory, and reasoning.* Hillsdale, NJ: Lawrence Erlbaum.

CHAPTER 12

Designing Good Experiments
To Test Bad Hypotheses

DAVID KLAHR

KEVIN DUNBAR

ANNE L. FAY

1. Introduction

All the contributors to this volume share the ultimate goal of producing a computational model of the scientific discovery process. Thagard and Nowak (this volume) distinguish between nonpsychological and psychological approaches to this goal. The former approaches rely heavily on AI techniques, including "computational techniques different from those available to humans," whereas the latter approaches involve detailed analysis of the behavior of humans actually engaged in different aspects of scientific discovery. The psychological approaches can be further divided into two paths. One involves analyses of the scientific record of real scientists making real scientific discoveries (Darden, 1987; Kulkarni & Simon, 1988; Langley, Simon, Bradshaw, & Zytkow, 1987; Thagard & Nowak, this volume). This path is necessarily coarse-grained because the mental processes of the scientists must be inferred from historical analyses, retrospective reports, or laboratory notebooks. However, the face validity of such a data base is extremely high because it has been deliberately selected as a consequence of having produced important scientific discoveries.

The other path to a psychological understanding of the scientific discovery process—and the one we follow in this chapter—involves the cre-

ation by the analyst of simulated contexts for scientific discovery. These
contexts can vary from very sparse (such as the well-known "2-4-6" task
invented by Wason, 1960)[1] to highly complex (see Mynatt, Doherty, &
Tweney, 1977). These simulated contexts enable researchers to perform
detailed analysis of the moment-to-moment behavior of subjects as they
work on a discovery task—typically ordinary college students but also
scientists (Tweney & Yachanin, 1985) and young children (Dunbar &
Klahr, 1989; Kuhn, Amsel, & O'Loughlin, 1988; Schauble, 1990). Although this approach has the shortcoming that the discovery task itself
is only analogous to science rather than being real science, it has two
important advantages. It enables us to precisely control the context of
the discovery process and to obtain fine-grained, on-line observations of
the thinking processes surrounding that discovery.[2]

Scientific discovery can be characterized as a process involving search
in two primary problem spaces—a space of hypotheses and a space of
experiments—with additional searches of subsidiary problem spaces, including an observation space, an instrumentation space, a data analysis
space, and a prior literature space (Newell, 1989). In our previous work
(Klahr & Dunbar, 1988), we extended Simon and Lea's (1974) dual
search notion to a general framework (called SDDS for Scientific Discovery as Dual Search) for the processes that coordinate and implement
dual search in the experiment and hypothesis spaces. The framework
was based on our empirical observations of subjects' behavior as they
formulated hypotheses and designed experiments to evaluate them. In
those studies, subjects were unconstrained with respect to both hypotheses and experiments. In this chapter, we narrow our focus to ask
how people search the experiment space when provided with a particular hypothesis to evaluate. By controlling both the hypothesis being

---

1. For recent discussions of much of the research spawned by Wason's original study,
   see Gorman and Carlson (1989), Klahr and Dunbar (1988), and Klayman and Ha
   (1987). Wason's ingenious task continues to intrigue researchers three decades
   after its invention (see Farris & Revlin, 1989; Gorman, 1989).
2. There is an interesting third alternative that combines the features of each of the
   psychological approaches. It involves the presentation of the essential knowledge
   context faced by major scientists at the time of their discoveries to "ordinary" but
   technically trained subjects to see if they can make the same discovery, given the
   same information and the same goals as the original discoverer (Dunbar, 1989;
   Qin & Simon, 1989). In the few instances in which this has been done, some
   subjects were able to rediscover important scientific laws.

evaluated and the extent to which it is correct, we are able to examine this search process in detail.

2. Designing Experiments

What does it take to design a good experiment? Given an hypothesis to be evaluated—either on its own merits, or in competition with alternative hypotheses—what formal rules, heuristics, and pragmatic constraints combine to yield a potentially informative experiment? How do subjects' expectations about the likelihood of an hypothesis being true or false affect the kinds of experiments that they design and their responses to information that is consistent or inconsistent with the hypothesis? In this chapter we describe a study in which we addressed these questions by presenting subjects with "bad" (incorrect) hypotheses and asking them to design a series of "good" experiments to test those hypotheses.

2.1 Previous Studies

In our earlier studies (Dunbar & Klahr, 1989; Klahr & Dunbar, 1988), we instructed subjects about all the basic features of a programmable robot and then asked them to extend that knowledge by experimentation. This training was intended to provide a rich context analogous to a scientist's partial knowledge about a domain in which further information can be obtained by experimentation. Our analyses focused on subjects' attempts to discover how a new function operates—that is, to extend their understanding about how the device works by formulating hypotheses and then designing experiments to evaluate those hypotheses; the cycle terminated when they believed that they had discovered how to predict and control the behavior of the device. To provide substantive background for the studies reported in this chapter, we start by summarizing one of our earlier studies.

2.1.1 The BigTrak Device

We used a computer-controlled robot tank (called *BigTrak*) that is programmed using a Logo-like language.[3] The device is operated by press-

---

3. This device was first used by Shrager (1985) in his investigation of "instructionless learning" (Shrager & Klahr, 1986).

ing various command keys on a keypad. BigTrak is programmed by first clearing the memory with the CLR key and then entering a series of up to sixteen instructions, each consisting of one of its six function keys (the command) and a one- or two-digit number (the argument). When the GO key is pressed BigTrak then executes the program. To illustrate, one might press the following series of keys: CLR ↑ 5 ← 7 ↑ 3 → 15 HOLD 50 FIRE 2 ↓ 8 GO. BigTrak would then do the following: move forward five feet, rotate counterclockwise 42 degrees (corresponding to 7 minutes on an ordinary clock face), move forward 3 feet, rotate clockwise 90 degrees, pause for 5 seconds, fire twice, and backup eight feet.

2.1.2 GENERAL METHOD OF PREVIOUS WORK

First, we established a common knowledge base about the device for all subjects prior to the discovery phase. We instructed subjects about how to use each of the basic function keys. Then the discovery phase started. Subjects were told that there is a Repeat key (RPT), that it takes a numerical parameter, and that there can be only one RPT in a program. They were asked to discover how RPT works by proposing hypotheses and running programs with RPT in them to test their hypotheses.[4] Subjects generated a concurrent verbal protocol that included hypotheses, experiments (programs), observations, evaluations, and revised hypotheses.

2.1.3 RESULTS

Nineteen of the 20 adult subjects in our first study (Klahr & Dunbar, 1988) discovered how the RPT key works within the allotted 45 minutes. The mean time to solution was 19.8 minutes. Subjects generated, on average, 18.2 programs.

Protocols were encoded in terms of the hypotheses listed in Table 1. We defined a "common hypothesis" as a fully specified hypothesis that was proposed by at least two different subjects. Across all subjects, there were eight distinct common hypotheses. Subjects did not always

---

4. On the original BigTrak, there was only one way that RPT worked: It repeated the previous N instructions once. In the studies reported here, we used several different rules for how RPT works.

Table 1. Common hypotheses and percentage of experiments conducted under each. Hypotheses are labeled according to the role of N: HS – Selector; HN – nil; HC – Counter.

Hypothesis	Current Designation*	Percent of Experiments Under Each Hypothesis
HS1: One repeat of last N instructions.	D	02
HS2: One repeat of first N instructions.		04
HS3: One repeat of the Nth instruction.	C	03
HN1: One repeat of entire program.		06
HN2: One repeat of the last instruction.		04
HC1: N repeats of entire program.	A	14
HC2: N repeats of the last instruction.	B	20
HC3: N repeats of subsequent steps.		02
Partially specified		03
Idiosyncratic		14
No Hypothesis		28

*Entries in this column show the labels for the four key hypotheses to be used in the present study. Note that they include the two most "popular" (A and B) and the two least popular (D and C).

express their hypotheses in exactly this form, but there was usually little ambiguity about what the current hypothesis was. We coded each experiment in terms of the hypothesis held by the subject at the time of the experiment. Table 1 shows the proportion of all experiments that were run while an hypothesis was held. (As noted earlier, HS1 in Table 1 is the way that BigTrak actually operated.)

Subjects proposed, on average, 4.6 different hypotheses (including the correct one). Fifty-five percent of the experiments were conducted under one of the eight common hypotheses listed in Table 1. Partially specified hypotheses, which accounted for 3% of the experiments, were defined as those in which only some attributes of the common hypotheses were stated by the subject (for example, "It will repeat it N times"). An idiosyncratic hypothesis was defined as one that was generated by only one subject. Such hypotheses are not listed separately in Table 1. For 28% of the experiments, there were no stated hypotheses.

2.2 The Hypothesis Space

The eight common hypotheses—which account for over half of the experiments—can be represented in a space of "frames" (Minsky, 1975). The basic frame for discovering how RPT works is depicted at the top of Figure 1. It consists of four slots, corresponding to four key attributes:

(1) The role of *N*: does it *count* a number of repetitions, or does it *select* some segment of the program to be repeated? (2) The unit of repetition: step, program, or group of steps? (3) Number of repetitions: 1, *N*, some other function of *N*, or no role at all? (4) Boundaries of repeated segment: beginning of program, end of program, *N*th step from beginning or end? A fully instantiated frame corresponds to a fully specified hypothesis, several of which are shown in Figure 1. There are two principal subsidiary frames for RPT—*N-role:Counter* and *N-role:Selector*. Within each of these frames, hypotheses differing along only a single attribute are shown with arrows between them. All other pairs of hypotheses differ by more than one attribute. Note that the hypotheses are clustered according to the *N*-role frame in which they fall. No arrows appear between hypotheses in one group and the other because a change in *N*-role requires a simultaneous change in more than one attribute. This is because the values of some attributes are linked to the values of others. For example, if *N*-role is Counter, the number of repetitions is *N*, whereas if *N*-role is Selector, then the number of repetitions is *1*.

This frame representation is a convenient way of capturing a number of aspects of the scientific reasoning process. First, it characterizes the relative importance that subjects give to different aspects of an hypothesis. Once a particular frame is constructed, the task becomes one of filling in or verifying "slots" in that frame. The current frame will determine the relevant attributes. That is, the choice of a particular role for *N* (such as *N-role:Counter*) also determines what slots remain to be filled (such as number of repetitions:*N*), and it constrains the focus of experimentation. Furthermore, frames enable us to represent the differential importance of attributes as the "frame type" becomes the most important attribute and its "slots" become subordinate attributes. This is consistent with Klayman and Ha's (1989, p. 11) suggestion that "some features of a rule are naturally more 'salient,' that is, more prone to occur to a hypothesis tester as something to be considered." In our context, a frame is constructed according to those features of prior knowledge that are most strongly activated, such as knowledge about the device or linguistic knowledge about "repeat." When a frame is constructed, slot values are set to their default values. For example, having selected the *N*-role:*Counter* frame, values for number of repetitions, units, and boundary might be chosen so as to produce HC1 (see Figure 1).

HYPOTHESIS
SPACE

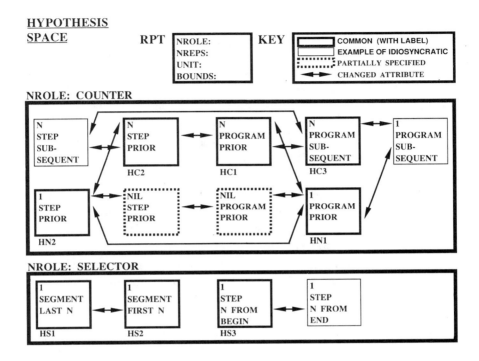

Figure 1. Frames for hypotheses about how RPT N works. Heavy borders correspond to common hypotheses from Table 1; dashed borders correspond to partially specified hypotheses; arrows indicate that adjacent hypotheses differ along a single attribute shown on the arrow; all possible hypotheses are not shown.

2.3   The Experiment Space

Subjects tested their hypotheses by conducting experiments, that is, by writing programs that included RPT and observing BigTrak's behavior. But it is not immediately obvious what constitutes a "good" or "informative" experiment. In constructing experiments, subjects are faced with a problem-solving task that parallels their effort to discover the correct hypothesis, except that in this case their search is not in a space of hypotheses but in a space of experiments.

This space can be characterized in many ways—the total number of commands in a program, the location of RPT in a program, the value of N, the specific commands in a program, the numerical arguments of specific commands, and so on. However, in this study, as in all of our previous investigations, subjects' verbal protocols suggest that they quickly realized that there are only two key features to their experiments. The first is λ, the length of the program preceding the RPT. The second is the value of N, the argument that RPT takes. Within the λ—N space, we identify three distinct regions according to the relative values of λ and N and their limiting values.[5] Region 1 includes all programs with RPT 1. Region 2 includes all programs in which the value of N is greater than 1 but less than λ. Region 3 includes all programs in which N is equal to or greater than λ. The regions are depicted in Figure 2, together with illustrative programs from the (4,1) cell in region 1, the (3,2) cell in region 2, and the (1,4) cell in region 3.

Programs from different regions of the experiment space vary widely in how effective they are in supporting or refuting different hypotheses. Table 2 shows how BigTrak would behave under different rules when executing programs from different regions of the experiment space. The programs are depicted on the left by generic commands (e.g., X, Y, Z) and the device behavior is shown under the column corresponding to each of the four rules used in this study. To illustrate, the second example shows a two-step program with $N = 1$. This is a region 1 experiment. Under rule A, the two-step program would be executed once and then repeated one more time. Under rule B, only the last step (Y) would be repeated one additional time. Under rule C, the first step would be repeated once, and under rule D, the last N steps (in this case, the last step, since $N = 1$) would be repeated once. This program cannot discriminate between rules B and D.

The generic examples in Table 2 can represent a continuum of discriminating power. The most highly differentiated programs are obtained if we substitute distinct commands for X, Y, and Z (for example, X = ↑ 2, Y = FIRE 1). The least informative programs are those in which both the command and the parameter are the same (e.g, X = Y = Z = → 15.) Intermediate between these two extremes are programs in which the commands are the same but the parameters are different, such as FIRE 1 FIRE 2 FIRE 3. For many such programs, behavior under the different rules is in fact distinct, but it is extremely difficult to keep

---

5. In previous analyses (Dunbar & Klahr, 1989; Klahr & Dunbar, 1988), we used a finer-grained categorization of the Experiment Space into six regions. The mapping from the earlier to the current regions is as follows: I & II → 1, III → 2, IV & V & VI → 3. The $\lambda = 1$, $N = 2$ cell from the old region 1 goes into the new region 3.

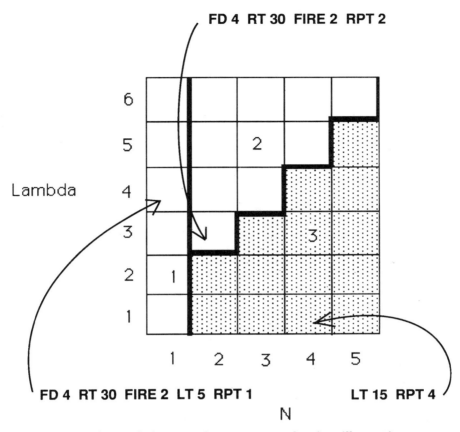

FD 4 RT 30 FIRE 2 RPT 2

Lambda

FD 4 RT 30 FIRE 2 LT 5 RPT 1   LT 15 RPT 4

N

Figure 2. Regions of the experiment space, showing illustrative programs
(Shown here is only the 6×5 subspace of the full 15×15 space.)

track of the BigTrak's behavior. We will present one such example in
Section 4.1.

Two important and subtle features of the rules are included in the
notation in Table 2. The first potentially confusing feature has to do
with the ambiguity inherent in the phrase "repeat it N times." Does it
mean N or $N + 1$ total executions of the repeated entity? That is, if a
program is supposed to repeat something twice, a subject might expect
to observe either two or three occurrences of that item or segment. The
underlined segments in Table 2 show the behavior generated by the
$N + 1$ interpretation (which is the one we use in our simulations). If
subjects use the N interpretation, then they would not expect to see
these extra segments. The second feature involves rules C and D when

Table 2. Behavior of BigTrak under four different rules and programs from each of the experiment space regions. Each row shows a generic program and how BigTrak would behave under each of the four rules used in this study. For each entry, executions under control of RPT are shown in boldface. (See text for further explanation.)

| | | | COUNTERS | | SELECTORS | |
No.	Region	Program	A: program N times	B: last step N times	C: Nth step once	D: last N steps once
1	1	X R1	XX	XX	XX	XX
2	1	X Y R1	XY**XY**	XY**Y**	XYX	XYY
3	1	X Y Z R1	XYZ**XYZ**	XYZ**Z**	XYZX	XYZZ
4	2	X Y Z R2	XYZXYZ**XYZ**	XYZZ**Z**	XYZY	XYZYZ
5	3	X Y Z R3	XYZXYZXYZ**XYZ**	XYZZZZ**Z**	XYZZ	XYZXYZ
6	3	X Y R3	XYXYXY**XY**	XYYY**Y**	XYY*	XYXY*
7	3	X R3	XX**X**	XX**X**	XX*	XX*
8	3	X Y Z R4	XYZXYZXYZXYZXYZ**XYZ**	XYZZZZZ**Z**	XYZZ*	XYZXYZ*

$N > \lambda$. For these programs (indicated by an asterisk), N is set equal to λ. Experiments in the three regions interact with hypotheses as follows:

1. Programs in region 1 have poor discriminating power. We have already described example 2, and example 3 similarly fails to discriminate B from D. Example 1—a minimalist program with $\lambda = N = 1$—has no discriminating power whatsoever.

2. Region 2 provides maximal information about all the most common hypotheses because it can distinguish between Counters and Selectors, and it can distinguish *which* Selector or Counter is operative. It produces different behavior under all four rules for any program in the region, and varying N in a series of experiments in this region always produces different outcomes.

3. Under rules C and D, results of experiments in region 3 may be confusing because they are executed under the subtle additional rule that values of N greater than λ are truncated to $N = \lambda$. Therefore, varying N in this region will give the impression that N has no effect on the behavior of the device (compare examples 1 and 7 or 5 and 8). Although some of the programs in this region are discriminating (such as example 5, with $\lambda = N = 3$), others either do not discriminate at all (C versus D in example 7) or depend on the truncation assumption to be fully understood (such as examples 6 to 8).

3. An Empirical Study of Testing an Incorrect Hypothesis

The brief summary of our earlier studies provides a context for the following description of a new study designed to explore subjects' responses to negative feedback. In these new studies, we always provided subjects with an initial hypothesis about how RPT might work. It was always wrong. In some conditions, it was only "somewhat" wrong, in that it was from the same frame as the way that RPT actually works. In others, it was "very" wrong, in that the suggested hypothesis came from a different frame than the actual rule.

3.1 Subjects

Thirty-six Carnegie Mellon undergraduates (27 males and 9 females) participated in the experiment for course credit. Most were science or engineering majors. All the subjects completed a questionnaire about their programming experience and skill and a self-rating of their skill in math, science, and mechanical reasoning. Subjects reported having taken between zero and five programming courses (mean 2.1, sd 1.4), and they tended to rate themselves between average and above average on all the technical and scientific scales.

3.2   Procedure

All subjects worked with a simulated version of the BigTrak on a Xerox Dandelion workstation. The simulator included five command keys ($\uparrow$, $\downarrow$, $\leftarrow$, $\rightarrow$, and FIRE). The workstation enabled us to control the way that RPT actually functioned and facilitated the recording of the programs that subjects wrote to evaluate their hypotheses.

There were three phases to the study. In the first, subjects were introduced to the (simulated) BigTrak and were trained to criterion on all its basic commands. In the second phase, subjects were told that there was a RPT key, that it required a numeric parameter, and that there could be only one RPT in a program. They were told that their task was to find out how RPT worked by writing programs to test a particular hypothesis. At this point, the experimenter suggested one possible way that RPT might work and instructed the subject as follows:

> Write down three good programs that will allow you to see if the Repeat key really does work this way. Think carefully about your program, and then write the program down on the sheet of paper... Once you have written your program down, I will type it in for you, and then I will run it. You can observe what happens, and then you can write down your next program. So you write down a program, then I will type it in, and then you will watch what the program does. I want you to write three programs in this way.

Next, the third phase began. Subjects wrote programs (experiments) to evaluate the given hypothesis. Although subjects did not have access to a record of the behavior of the device in earlier experiments, they did have access to the list of programs that they had written, and they often referred to them in commenting on differences among the most recent outcome and previous ones. Subjects were instructed to give verbal protocols. This gave us a record of (1) what they thought about

Table 3. Percentage of experiments run in previous studies for each Repeat rule used in the present study.

	STUDY 1	STUDY 2
COUNTERS		
A: REPEAT THE ENTIRE PROGRAM N TIMES	14	13
B: REPEAT THE LAST STEP N TIMES	20	26
SELECTORS		
C: REPEAT THE NTH STEP ONCE	3	5
D: REPEAT THE LAST N STEPS ONCE	2	5

the kinds of programs they were writing while testing their hypotheses, (2) what they observed and inferred from the device's behavior, and (3) what their hypothesis was about how RPT actually worked. When subjects had written, run, and evaluated three experiments, they were given the option of writing additional experiments if they were still uncertain about how RPT worked.

3.3 Design

The BigTrak simulator was programmed so that each subject worked with a RPT command obeying one of the four rules listed in Table 3. Note that there are two Counter rules and two Selector rules. Table 3 also summarizes the results from two earlier studies showing the relative frequency with which subjects ran experiments to test four specific hypotheses. Two of them (A and B) accounted for approximately one-third of all experiments, while the other two (C and D) were tested on only 10% of all experiments. More generally, all of our earlier studies showed that Counter hypotheses were regarded as highly probable and Selector hypotheses were regarded as improbable. This consistent preference made it possible for us to investigate the extent to which the a priori belief in particular hypotheses affected the types of experiments designed and the interpretation of results.

The key feature of this study is that *RPT never worked in the way that was suggested.* The design is shown in Table 4. The Given hypothesis

Table 4. Design of given-actual conditions.

GIVEN	ACTUAL	
	COUNTER	SELECTOR
COUNTER	A → B B → A	A → C A → D B → C B → D
SELECTOR	C → A C → B D → A D → B	C → D D → C

is the one that was suggested by the experimenter, and the Actual hypothesis is the way that BigTrak was programmed to work.[6] We used a between-subjects design, with three subjects in each of the given-actual conditions ($N = 36$). This yielded 12 subjects in within-frame conditions; for example, Repeat the program N times (A) → Repeat the last step N times (B); Repeat last N steps once (D) → Repeat Nth step once (C). There were 24 subjects in the between-frame conditions; for example, Repeat the program N times (A) → Repeat last N steps once (D); Repeat Nth step once (C) → Repeat the program N times (A).

3.4   Questions About Searching the Experiment Space

Now that we have described the details of our design, we can pose some specific questions. First, with respect to overall effort and success rates:

1. Will subjects have more difficulty when they have to change frames in order to discover the Actual rule than when they can remain within the same frame as the Given hypothesis?

2. Will subjects find it easier to discover rules from the preferred frame (Counters) than from the nonpreferred frame (Selectors)?

3. Will the difficulty of crossing frame boundaries interact with preferences for hypotheses? That is, will it be easier to discover a Counter

---

6. In our discussion, we will distinguish among three categories of hypotheses and rules. *Given* hypotheses are the ones initially suggested by the experimenter, *Current* hypotheses are the ones currently being evaluated by the subject, and *Actual* rules are the ways that RPT actually works in a particular condition. Ideally, a subject would start with Current = Given and end with Current = Actual.

when given a Selector hypothesis than to discover a Selector when given a Counter hypothesis?

4. Will the extent of "rational" search of the hypothesis space depend on experimental conditions? To what extent will subjects propose hypotheses that are consistent with the evidence available to them?

5. Will subjects find it easier to reject hypotheses that have been given to them rather than hypotheses that they have generated themselves? In our prior research, most subjects began with hypotheses A and B and needed a considerable amount of disconfirming evidence before abandoning their hypotheses. When subjects are actually given hypotheses to test, they may more readily abandon their hypotheses.

Second, with respect to search in the experiment space:

1. Will subjects' interpretation of what a "good experiment" is vary according to the Given-Actual condition in which they find themselves? That is, will experiments for favored hypotheses tend to demonstrate the presumed effect of RPT and experiments for unfavored hypothesis tend to have the power to discriminate between alternative hypotheses?

2. To what extent will subjects adopt the same experiment space that we have presented here? Will their choice of experiments reflect an implicit understanding of the interactions shown in Table 2?

3. What kinds of pragmatic rules will subjects apply to their search of the experiment space? Will they design programs that are easily observable, discriminatory, and memorable?

Finally, with respect to the observation and encoding of experimental outcomes:

1. Given that the BigTrak never works the same way as the Given hypothesis, how will subjects interpret the disconfirming evidence?

2. Will disconfirmation result in subjects searching a new region of the experiment space?

3. Will hypothesis preference also influence subjects' encoding and evaluation of experimental outcomes as well as overall success rates? That is, will subjects tend to distort their encoding of evidence in the direction of confirming favored hypotheses?

4. Experimental Results

The raw data are comprised of subjects' written programs as well as transcriptions of subjects' protocols (verbalizations) during the experimental phase. The protocols provided the basis for all of our measures of hypotheses changes and search in the experiment space. In Section 4.1, we informally describe two characteristic protocols, and then in subsequent sections we provide a quantitative analysis based on the full set of protocols.

4.1 Complete Protocols

The subject protocols are extremely rich, and in this section our aim is only to convey a general sense of the kind of encodings and inferences that we make from them. In the following two summaries, we focus on the ease with which subjects coordinate their search in the hypothesis and experiment spaces. The complete protocols are listed in Appendix A and Appendix B. Line numbers correspond roughly to major clauses. For each experiment, the commands used in the program are on the left side of the listing, and the actual behavior of BigTrak is shown in boldface type on the right side. Experimenter comments are shown in uppercase.

4.1.1 SUBJECT DP

Subject DP had experience with several programming languages (LOGO, LISP, PASCAL) and reported between 100 and 500 hours of programming experience. He rated himself as "above average" in mathematics and science, and average in "handling new gadgets." DP was in the Counter → Selector condition; he was given rule A: *Repeat entire program N times.* The actual rule was rule C: *Repeat Nth step once.* DP discovered the correct rule after five experiments.

Several general characteristics of DP's protocol make it interesting (but not unusual). First, even before the first experiment, DP rejected the given hypothesis and proposed an alternative (003: "I want to test to see if Repeat repeats the statement before it"; for example, this is rule B, not rule A.) Second, throughout the experimental phase, DP made many explicit comments about the attributes of the experiment space. He clearly attended to the properties of a "good" experiment.

Third, DP operated in an experiment space that included a feature that we have ignored so far—whether the range of influence of RPT extends to commands that precede it, follow it, or both. (We have included only the first of these in our analysis so far.) Several of our subjects explored this possibility, but it was not a dominant focus for most experiments.

DP first focused on the question of the before/after range of RPT, and he wrote a minimal program with one step on each side of RPT. Note that he used easily discriminated commands (left and right turns) so that, if RPT was having an effect on either side of its location in the program, it would be unambiguously evident. (This ability to write programs that contain useful "markers" is an important feature of our subjects' behavior, and we will return to it later.) DP was very clear about his intentions in his first experiment (003–010): to determine whether RPT acts on instructions before or after the RPT command. To resolve this question, DP conducted an experiment with commands both before and after the RPT key. This experiment was appropriate as it allowed DP to discriminate between these two rival hypotheses. However, with respect to being able to discriminate between the Given hypothesis (A), the Current hypothesis (B), and the Actual hypothesis (C), the program yields ambiguous results. DP extracted from the first experiment the information he sought (017): "It appears that the Repeat doesn't have any effect on any statements that come after it."

For the second experiment DP returned to the question of whether the Given hypothesis (A) or the Current hypothesis (B) was correct, and he decided to increase λ from 1 to 2. He also decided to include one step following the RPT "just to check" that RPT had no effect on instructions that followed it (022-023). Thus, DP was in fact testing three hypotheses; A, B, and "after." Once again, he used commands that could be easily discriminated. He continued to write a program from region 3 of the experiment space ($\lambda = 2$, $N = 2$). DP observed that there were two executions of the ↑ 2 instruction, and he concluded (028) that "it only repeats the statement immediately in front of it." This conclusion is consistent with the data that DP had collected so far, but the hypothesis (B) was not in fact how the RPT key worked.

For the third experiment, DP continued to put commands after RPT just to be sure they were not affected. However, given that his Current hypothesis was confirmed in the previous experiment, he next decided to write a program that further increases the length of the program. This was his first experiment in region 2. The goal of this experiment was to "see what statements are repeated" (032). He realized that the outcome of this experiment was inconsistent with his Current hypothesis (B), whereas the outcome of the previous experiment was consistent with B (050): "It seemed to act differently in number 2 and number 3." The unexpected result led DP to abandon hypothesis B, and he decided to continue beyond the mandatory three experiments.

For the fourth experiment, DP used a different value of N (055): "just to see if that (a value of 3 instead of 2) has anything to do with it." Here, too, DP demonstrated another important characteristic of many of our subjects' general approach to experimentation. He used a very conservative incremental strategy, similar to the VOTAT (vary one thing at a time) experimental strategies described by Tschirgi (1980) and the Conservative Focusing strategy described by Bruner, Goodnow, and Austin (1956). This approach still led him to put commands after the RPT, even though he seemed confident that RPT had no effect on them and even though they placed greater demands on his observational and recall processes. At the $\lambda - N$ level, DP executed VOTAT consistently throughout his series of five experiments. The $\lambda - N$ pairs are: 1–2, 2–2, 3–2, 3–3, 3–1. For the last three experiments, even the specific commands and their parameters remained the same, and only N varied. This moved him from region 2 into region 3. While analyzing the results of this experiment (061–071) in conjunction with earlier results, DP changed from the Counter frame to the Selector frame. First he noticed that "the number three" statement (the $\downarrow$ 1) was repeated twice in this case but that "the turning statement" was repeated (executed) only once (061–063). The implied comparison was with the previous experiment, in which the turning statement ("the right 15 command" [064]) was the command that got repeated. The next sentence is of particular interest:

> ...because when I change the number not only did it change ...it didn't change the uh ...the number that it repeated but it changed the uh...the actual instruction (066–069).

We believe that DP was attempting to articulate a change from the Counter frame to the Selector frame, as the following paraphrase of his comments indicates:

> When I changed the value of N, it didn't change the *number* of repetitions, but it did change *which* commands got repeated.

DP went on to clearly state two instantiated versions of the correct rule by referring to previous results with $N = 2$ and $N = 3$, and he designed his fifth experiment to test his prediction with $N = 1$. The outcome of this final experiment, from region 1, in conjunction with earlier results, was sufficient to convince him that he knew how RPT worked.

4.1.2 SUBJECT JS

JS rated himself as above average in mathematics and science as well as in "handling new gadgets." He reported having between 50 and 100 hours of programming experience. This subject was also in the $A \rightarrow C$ condition. JS's protocol had two interesting features. First, he never fully accepted the Given hypotheses (A: Repeat entire program N times), and at the very outset he proposed a few alternatives. Second, he was very articulate about several aspects of his experimental strategy, not only with respect to both the $\lambda - N$ space but also in terms of the logic of a disconfirming strategy and pragmatic constraints, such as designing programs that are easy to observe and encode.

JS started by expressing doubt about the Given hypothesis and setting out to disconfirm it (002–005) using a "simple program" (006) with "distinct steps" (009) that could be "distinguished" (012). As he developed the program, he proposed two alternative hypotheses and reasoned about them on the basis of plausibility and functionality (013–017). As he developed his first program, JS described its predicted behavior as if his Current hypothesis were Repeat Nth step once, which was the Actual rule. That is, he expected the RPT 1 to execute the ↑ 1 after the ↓ 1, which would "bring it back to its original position" (022). JS also added a command following the RPT just to see if RPT had any effect on subsequent commands, although he did not seem to expect it.

The experimenter now asked JS to make a prediction before running the program (032), and JS gave two possible outcomes. He predicted that, if the Given hypothesis was correct, then, after the program was executed the first time, it would be executed again in its entirety: "It will continue with the rest of the program" (037). However, if his alternative hypothesis (C) was correct, then "the only thing I'm thinking it might do is I think it might just move forward 1 (repeat the first step only), and then it'll end up turning to the left 30" (038–040).

The program ran, and JS correctly observed and interpreted its behavior as disconfirming the Given but confirming his Current (and the
Actual) hypothesis (042–049). However, JS then realized that a region
1 program did not rule out another plausible hypothesis: Repeat first N
steps once. He deliberated for a bit on what kind of experiment would
best discriminate between the two possibilities, and for his second experiment he constructed a region 2 program with $\lambda = 4$ and $N = 3$
and four highly discriminable commands. He also articulated a VOTAT
strategy (061–063): "I want to run the same program because I know
what it does. I just want to change the condition of the repeat."

At this point, JS stated the correct rule as well as his now-disconfirmed
hypothesis (007–082):

> So it's just repeating the step number of the...the number you
> put after the Repeat it repeats that sequence,...it doesn't repeat
> first, second, and third like I thought it might, it just repeats the
> third step.

Having discovered the correct rule, JS went on to explore the effect of
having $N > \lambda$ and wrote one more experiment to attempt to resolve that
question. He appeared to end somewhat unsure of this subtle feature.

4.1.3   GENERAL FEATURES OF THE SUBJECTS' BEHAVIOR

We have presented only two of our 36 protocols, but they suffice to illustrate several general features of the subjects' approach to this task.
First, subjects did not always accept the Given hypotheses, even before
running their first experiment. (Recall that both JS and DP expressed
doubt about the Given hypothesis *prior* to running their first experiment and proposed an alternative.) This initial skepticism varied in its
degree and in the conditions under which it occurred. We can define
"mild skepticism" as the consideration of an alternative hypothesis from
the same frame as the Given hypothesis and "extreme skepticism" as the
consideration of an alternative from a different frame. For both Counter
and Selector subjects, nearly two-thirds (17/26) of the hypotheses considered in experiment 1 (for all 18 subjects in each condition) were,
in fact, the Given hypotheses. However, which *additional* hypotheses
were generated by subjects depended on the Given condition. Whereas
78% of all non-Given hypotheses suggested by Selector subjects were
Counters, only 22% of all non-Given hypotheses suggested by Counter

subjects were Selectors. Extreme skepticism occurred mainly among subjects in the Given = Selector group.

Most subjects showed a clear understanding of the two principal dimensions of the $\lambda - N$ space. Their protocols are filled with comments about "using longer programs," "using a different value of N," and so forth. At a finer grain of analysis, subjects were also aware of the importance of what might be called "good instrumentation"—designing programs that have identifiable markers in them. We already saw one such example in subject JS (Appendix B, lines 007–009). The following statements by other subjects are typical (emphasis added):

I don't want to have two of the same move in there yet, *I might not be able to tell if it was repeating the first one or if it was doing the next part of my sequence* (AD03).

I'm just going to make up some random but different directions *so that I'll know which ones get executed* (RS22).

I'm going to use a series of commands that will ... *that are easily distinguished from one another* and won't run it off the screen (GM27).

... so I'm going to pick two (commands) that are the direct opposite of each other, to see if ... they don't really have to be direct opposites but ... anyhow, I'm just going to write a program that consists of two steps, *that I could see easily* (BB04).

In addition to working in both the $\lambda - N$ space and the instrumentation space, subjects were generally sensitive to pragmatic constraints, such as using small values of N on commands, so that BigTrak's behavior could be easily observed and remembered.[7]

Although many subjects could articulate these general strategies, they could not always carry them out, as the following selection from subject MA indicates. MA was in the most difficult Selector → Counter condition. He was given D: Repeat last N steps, and the Actual rule was B: Repeat last step N times. MA expressed some doubt about the Given hypothesis and articulated a good experimental strategy:

---

7. Although it is not shown in this chapter, such awareness of pragmatic constraints contrasts markedly with the behavior of middle-school children in the same situation (Dunbar, Klahr, & Fay, 1989).

> OK, if it repeats the last N steps—which we are presuming, it may or may not do that—if it does, then you'd want to write a program which would have a certain amount of steps before the Repeat key, and not repeat all of them, so that you could see if it actually does that. I'm also going to add steps after it so that, if it repeats the steps after it, you'll be able to see that. So the problem is just making up steps that you can differentiate between, so that's what I'm going to do.

Unfortunately, he decided to write a program that included only FIRE commands, and, although he expressed some doubt about whether he would be able to interpret its behavior, he proceeded as planned:

> I'll have it, OK, well, I was thinking of having it not move anywhere, just fire, but I don't know if I'll be able to tell those apart on the screen. So why don't I do that anyway? So I'm just going to fire once, I'll fire twice, I'll fire 3 times, and then I'll repeat the previous 2 steps, then I'll have it fire four times, then fire 5 times. [FIRE 1, FIRE 2, FIRE 3, REPEAT 2, FIRE 4, FIRE 5].

At this point BigTrak fired 21 times, but MA counted 23 FIREs, got confused, and abandoned the all-FIRE approach to experimentation.[8]

In addition to these general characteristics of individual programs, many subjects were systematic in the *sequence* of programs that they wrote, following, as suggested earlier, a strategy of varying only one thing at a time (such as changing either λ or N but not both from one experiment to the next). We will present more data on this issue in a later section.

4.2   Overall Difficulty

As predicted, subjects were less successful at discovering Selector rules than Counter rules. For the two Given conditions, only 1 of the 18 subjects with Actual-Counter rules failed to discover the rule, whereas 5 of 18 failed to discover Selector rules. The proportion of successful subjects in each condition was: Counter → Counter, 100%; Selector → Counter, 92%, Counter → Selector, 67%, Selector → Selector, 83%. Thus, discovering a Counter rule was easiest, whereas discovering a Se-

---

8. In fact, all four rules are distinguishable under this program. Rules A, B, C, and D would FIRE 27, 21, 17, and 20 times, respectively. However, this is extremely difficult for the subject to figure out under these circumstances.

lector rule was more difficult. Also, switching from the Selector frame to the Counter frame was much easier than switching from the Counter frame to the Selector frame. Given that we already knew that subjects regard Counter hypotheses as more likely than Selectors (Klahr & Dunbar, 1988), the results of this study suggest that it is the a priori strength of belief in hypotheses that will determine how difficult it is to switch frames.

4.2.1   TRIAL OF THE CORRECT HYPOTHESIS

Another aggregate measure of the relative difficulty of the four conditions is the trial on which subjects arrived at the correct rule, that is, the point when the Current and Actual hypotheses became the same. As shown in the protocol listed in Appendix A, subjects usually stated the Current hypothesis just before they wrote an experiment to test it. Thus, we can compute the proportion of subjects who arrived at the correct hypothesis prior to each experiment. Figure 3 shows the cumulative proportion of subjects in each condition who stated the correct hypotheses prior to the Nth experiment in their series. The effects of condition are very clear. Although a few subjects immediately rejected the Given hypothesis and luckily guessed the Actual rule prior to the first experiment, there is no reliable effect for such correct anticipations. By the second experiment, half of the subjects in both Actual = Counter groups had proposed the Actual rule, but none of the subjects in the Counter → Selector group had.

Recall that subjects were asked to "write three good experiments" to discover how RPT worked. Thus, the proportion of subjects correctly identifying the rule by experiment 3 provides a measure of success on the task as initially presented. As Figure 3 shows, all Counter → Counter subjects could make the minor revision in the preferred hypothesis necessary to go from Rule A to B or B to A by their third experiment. However, when subjects had to change from a Counter to a Selector only 42% of them were able to abandon a preferred Counter for a Selector by experiment 3. As noted above, the difficulty of frame change was asymmetric, as all but one-quarter of the Selector → Counter subjects discovered that the unpreferred Selector was wrong and discovered the correct counter. Finally, even though no frame change was required, subjects had difficulty making the minor within-frame revision necessary in the Selector → Selector condition, and 33% of them failed to do

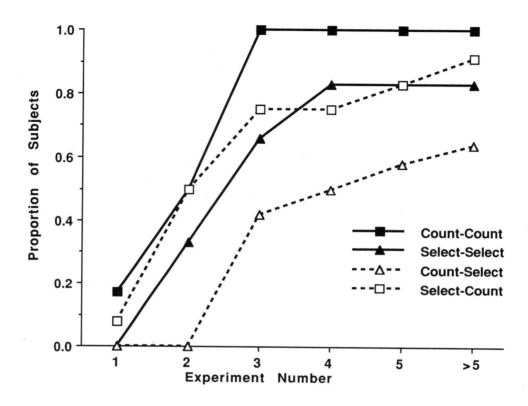

Figure 3. Proportion of subjects generating correct hypothesis by the Nth experiment.

so by the third experiment. This relative order of difficulty remained beyond experiment 3: Counter → Counter was relatively easy, Counter → Selector was relatively difficult, and the two Given = Selector conditions were roughly equivalent and of intermediate difficulty.

4.2.2   THE NUMBER OF EXPERIMENTS

The success-rate measures indicate that the frame change required by the Counter → Selector condition was particularly difficult. Another sensitive measure of difficulty is the number of experiments run. Recall that, once subjects completed the three mandatory experiments, they were free to run additional experiments until they were satisfied that

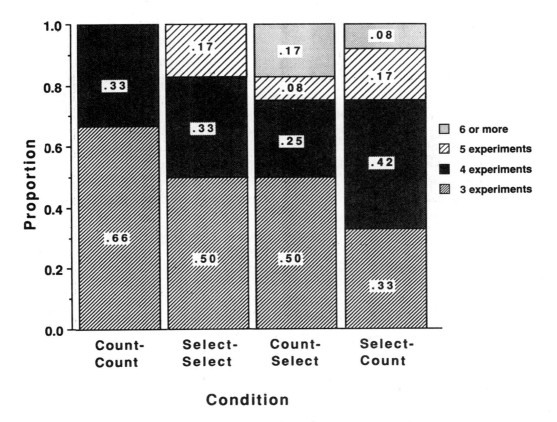

Figure 4. Proportion of subjects running N experiments.

they had discovered the correct rule for RPT. As shown in Figure 4, only one-third of the Counter → Counter subjects chose to run a fourth experiment, and none ran more than four. Half of the subjects in both the Counter → Selector and Selector → Selector conditions, and two-thirds of the Selector → Counter subjects ran four or more experiments. More of the subjects in between-frame conditions ran extra experiments than did subjects in within-frame conditions. The mean number of extra experiments per subject was 0.5 for the within-frame conditions and 1.3 for the between-frame conditions.

4.2.3 IDENTICAL EXPERIMENTS

When subjects were particularly surprised or confused by an experimental outcome, they occasionally repeated an experiment, that is, wrote a program with the same $\lambda - N$ combination as an earlier (usually immediately preceding) program. Although this was an uncommon event, it provided another sensitive index of the relative difficulty of our experimental conditions. Of the 150 total experiments, we observed 14 such pairs of identical experiments, and they occurred only in the frame-change conditions. For both Selector $\rightarrow$ Counter and Counter $\rightarrow$ Selector, there were seven repeats. In most of these cases, the problem was that subjects misencoded the outcome of the first experiment, not because it was particularly complex but because their expectations at some crucial point left them unprepared to notice an essential piece of behavior of the device.

4.3   Search in the Experiment Space

Although the legal range of values for both λ and N was from 1 to 15, subjects tended to be conservative in both the length of program they ran and the value of N. Over 90% of the experiments were within the $\lambda \leq 6$ by $N \leq 5$ experiment space depicted in Figure 2, and more than 60% were within a 4 by 3 subset of that range, even though it represents only 5% of the full space. Each experiment was classified according to its location in the $\lambda - N$ space shown in Figure 2. If subjects were selecting values of λ and N at random, then the expected relative frequency of experiments in each of the experiment space regions would be proportional to the size of that region in the 6×5 experiment space (region 1, 6/30; region 2, 10/30; region 3, 14/30) and would be the same for all conditions. If subjects were sensitive to the interaction between the potential informativeness of different regions of the experiment space and the hypothesis being tested, then we would expect to see an effect of frame-type and experiment space region. More specifically, when the goal of hypothesis testing is to demonstrate an effect, subjects should design experiments that will highlight that feature. For Counter hypotheses, this focus would lead to an attempt to demonstrate that N controls the number of repetitions, which is best demonstrated by larger values of N. For small values of λ, this tends to produce programs in region 3. On the other hand, the clearest way to demonstrate a Selector hypothesis is to use a value of N that disambiguates the selected

segment or step from first, last, or all steps in a program. Region 2 is
the preferred region for such demonstrations.

4.3.1 Experiment Space Distributions

Figure 5 shows the distribution of first and third experiments under two
different aggregations. The upper panels show the percentage of subjects whose first experiments were in each cell of the experiment space,
as a function of the frame of the Given hypothesis. The lower panels
show the distributions of the third experiment as a function of whether
the subjects were in frame-change or same-frame conditions. For each
of the upper panels there were 18 subjects and for the lower panels
there were 24 and 11.[9] In the first experiment, region 2 is underrepre-
sented for subjects testing Counter hypotheses and overrepresented for
Selector hypotheses, and the opposite is true for region 3. Collapsing
over all cells in a region, the two distributions are significantly different from each other ($\chi^2 = 13.6, p < .005$) and from a random model.
These results show that, even in their first experiments, the subjects
were sensitive both to the properties of the experiment space and to
the plausibility (to the subjects) of the Given hypothesis. When given
a plausible Counter, subjects focused on number of repetitions rather
than on what was to be repeated. This produced programs with large
values of N (for Counters, 39% of first experiments had $N > 2$, versus
only 17% for Selectors) and relatively short programs (one-third of the
Counters but none of the Selectors had $\lambda = 1$ on their first program.)
On the other hand, when given an implausible Selector, subjects were
likely to start with an experiment that could clearly discriminate between Counters and Selectors. These are longer programs with small
values of N (region 2 experiments), and they were used by 55% of the
Selector subjects but only by 6% of the Counter subjects.

Recall that, for half of the subjects, the Given hypothesis was from a
different frame than the Actual hypothesis. Consequently, classification
of experiments by frame of Given hypothesis became increasingly invalid
as subjects started to discover that the Given hypothesis was incorrect
and designed experiments to test a Current hypothesis from the Actual
frame. Therefore, for the third experiment, programs were classified ac-

---

9. One of the same-frame subjects wrote a program that *began* with RPT, so only
two of his experiments are included here.

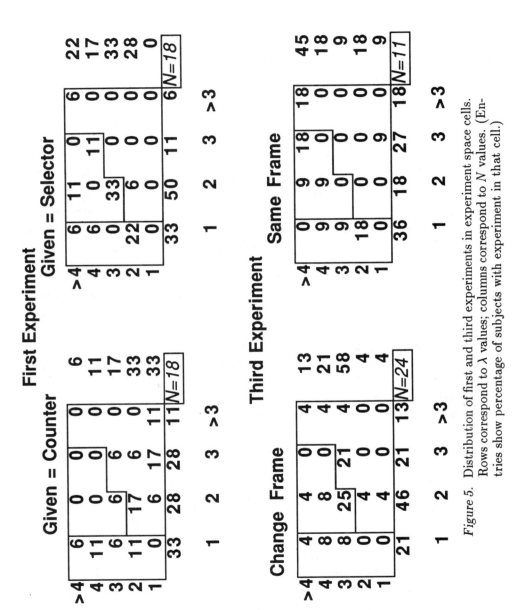

Figure 5. Distribution of first and third experiments in experiment space cells. Rows correspond to λ values; columns correspond to *N* values. (Entries show percentage of subjects with experiment in that cell.)

cording to whether or not a frame change was necessary. If we aggregate over all the cells in a region, then by the third experiment the frame change by region distributions were not significantly different from one another or from a random distribution of experiments in experiment space regions. However, a cell-by-cell analysis reveals a strong effect of frame change. For the third experiment, 46% of the frame-change subjects, but none of the same-frame subjects, had experiments in cells 3,2 or 3,3. These cells tended to be selected as a consequence of the high discriminability of 3,2 and the incremental VOTAT strategy described in the next section. By experiment 3, subjects in the relatively difficult change-frame conditions were avoiding short programs (only 8% had $\lambda < 3$), and the same-frame subjects show a bimodal distribution for λ. Change-frame subjects are also more consistent in their choice of N (nearly 70% with N equal to 2 or 3).

To determine the effect of experiment space region on overall success rate, we analyzed the data according to whether subjects ever went into region 2 and what region subjects were in just prior to their announcement of the correct hypothesis. There were two kinds of very clear regional effects. First, of the 30 subjects who were successful, 28 went into region 2 at least once (93%), and 4 of the 6 subjects who failed to reach the correct hypothesis never went into region 2 (67%). The two who did go into region 2 wrote programs that did not discriminate between the Actual hypothesis and an idiosyncratic hypothesis that they held. Second, with respect to the region preceding the correct hypothesis, Actual = Counter subjects were in region 2 55% of the time, and Actual = Selector subjects were there 71% of the time. Only 4 of the 30 successful subjects were in region 3 immediately prior to announcing the correct hypotheses. Of these 4, 3 were in Actual = Counter groups where an experiment in region 3 would be sensitive to variations in N, and therefore highly informative.

4.3.2   Incremental Search in the Experiment Space

The analysis of experiment space regions gives a picture of the properties of experiments in isolation, but it does not reflect the nature of the incremental paths followed by subjects as they moved from one experiment to the next. The VOTAT strategy mentioned earlier would lead to conservative moves in the experiment space that do not vary both λ and N at the same time (including moves that vary neither). Overall,

Table 5. Proportion of conservative transitions in experiment space.

Condition	First Transition	All Transitions
Counter-Counter	0.33	0.43
Selector-Selector	0.50	0.44
Counter-Selector	0.67	0.64
Selector-Counter	0.67	0.64
Mean	0.54	0.53

about half of the experiment space moves were conservative, but they were more conservative in the frame-change conditions. Table 5 shows the proportion of conservative moves for each condition. The first column shows the proportion only for the first transition (between the first and second experiments), and the second column shows the proportion of all transitions that were conservative. It is clear that, for the Counter → Counter condition, when both the Actual and the Given hypotheses were from the preferred frame, subjects were relatively bold in proposing their second experiment, and two-thirds of them changed both λ and N. However, in frame-change conditions, where the outcome of subjects' first experiment was highly discrepant with their expectations based on the Given hypothesis, subjects were much more conservative in moving about the experiment space; only one-third of them changed λ and N simultaneously.

4.3.3   The Discriminating Power of Experiments

In Section 2.3, we presented a formal analysis of the discriminating power of the different regions of the experiment space. In this section, we summarize the discriminating power of the experiments actually run by the subjects. Each experimental outcome was coded in terms of how many hypotheses were consistent with it. For each subject on each experiment, we considered only the four hypotheses used in this study plus any idiosyncratic hypotheses that the subject might have mentioned. Then we computed, for each condition, the mean number of hypotheses that would be consistent with each experimental outcome (averaged over all the subjects in the condition.) The results are listed in Table 6.

Table 6. Mean number of hypotheses consistent with experimental outcomes.

	EXPERIMENT NUMBER			
CONDITION	1	2	3	4
COUNTER-COUNTER	1.5	1.0	1.2	1.0
SELECTOR-SELECTOR	1.7	1.3	1.2	1.0
COUNTER-SELECTOR	2.8	1.8	1.6	1.5
SELECTOR-COUNTER	1.6	1.3	1.1	1.1

Whereas three of the groups were able to start with programs consistent with only one or two hypotheses, the Counter → Selector subjects designed experiments at the outset whose outcomes were consistent with between two and three hypotheses, and even by their third experiment they were just approaching the first experiment mean of the other three groups.

Another way of describing the discriminating power of subjects' search of the experiment space is in terms of the regions that were avoided while testing particular hypotheses. For first experiments, all subjects avoided an $N = 1$, $\lambda = 1$ experiment as it would not discriminate among any of the hypotheses. Two-thirds of the Given = Counter subjects conducted experiments that could distinguish between the other Counter hypothesis, suggesting that they were testing more than one hypothesis at a time and were avoiding undiscriminating regions of the experiment space. All the Given = Selector subjects conducted first experiments in regions that would discriminate between one selector hypothesis and another.

These results suggest that, when given an hypothesis to test, subjects did consider other hypotheses within that frame, and they wrote programs that would allow them to discriminate between same-frame alternatives. If subjects were only considering hypotheses within the frame of the given hypothesis, then we should expect to see many experiments that would not distinguish between hypotheses from different frames. In fact, 47% of first experiments cannot rule out specific hypotheses from the alternate frame. If we break this down further, we find that, when given Counters, only 33% of first programs can rule out (or confirm) Selectors, whereas, when given Selectors, 66% of programs could rule out (or confirm) Counters. Again, this reflects the a priori belief that RPT works like a Counter, an important factor in determining what parts of the experiment space to search.

4.4   Response to Discrepancies

One rough measure of the extent to which an hypothesis is incorrect is the difference (Δ) between the number of commands that actually get executed and the number that were expected to be executed under the Current hypothesis. For example, consider program 5 in Table 2. If the Current hypothesis is A and the Actual is B, then $\Delta = 6$. If the Current-Actual pair is B-D, then $\Delta = 0$, and if it is C-D, then $\Delta = 1$. When $\Delta = 0$, there remains the possibility that the *content* of the experimental outcome is discrepant with the prediction, but when Δ is nonzero there is no uncertainty—the prediction is not supported by the outcome. Subjects appear to be sensitive to the size of Δ, even though it abstracts over particular program content. In 69% of the experiments where $\Delta = 0$, subjects changed hypotheses, but in 82% of the cases where $0 < \Delta \leq 2$, and on 100% of the cases where $2 < \Delta$, subjects changed their hypotheses.

5.   Discussion

Our subjects were remarkably adept at designing and interpreting experiments in a novel domain. When given a plausible hypothesis, they tended to design experiments that demonstrated the effect that was to be expected. When given implausible hypotheses, they wrote programs that were good discriminators. When the discrepancy between the Given and the Actual hypothesis was very great, subjects were conservative in moving from one experiment to the next. The fundamental question for builders of computational models of the experimental design process is how subjects bring to bear general heuristics for "good experiments" in this novel domain.

5.1   Hypothesis-Generation Heuristics

Any scientific enterprise is conducted in the context of the currently available knowledge of the domain, and initial hypotheses are determined by the knowledge of the domain. In the case of the BigTrak domain, almost all the commands that were learned in the initial phase work by executing a command N times. As a result, subjects were initially predisposed toward hypotheses that are Counters. This is evident in the results of this study and our previous work (Klahr & Dunbar, 1988). The study discussed in this chapter also suggests that subjects considered more than one hypothesis at a time—both the subject protocols and the types of experiments conducted suggest that the subjects considered various hypotheses within a frame. Thus, one heuristic used is that of generating a frame and then generating various slot values within that frame. Then experiments are conducted that will discriminate between rival hypotheses within the frame. The data also suggest that it is easy to think of hypotheses from an alternate frame, but only when the strength of belief in the current frame is less than that of the alternate frame. Thus, subjects in our Selector → Counter group were much more successful than in the Counter → Selector group.

These findings suggest that a useful heuristic in a computational model of experiment generation would be to initially generate different frames and conduct experiments that distinguish between frames, rather than designing experiments that discriminate between rival hypotheses that are all from the same frame. This heuristic is slightly different from the one that is usually used in discussions of scientific methodology. The usual claim is that multiple hypotheses should be considered when designing an experiment, but here we are arguing that this is most effective when the alternate hypotheses come from different frames. Once the frame is established, then the correct slot values of the frame can be determined. Essentially, we are advocating a form of breadth first search.

One interesting and unexpected result of this study is the fact that subjects tended to test multiple hypotheses, whereas subjects in our previous work and in the work of others (such as Mynatt, Doherty, & Tweney, 1977) generally avoided testing multiple hypotheses. In this study, subjects were given hypotheses to test, whereas in most other studies subjects must generate their own initial hypotheses. This difference in procedure had two effects. First, subjects almost always generated hypotheses other than the one given, resulting in the testing of multiple hypotheses. Second, subjects abandoned the given hypothesis much more readily than if they had generated the hypotheses themselves. In the Klahr and Dunbar (1988) study, most subjects' initial (self-generated) hypothesis was A. They only discovered that RPT worked according to rule D after 15 experiments. In the present study, two of the three subjects in the A → D group discovered that RPT worked according to rule D after only four experiments. These results suggest that self-generated hypotheses are given higher strength values than externally generated hypotheses—a fact that becomes apparent when articles are submitted for publication!

5.2   Experiment-Generation Heuristics

The BigTrak domain may appear relatively simple in comparison with that faced by a scientist in a laboratory, but the size of the BigTrak experiment space is surprisingly large. Counting only commands and not their numerical arguments as distinct, there are over 30 billion distinct programs (5^{15}) that subjects could choose from for each experiment. Even if we limit the space to programs of length less than or equal to 4, there are nearly 800 different experiments to choose from ($5^4 + 5^3 + 5^2 + 5$). Most subjects appear to understand immediately that specific instructions are not important and that only the $N - \lambda$ space is relevant, but even it can be as large as 225 cells (15×15). Thus, when asked to write only three experiments, subjects must prune this space effectively. There is clear evidence that subjects do manage to drastically prune the space. As noted earlier, 60% of the experiments occurred within a $\lambda \leq 4$, $N \leq 3$ area of the experiment space, although it represents only 5% of the 15×15 experiment space. Even within this preferred area, experiments were not uniformly distributed. The 1, 1 cell was never used, presumably because subjects realize that it provides no information. Conversely, the 3, 2 cell was disproportionately selected 18 times out of 106 total programs in the first three experiments. This is five times more than expected in a random selection from a 6×5 space, and twice the expected frequency in a 4×3 space. This cell represents the minimum values of λ and N in the maximally informative region 2.

What enables subjects to be so effective in constraining their search in the experiment space? We believe that the following heuristics are operating:

1. *Maintain observability.* Given that the BigTrak moves along the screen from one location to another, there is no permanent record of behavior, and subjects must remember what the device actually did. Thus, one heuristic is to write short programs, making it possible

to remember what happened and to compare the results with those predicted by the Current hypotheses. Other uses of this heuristic are to use small values of N to move forward or backward (this is easy to see, and the BigTrak does not go off the screen) and to make turns that are easy to see, such as right-angle, and 180-degree turns.

2. *Design experiments giving "characteristic" results.* In the BigTrak domain, this translates into "use distinct commands." When all the commands in a program are the same, it is extremely difficult to discriminate between rival hypotheses (see the protocol from subject MA quoted earlier). Almost all subjects attempted to write programs where every command was different. This makes it possible to determine what specific commands were repeated as well as the order in which they were repeated. This heuristic substantially reduces the size of the experiment space while maximizing the observability of the programs.

3. *Focus on one dimension of an hypothesis.* Most hypotheses are complex entities and have many aspects that can be focused on. Auxillary hypotheses, ancillary hypotheses, and additional assumptions that are not tested must be made (see Lakatos & Musgrave, 1970). That is, in going from an hypothesis to an experiment, what is thought to be crucial will be focused on. Our results show that in the BigTrak domain subjects tend to focus on one dimension of an hypothesis at a time. For example, when given a Counter hypothesis, subjects initially focused on the number of times something was repeated rather than what is repeated. This heuristic means that subjects miss some of the features of an experimental result as they are considering the result only in terms of the current dimension of the hypothesis that is being focused on. Furthermore, the finding that many experiments changed only one feature of the experiment at a time suggests that the focus was not on only one aspect of an hypothesis but also on one aspect of an experiment. As we mentioned previously, this strategy has been often discussed in the concept attainment literature (Bruner, Goodnow, & Austin, 1956; Tschirgi, 1980).

4. *Exploit surprising results.* Kulkarni and Simon (1988) argue that this heuristic was used by Krebbs in his discovery of the Ornithine cycle, and they have instantiated the heuristic in their program KEKADA. Holland, Holyoak, Nisbett, and Thagard (1986) also note that the generation of new hypotheses from surprising findings (abduction) is

a useful inductive procedure, and they have instantiated it in their PI program. Our results suggest that our subjects also used such a heuristic, but solid evidence for it remains elusive.

One problem is that "surprise" itself is not well defined. If it is defined as *any* discrepancy between expected and observed outcomes, then the heuristic loses much of its power. We have already described one rough measure that could be used to define surprise, Δ, the difference between the expected and actual number of commands executed by a program (see Section 4.4). It is clear that subjects do not ignore Δ, once it is big enough. The identical experiments (Section 4.2.3) provide additional support for subjects' tendency to respond to surprising results. Recall that identical experiments occurred only in the frame-change conditions, which is where subjects are most surprised by the qualitative nature of the discrepancy between the expected and the actual outcome.

Subjects' verbal protocols suggest that they respond to surprise by setting up a new goal of tracking down its source. The successful subjects in the Counter $\rightarrow$ Selector condition used this heuristic. They focused on why the program or step was not repeated N times and changed their goal from trying to fit the result into a Counter frame to using the surprising experimental result to induce new hypotheses. Subjects in this condition who did not use this strategy continued to focus on how many times things were repeated rather than focusing on the surprising result. Dunbar (1989), in a study that simulated a discovery in a genetics experiment, also found that only subjects who used the strategy of generating a new goal of explaining surprising results were able to discover the mechanism underlying genetic control. This shift of focus usually produces a shift to a new region of the experiment space. The outcome of the next experiment, in turn, leads to the generation of new hypotheses. We have also discussed a group of subjects that use this heuristic (Experimenters) in Klahr and Dunbar (1988).

5. *Use the a priori strength of an hypothesis to choose an experimental strategy.* One of the most often discussed issues in the literature on scientific reasoning has been that subjects tend to attempt to confirm rather than disconfirm their current hypothesis (see Klayman & Ha, 1987). Our study reveals that the strategies of confirmation and disconfirmation varied with the strength of the belief in the currently held hypothesis. When the hypothesis was thought to be

highly likely, subjects often set themselves the goal of demonstrating the key features of the given hypothesis rather than conducting experiments that could discriminate between a large number of hypotheses. A less common strategy for highly likely hypotheses was to use the RPT key as a subgoal to perform an action, for example, drawing a square. In another study (Dunbar, Klahr, & Fay, 1989), we found that young children frequently use this strategy. For hypotheses with low a priori strength, subjects usually propose hypotheses from frames other than the Given frame and conduct experiments that will discriminate between rival hypotheses. Subjects search the hypothesis space before conducting any experiments, and, when they design an experiment, they select an experiment that is in a region of the experiment space that can potentially disconfirm the hypothesis that they are testing.

Not only do subjects appear to use these heuristics, but they also appear to be able to deal with their inherent contradictions. As we noted earlier, no subject ever used the 1, 1 cell, even though it would yield the easiest to observe behavior, because it is so uninformative with respect to discriminating among rival hypotheses. The frequent use of the 3, 2 cell represents a "minimax" solution to the conflicting heuristics of minimizing cognitive load and maximizing discriminability. We are not suggesting that subjects are able to carry out an optimization algorithm that selects this solution. Instead, we believe that the interaction of multiple heuristics produces in our subjects the same kind of behavior that Giere (1988) describes in terms of "the scientist as satisficer."

6. Conclusion

As we stated at the beginning of this chapter, our work falls into the category of psychological studies of scientific discovery in simulated contexts. Our simulated situation is not designed to mimic any particular real scientific discovery, but rather to create a situation in which the thinking processes of subjects are similar to those of scientists when working on real problems. Clearly, a subject's discovery of how RPT works is of scant scientific import. However, we believe that the BigTrak context does give us some insight into the psychology of scientific discovery. In particular, the study described here was designed to contrast high and low plausibility hypotheses (Counters versus Selectors) and minor versus major theory changes (same-frame versus frame-change conditions). Our results suggest a number of powerful heuristics that can be used to design experiments and formulate new hypotheses. Some of these heuristics are very successful and lead toward discovery. For example, generating hypotheses from alternative frames and setting new goals of explaining surprising results led toward the discovery of the correct hypothesis and resulted in fewer experiments. Other heuristics that tended to be less effective were searching for confirmation and focusing on hypotheses within one frame.

Some of the "good" heuristics that we have discovered are similar to those that have been discovered in other approaches to scientific reasoning that we mentioned earlier—historical analyses of scientific discovery (Darden, 1987) and computational models (Holland, Holyoak, Nisbett, & Thagard, 1986; Kulkarni & Simon, 1988; Langley, Simon, Bradshaw, & Zytkow, 1987). This is encouraging as it suggests that we are coming closer to an understanding of the processes underlying scientific discovery. However, as Klayman and Ha (1987) have noted, certain hypothesis-testing methods that are useful in one context may be totally inappropriate in other contexts. Thus, a further goal for our research is to discover the contexts under which heuristics should and should not be used.

Acknowledgements

This research was supported in part by the Personnel and Training Research Programs, Psychological Sciences Division, Office of Naval Research, under Contract No. N00014-86K-0349, in part by the A. W. Mellon Foundation, and in part by grant number OGP0037356 from the National Sciences and Engineering Research Council of Canada. We thank Robert Siegler for convincing us that our previous draft was not the final draft.

References

Bruner, J. S., Goodnow, J. J., & Austin, G. A. (1956). *A study of thinking.* New York: Science Editions, Inc.

Darden, L. (1987). Viewing the history of science as compiled hindsight. *Artificial Intelligence, 8,* 33–41.

Dunbar, K. (1989). Scientific reasoning strategies in a simulated molecular genetics environment. *Proceedings of the Eleventh Annual Meeting of the Cognitive Science Society* (pp. 426–433). Ann Arbor, MI: Lawrence Erlbaum.

Dunbar, K., & Klahr, D. (1989). Developmental differences in scientific discovery strategies. In D. Klahr & K. Kotovsky (Eds.), *Complex information processing: The impact of Herbert A. Simon.* Hillsdale, NJ: Lawrence Erlbaum.

Dunbar, K., Klahr, D., & Fay, A. L. (1989, April). Developmental differences in scientific reasoning processes. Paper presented at the biennial meeting of the Society for Research in Child Development, Kansas City, MO.

Farris, H., & Revlin, R. (1989). The discovery process: A counterfactual strategy. *Social Studies of Science, 19,* 497–513.

Giere, R. N. (1988). *Explaining science: A cognitive approach.* Chicago, IL: University of Chicago Press.

Gorman, M. E. (1989). Error, falsification and scientific inference: An experimental investigation. *The Quarterly Journal of Experimental Psychology, 41A(2),* 385–412.

Gorman, M. E., & Carlson, B. (1989). Can experiments be used to study science? *Social Epistemology, 3,* 89–106.

Holland, J., Holyoak, K., Nisbett, R. E., & Thagard, P. (1986). *Induction: Processes of inference, learning, and discovery.* Cambridge, MA: MIT Press.

Klahr, D., & Dunbar, K. (1988). Dual space search during scientific reasoning. *Cognitive Science, 12,* 1–55.

Klayman, J., & Ha, Y. (1987). Confirmation, disconfirmation and information in hypothesis testing. *Psychological Review, 94,* 211–228.

Klayman, J., & Ha, Y. (1989). Hypothesis testing in rule discovery: Strategy, structure, and content. *Journal of Experimental Psychology: Learning, Memory, and Cognition, 15,* 596–604.

Kuhn, D., Amsel, E., & O'Loughlin, M. (1988). *The development of scientific thinking skills.* New York: Academic Press.

Kulkarni, D., & Simon, H. A. (1988). The processes of scientific discovery: The strategy of experimentation. *Cognitive Science, 12,* 139–175.

Lakatos, I., & Musgrave, A. (Eds.). (1970). *Criticism and the growth of knowledge.* New York: Cambridge University Press.

Langley, P., Simon, H. A., Bradshaw, G. L., & Zytkow, J. M. (1987). *Scientific discovery: Computational explorations of the creative processes.* Cambridge, MA: MIT Press.

Minsky, M. (1975). A framework for representing knowledge. In P. H. Winston (Ed.), *The psychology of computer vision* (pp. 211–277). New York: McGraw-Hill.

Mynatt, C. R., Doherty, M. E. & Tweney, R. D. (1977). Confirmation bias in a simulated research environment: An experimental study of scientific inference. *Quarterly Journal of Experimental Psychology, 29*, 85–95.

Newell, A. (1989). Putting it all together: Final comments. In D. Klahr & K. Kotovsky (Eds.), *Complex information processing: The impact of Herbert A. Simon.* Hillsdale, NJ: Lawrence Erlbaum.

Qin, Y., & Simon, H. A. (in press). Laboratory replication of scientific discovery processes. *Cognitive Science.*

Schauble, L. (1990). Belief revision in children: The role of prior knowledge and strategies for generating evidence. *Journal of Experimental Child Psychology, 49*, 31–57.

Shrager, J. (1985). *Instructionless learning: Discovery of the mental model of a complex device.* Doctoral dissertation, Department of Psychology, Carnegie Mellon University, Pittsburgh, PA.

Shrager, J., & Klahr, D. (1986). Instructionless learning about a complex device. *International Journal of Man-Machine Studies, 25*, 153–189.

Simon, H. A., & Lea, G. (1974). Problem solving and rule induction: A unified view. In L. W. Gregg (Ed.), *Knowledge and cognition* (pp. 105–128). Hillsdale, NJ: Lawrence Erlbaum.

Tschirgi, J. E. (1980). Sensible reasoning: A hypothesis about hypotheses. *Child Development, 51*, 1–10.

Tweney, R. D., & Yachanin, S. A. (1985). Can scientists assess conditional inferences? *Social Studies of Science, 15*, 155–173.

Wason, P. C. (1960). On the failure to eliminate hypotheses in a conceptual task. *Quarterly Journal of Experimental Psychology, 12*, 129–140.

APPENDIX A

A Complete Protocol for Subject DP26

CLR and GO commands have been deleted.

BigTrak's Behavior is shown in boldface type.

Given: A — Repeat entire program N times.

Actual: C — Repeat Nth step once.

002 What I want to do is um...

003 I want to test to see if repeat repeats the statement before it,

004 so I'll write just a little program that does something,

005 say turns left ten, and then put the repeat statement in

006 and we'll just repeat it twice 'cause it's not really important

007 for the first program.

008 And then just to check I'll put in a statement after the repeat

009 to see if it has any effect on that,

010 so it's just left 10, repeat it twice and then right ten.

011 EXP: OK, SO WHAT ARE THE COMMANDS?

012 Left ten, repeat two, and right ten.

013 Experiment 1: ← 10 RPT 2 → 10

← 20 → 10

014 It did what you would expect it would do,

015 it did repeat the left ten twice so it turned left twenty

016 and then back to the right ten,

017 so it appears that the repeat doesn't have any effect on any statements

018 that come after it,

019 so I guess now we'll try putting more than 1 statement

020 in front of the repeat and see what happens

021 we'll do left ten and forward two, repeat twice,

022 and just to check it again at the end we'll have it fire once

023 and see if it repeats or not

024 It's left ten, forward two, repeat twice, fire one.

025 Experiment 2: ← 10 ↑ 2 RPT 2 FIRE 1

← 10 ↑ 4FIRE1

026 Ok, it only turned left once,

027 but it went forward four,

028 so it looks like it only repeats the statement immediately in front
029 of it.
030 I guess for the last test we'll just try a more complicated program
031 with the repeat in the middle of it
032 and see what statements are repeated
034 So this is just a long program with a repeat in the middle of it
035 to see what statements are repeated.
036 EXP: OK, SO WHAT ARE THE COMMANDS?
037 Forward one, right fifteen, back one, repeat two, left ten,
038 forward 1.
039 Experiment 3: ↑ 1 → 15 ↓ 1 RPT 2 ← 10 ↑ 1
040

$$↑ 1 → 15 ↓ 1 → 15 ← 10 ↑ 1$$

046 Ok, that time it repeated the uh.. the right fifteen command
047 EXP: SO DO YOU KNOW HOW IT WORKS,
048 OR WOULD YOU LIKE TO WRITE MORE PROGRAMS?
049 I'm not really sure how it works,
050 because it seemed to act differently in number two, and number 3.
051 It would be best to write more.
052 Should I just go ahead?
053 EXP: YES JUST WRITE NUMBER FOUR.
054 Let's just try the same program as before,
055 except we'll put a three in, repeat three instead of repeat 2,
056 and see if that has anything to do with it.
057 Forward one, right fifteen, back one, repeat three, left ten,
058 forward one
059 Experiment 4: ↑ 1 → 15 ↓ 1 RPT 3 ← 10 ↑ 1
060

$$↑ 1 → 15 ↓ 2 ← 10 ↑ 1$$

061 That time it repeated the statement here but it repeated it twice,
062 the number four.. the number three..
063 it repeated the turning statement once.
064 EXP: WOULD YOU LIKE TO TRY ANOTHER PROGRAM?
065 I guess, I don't really have any idea of what it's doing
066 because when I change the number not only did it change..
067 it didn't change the uh..
068 the number that it repeated
069 but it changed the uh.. the actual instruction.

070 I guess my conjecture is, right now, that it says repeat two
071 so it repeats the second instruction,
072 and here it repeats three and it repeats the third instruction.
073 So we'll try the same thing with repeat one,
074 and see if it repeats the first instruction.
075 Forward one, right fifteen, back one, repeat one, left 10, forward one
076 Experiment 5: ↑ 1 → 15 ↓ 1 RPT 1 ← 10 ↑ 1
077

                                    ↑ 1 → 15 ↓ 1 ↑ 1 ← 10 ↑ 1

078 Ok, I think I know what it does now.
079 EXP: OK..
080 When it hits the repeat statement..
081 when it says repeat one it means at this point repeat statement
082 number one
083 and in this case it went forward and it turned and it went back
084 and then it came forward again, which is the first statement.
085 and it did something similar, I mean it went forward one, turned right
086 went back, and it hit repeat three and this is the third statement
087 so it went back to here
088 EXP: OK, SO HOW, IN GENERAL, DOES THE REPEAT KEY WORK?
089 If you type, it looks, when it hits the repeat statement,
090 if you look through the program when there's like repeat six
091 it takes the sixth statement and does that,
092 when it hits the repeat statement it'll repeat the sixth statement.
093 EXP: OK, GREAT.

APPENDIX B

A Complete Protocol for Subject JS02

CLR and GO commands have been deleted.

BigTrak's Behavior is shown in boldface type.

Given: A — Repeat entire program N times.

Actual: C — Repeat Nth step once.

002 **Alright, Program 1, if that is the hypothesis,**
003 **which I'm not so sure, if it's not the hypothesis**
004 **I'm going to design a program**
005 **that's going to prove that it's not the hypothesis,**
006 **and I think a good way of doing that would be a simple program,**
007 **so, uh.. I'm going to put in first move forward one**
008 **and uh.. that's just a good way to start off**
009 **and I want distinct steps here to see if it is repeating it**
010 **so I will have a right turn, fifteen degrees**
011 **then I think a good maneuver here would be just to have it fire once**
012 **it's just something that's distinguished.**
013 **Then to see if this thing moves like this**
014 **it might go in a reverse order**
015 **or it might just repeat the step number**
016 **but I sort of doubt that**
017 **because there's no numbered lines to these programs**
018 **I'm going to have it move backwards one**
019 **and that will put it back to the left facing forward**
020 **and then we will try a repeat which will..**
021 **we'll try to repeat one**
022 **repeat one will bring it back to its original position**
023 **but it will be facing the opposite direction**
024 **so after the repeat one**
025 **to see what happens to the instructions that happen afterwards**
026 **we will put a turn left, thirty degrees**
027 EXP: I'M GOING TO HOME IT AND CLEAR IT, NOW YOU TELL ME WHAT TO PRESS
028 **Ok, up one, to the right fifteen, fire one, backwards one, repeat one**
030 **left thirty.**
031 Experiment 1: ↑ 1 → 15 FIRE 1 ↓ 1 RPT 1 ← 30

032 EXP: NOW WHAT DO YOU THINK MIGHT HAPPEN WHEN I PRESS GO

033 Um, well I think it's definitely going to execute the first part of it

034 it's going to end up facing to the right

035 but over one block to the left of the position it's in now

036 and, uh, then if the hypothesis for the repeat is correct

037 then it will continue with the rest of the program

038 if it's not, the only thing I'm thinking it might do

039 is I think it might just move forward 1

040 and then it'll end up turning to the left 30, reversing it's direction

041 EXP:OK, I'M PRESSING GO

$$\uparrow 1 \rightarrow 15\text{FIRE}1 \downarrow 1 \uparrow 1 \leftarrow 30$$

042 Aha, that's what I thought it would do

043 but that's not what the hypothesis said.

044 EXP: SO WHAT ARE YOU THINKING?

045 Well it's the original idea,

046 it's uh.. if I ran this same program and I said repeat two

047 it would repeat the second step.

048 if I said repeat three, it's going to fire again.

049 It's repeating the order of the steps that I put in,

050 I think.

051 Or, it might,

052 I want to try something here,

053 EXP: WHAT ARE YOU THINKING?

054 Well I'm thinking it might also be..

055 It'll repeat the first step..

056 If I put two, it might repeat the first and the second step

057 so I'm going to try two

058 actually I'll try three.. no I'll try two.

059 Do I have to write this whole program in again?

060 EXP: WHATEVER PROGRAM WE'RE GOING TO DO, YOU NEED TO WRITE IT FOR ME

061 I want to run the same program,

062 because I know what it does,

063 I just want to change the condition of the repeat

064 because I want to see if it's going to repeat

065 the first two instructions

066 or it's just going to repeat the second instruction

067 so we will give it a.. actually we'll give it a three

068 because if it's the first condition

069 then, um.. if it's my first idea

070 it's going to repeat just the third step

071 then I'll have to worry about it turning fifteen degrees

072 it just, it'll be easier

073 EXP: HOME CLEAR NOW WHAT?

074 Up one, to the right fifteen, fire one, backwards one, repeat three,

075 to the left thirty

076 Experiment 2: ↑ 1 → 15 FIRE 1 ↓ 1 RPT 3 ← 30

$$↑ 1 → 15FIRE1 ↓ 1FIRE1 ← 30$$

077 So it's just repeating the step number of the..

078 the number you put after the repeat it repeats that sequence,

079 that third (unintelligible) the fire or the turn right

080 or the turn left

081 it doesn't repeat first second and third like I thought it might,

082 it just repeats the third step.

083 EXP: CAN YOU WRITE ONE MORE PROGRAM TO BE SURE?

084 Yeah I'll write one more program.

085 Ok, this has some interesting things,

086 we'll make it move backwards three,

087 we will make it turn to the right sixty,

088 and we'll make it turn to the left..

089 no we want it to go.. we'll have it go straight ahead two

090 and we'll have it fire.. fire five times

091 and then we'll have it, let's see what I want to repeat here..

092 and then we'll have it do a nice little spin,

093 I'm curious, one two three four

094 fifth instruction, we'll make it..

095 (unintelligible) the sixth instruction,

096 I doubt it but I'm curious

097 because that would be the one that's after this,

098 One two three four five, um..

099 and the sixth instruction will be um..

100 what could we make it do interesting..

101 We don't have any backward.. yes we do have a backwards

102 something different, we will make it turn left ten

103 EXP: HOME CLEAR

104 Backwards three, right sixty, forward two, fire five, repeat six,

105 left ten
106 Experiment 3: ↓ 3 → 60 ↑ 2 FIRE 5 RPT 6 ← 10

↓ 3 → 60 ↑ 2FIRE10 ← 10

107 I wonder if it did that because repeat six,
108 since six didn't occur yet,
109 I should have put another step in there,
110 because six didn't occur yet
111 it might not actually be repeating the sixth one,
112 it may just be going on,
113 but that doesn't disprove anything anyway, it's just a thought
114 EXP: SO WHAT ARE YOU THINKING?
115 Well I think that whatever number you put after
116 it repeats that instruction line,
117 I set up this third program to prove that again,
118 but what I was curious about when I designed this program,
119 is whether it would repeat something it actually hadn't done yet.
120 Do you know what I'm saying?
121 Because so far it had moved backwards and turned around
122 and gone forward, it had fired five times,
123 then I'm asking it to repeat the sixth step in the program,
124 but the sixth step hadn't occurred.
125 Now what I should have done,
126 is I should have included another instruction
127 I should have had it repeat the seventh step
128 and put in a sixth instruction that was different,
129 because I don't know, from my last program,
130 I don't know whether,
131 I know that the number six means it'll repeat the sixth instruction
132 but, since it hadn't done it yet,
133 I don't know whether it went to the sixth one..
134 because of the repeat six,
135 or it said repeat six is an illegal quantity to put in there,
136 therefore we go on to the next instruction
137 and it just did the sixth instruction anyway
138 EXP: ARE YOU REALLY SURE YOU KNOW HOW IT WORKS,
139 EXP: OR DO YOU WANT TO WRITE ANY OTHER PROGRAMS TO BE SURE?
140 I'm really sure..
141 I'm curious about the last thing

142 whether it will actually repeat something that hasn't occured yet
143 EXP: BUT YOU'RE FAIRLY SURE YOU KNOW HOW IT WORKS?
144 yes
145 EXP: OK WHY DON'T WE STOP THERE THEN

CHAPTER 13

Scientific Discovery
in the Layperson

MICHAEL J. PAZZANI
MARGOT FLOWERS

1. Introduction: Everyday Reasoning and Scientific Reasoning

Scientific discovery can be broken down into two interacting processes:

1. *Hypothesis generation*, which we view as a bottom-up, data-driven process in which one proposes an explanation for some unexplained, unusual, or unexpected occurrence in term, of some causal mechanism.

2. *Hypothesis evaluation*, which we view as a top-down, theory-driven analysis in which one evaluates a hypothesis against further data or against alternative hypotheses.

However, these component processes are not unique to science. In the course of learning to master the environment, the layperson also performs these same activities. Thus, we claim that progress in understanding the process of scientific discovery can be aided by understanding discovery in the layperson.

In our view, the primary difference between discovery in the layperson and the scientist is one of context—a scientist's discovery must be novel to both the scientist and society, whereas a discovery in the layperson need be novel only to the layperson. Our claim, simply stated, is that there are minimal differences between the processes used to invent the wheel and the processes used to reinvent the wheel (when one is previously unaware of wheels). The process of scientific discovery involves analysis of data, formation of new hypotheses, testing of hypotheses by experimentation, the assertion and retraction of hypotheses as they are confirmed or contradicted, and revision of incorrect hypotheses. These are precisely the same kinds of activities used in everyday situations by nonscientists. Similarly, both scientists and nonscientists are subject to similar sources of error—incorrect analysis of data, rejection of evidence, influence of self-interest, and influence of prior expectations. The processes used when a scientist interprets evidence and forms hypotheses differ only in degree and emphasis, and not kind, from those used by people in everyday life (Kuhn, Amsel, & O'Loughlin, 1988).

Our belief is that a model of the discovery processes used by the layperson will require only minor modification to model the scientific discovery process.[1] We will present examples of everyday discoveries by the layperson, compare these activities with those of scientists, and present examples of computational systems that model the reasoning processes of the layperson. Finally, we will discuss how these programs may be used to model the scientific discovery process.

In most instances, it is difficult to construct an accurate model of the cognitive processes of a human subject in a controlled experimental situation that occurs over a small period of time. Although we are encouraged that some are studying the reasoning processes of famous scientists over long periods of time (Hadamard, 1954; Hovey, 1962; Jenkins, 1983; Jenkins & Jeffrey, 1984; Osowski, 1986; Tweney, this volume; Wertheimer, 1959), it is an exceedingly difficult and time-consuming task. There are many advantages to studying discovery in laypeople:

- There are more laypeople than scientists, providing more examples to study. For example, Thagard (1989) presents the seven examples of revolutionary changes in the history of science. Yet each year, thousands of children learn there is no Santa Claus.

- Negative as well as positive examples can be studied. The mistakes that people make provide useful information about constraints on cognitive processes (e.g., Klahr & Dunbar, 1988). Scientists who fail to make discoveries are quickly lost to obscurity, and thus examples of their failures are hard to capture for study.

---

1. One obvious place for refinement is the amount of effort a scientist spends in confirming hypotheses.

● Repeatable experiments are possible. Subjects can be observed making discoveries under different conditions. Such experiments can reveal the utility of a piece of information for a particular discovery. Protocols of subjects making discoveries can be recorded and analyzed (e.g., Klahr & Dunbar, 1988).

In this chapter, we identify a number of relevant cognitive traits and biases used by the layperson. We support the claim that scientific reasoning can be viewed as a special case of everyday reasoning techniques, giving examples from the domain of science. We illustrate how these techniques can be incorporated into models of scientific reasoning and how these models better capture the processes involved in scientific discovery.

We first discuss the notion of the intuitive scientist, how laypersons can be viewed as scientists, and how the reasoning of scientists reflects everyday reasoning strategies. Next, we review the computational models of learning and argumentation used by OCCAM and ABDUL/ILANA, two programs that model commonsense reasoning tasks. Then, we examine a number of case studies of biases and argumentation in laypersons and scientists to show how the knowledge-intensive approach to learning can be applied to scientific discovery and how the evaluation of scientific hypotheses can be approached with techniques for understanding arguments.

2.   Background: The Layperson as Intuitive Scientist

Our claim bears some resemblance to the notion of the "intuitive scientist" (Kelley, 1971; Kelley, 1973; Nisbett & Ross, 1980). Kelley advocated the idea that the layperson makes causal inferences in a manner similar to that of the formally trained scientist, by detecting covariation between effects and potential causes. However, this work does not address the issue of how accurately people perceive covariation between events or how prior expectations influence the perception of covariation (Chapman & Chapman, 1969; Jennings, Amabile, & Ross, 1980; Kuhn, Amsel, & O'Loughlin, 1988).

One recent experiment demonstrates the dominant role of prior knowledge in the detection of a predictive relationship (Pazzani & Schulenburg, 1989). In this experiment, there were two different learning tasks that used the same stimuli. Subjects were shown photographs of a person with a balloon. The photographs varied in four ways:

- Age: Either an adult or a small child was performing an action.
- Action: The person either dipped the balloon in a bowl of water or stretched the balloon.
- Color: Either a yellow or purple balloon was shown.
- Size: The balloon was either very small or very large.

To study how prior knowledge can facilitate the prediction task, one group of subjects was instructed to predict whether or not the balloon would be inflated, whereas another group of subjects had to perform a classic concept acquisition task (see Bruner, Goodnow, & Austin, 1956) and was instructed to learn a way to identify which balloons belonged to an arbitrary category called *alpha*. These two groups were further subdivided into one that had to learn a conjunctive relationship (color = yellow and size = large) and one that had to learn a disjunctive relationship (age = adult or action = stretched). The latter relationship was designed to be consistent with a subject's prior knowledge about the ease of inflating balloons,[2] whereas the conjunctive relationship was not.

The subjects in this experiment were 88 undergraduates who received extra credit in an introductory psychology course for their participation. Each subject was shown a photograph, asked to make a prediction (or classification), and then was shown the correct answer. This process continued until the subject made six correct predictions in a row. For each subject, the number of the trial on which the last error was made was recorded.

If subjects relied only on the detection of covariation, one would expect that the result would be the same for the prediction and classification task. Instead, Figure 1 shows that the task of learning a predictive relationship is influenced by prior knowledge. This effect is so strong that it dominates the well-known finding that conjunctive concepts are easier to learn than disjunctive concepts (Bruner, Goodnow, & Austin, 1956).

---

2. Note that "consistency" does not imply "deducible from." In particular, when there are multiple interacting factors, the influence of each factor may not be obvious.

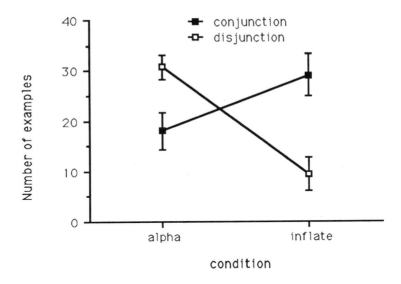

Figure 1. The ease of acquiring predictive (inflate) and descriptive (alpha) concepts. The disjunctive relationship is consistent with prior knowledge on the ease of inflating balloons, whereas the conjunctive relationship violates these beliefs.

3. The Scientist as Layperson

Kelley has revised the notion of the intuitive scientist to deal with two
related findings:

- The role of prior knowledge in the perception of covariation as described in the previous section.
- The perseverance of beliefs in the face of new evidence. Once a belief
becomes accepted, new evidence is not treated in the same manner
as prior evidence (Gorman, 1986; Klahr & Dunbar, 1988; Lord, Ross,
& Lepper, 1979; Nisbett & Ross, 1980; O'Brien, Costa, & Overton,
1986).

Early scientific discovery programs (e.g., Langley, Bradshaw, & Simon,
1983), which concentrated on empirical techniques for making scientific
discoveries, were not intended to account for the biases that people have
in learning and evaluating causal relationships. However, scientists exhibit similar biases (Chapman & Chapman, 1969; Eddy, 1982; Elstein,
Shulman, & Sprafka, 1978; Faust, 1984; Greenwald, Pratkanis, Lieppe,
& Baumgardner, 1986; Mahoney & Kimper, 1976). Rather than viewing these biases as non-normative shortcomings (Cohen, 1981; Nisbett

& Ross, 1980; Stitch & Nisbett, 1980), we view them as essential parts
of the scientific discovery process. In particular, prior knowledge provides an important constraint on the interpretation of the results of an
experiment.

It is not an accident that chemists make chemical discoveries. It is
not simply because chemists have beakers, perform chemistry experiments, and analyze the results of chemical experiments. Rather, it is
because they have prior knowledge of chemistry and the types of factors
that are likely to influence chemical processes. This knowledge enables
a chemist to focus on a small number of potentially relevant features.
Hypotheses that are consistent with the data but that are inconsistent
with knowledge of chemistry are not actively pursued. This focusing
effect is especially important in scientific discovery, in which novelty to
society is stressed. Reliance on prior knowledge serves as an effective
means of controlling distributed problem-solving techniques. The different background knowledge of different scientists causes each scientist
to focus on a different hypothesis.

Similarly, it is important that scientists be more skeptical of data that
disagree with their hypotheses and prior knowledge. Unexpected occurrences need explanations. When data are consistent with a hypothesis,
the hypothesis serves as an explanation. However, when data are not
consistent with a hypothesis, there are a number of possible explanations. The hypothesis may be incorrect, or there may be measurement
error, equipment error, self-deception, or fraud. Until other alternatives
are sufficiently substantiated, a hypothesis is not usually abandoned.
This bias causes scientists to focus their energies on discoveries within
an existing framework rather than continuously switching frameworks or
expending considerable effort triple-checking equipment that produced
the expected result.

4.   Using Causal Knowledge and Argument Structure

In order to model the effects of bias and subjective evaluation of evidence in scientific discovery, we propose the use of causal knowledge
and argument structure. As background, in this section we discuss two
process models that address these kinds of knowledge.

4.1   The OCCAM System

OCCAM (Pazzani, 1989; Pazzani, Dyer, & Flowers, 1986) can be viewed as a model of the intuitive scientist that incorporates a learning technique called *theory-driven learning* (TDL). The system addresses the problem of learning causal knowledge by observing examples of events and their consequences. The task domain involves acquiring simple rules of causation, such as those that describe the outcome of common events in the life of a small child (e.g., when a cup made of glass is dropped, it usually breaks).

A distinction can be made between a theory of *causality*—general principles that lead one to believe that a particular class of actions has a necessary consequence—and a theory of *causation*[3]—indicating specific inference rules that indicate the effects of a particular class of actions. OCCAM*'s objective is to construct a theory of causation, given a theory of causality and a number of observations.* The model of theory-driven learning was influenced by a number of experimental studies in cognitive, social, and developmental psychology that have explicated a number of principles which people use when learning causal relationships (Bullock, Gelman, & Baillargeon, 1982; Shultz & Kestenbaum, 1985).

OCCAM represents a theory of causality as a set of causal patterns. A total of approximately 25 causal patterns have been implemented. The causal patterns make use of temporal and spatial constraints to suggest causal relationships. One such causal pattern is:

> When similar actions on an object are followed by a state change, and a feature of the object correlates with the state change, then that feature is needed for the action to result in the state change.

In TDL, a causal pattern that matches a training example proposes a hypothesis, which is then tested against new data and either accepted or rejected, depending on the accuracy of the hypothesis. Theory-driven learning can be viewed as a form of explanation-based learning (DeJong & Mooney, 1986) in which the domain knowledge (the set of causal patterns) is known to be overly general. Since it is possible for the theory of causality to propose hypotheses that are not true, these hypotheses are evaluated against further examples.

---

3. A theory of causation is also termed a *domain theory* (Mitchell, Kedar-Cabelli, & Keller, 1986).

To gain an understanding of the role and scope of TDL, it is necessary to describe the overall learning architecture of which the process is a component. One important task for OCCAM is to determine what learning strategy is appropriate for each new example. The learning strategies are ordered in OCCAM by the amount of knowledge they require. Thus, OCCAM prefers explanation-based learning (EBL) if it can produce an explanation. If an explanation cannot be produced and the event fits a known causal pattern, then OCCAM uses TDL. As a last resort, OCCAM attempts similarity-based learning (SBL). Thus, TDL's role is restricted to those observations that cannot be explained by the current knowledge of causation (otherwise, EBL would be used) and that meet the constraints of a potential causal relationship (that is, that match a causal pattern). Regularities between observations that cannot be explained and that do not match a causal pattern (e.g., the opening of a garage door by pressing the button on a remote control) can be detected and generalized by SBL in OCCAM.

In OCCAM, the different learning strategies operate over the same memory. The memory that stores the results of SBL and TDL is the same memory that is used to create explanations for EBL. This lets the system use the results of data-intensive empirical strategies to enable later knowledge-intensive learning. OCCAM's memory is organized as a hierarchy of explanatory schemata similar to Schank's (1982) MOPs. The hierarchy of schemata serves as a discrimination net for making predictions and finding explanations. There are several advantages to this architecture for integrating empirical and explanation-based learning in OCCAM. First, the empirical-learning component can acquire the background knowledge required by the explanation-based component. Second, when there is sufficient background knowledge, only a small number of examples are required to create a set of schemata that make accurate predictions.

In spite of its utility, explanation-based learning alone cannot adequately explain many scientific discoveries. In discovery tasks, the goal is to produce a complete and correct background theory; thus such a theory does not exist ahead of time to be utilized as required by EBL methods. Nonetheless, as demonstrated by the experiment in Section 1.1, prior background knowledge can focus the hypothesis-generation process.

There is a major difference between a schema that encodes knowledge whose only support is a number of examples and a schema whose support also includes an underlying theory. The differences between relevance and simple correlation affect how OCCAM handles exceptions to schemata, consistent with OCCAM's preference for knowledge-intensive strategies over data-intensive strategies. Exceptions to schemata are treated differently if they have proven relevant—that is, if it has been previously explained why a certain regularity holds—than if they are merely correlational. When new data do not agree with a correlational schema, the schema itself is questioned. In contrast, when new data contradict a schema supported by an underlying theory, the data are questioned. Only when the data withstand such questioning or when the amount of data is overwhelming is a schema that is supported by an underlying theory abandoned. This is an indication that the underlying theory is incorrect and needs revision to account for new empirical findings.

4.2   The ABDUL/ILANA System

In science, the questioning and confirmation of data and theories can be viewed as a kind of argument. Thus, aspects of scientific creativity can be analyzed and modeled by incorporating techniques and heuristics of argument analysis and response formation. In this section, we illustrate our basic approach to arguments with an example drawn from the domain of political argumentation and present an overview of AB-DUL/ILANA (Birnbaum, Flowers, & McGuire, 1980; Flowers, McGuire, & Birnbaum, 1982; McGuire, Birnbaum, & Flowers, 1981). Later we will show how the process of creating, evaluating, rejecting, and updating scientific hypotheses can be viewed as a similar kind of argument and thus how argument structure analysis can be useful in analyzing the relationships between theory and data and in determining what research steps are sensible responses.

ABDUL/ILANA views an argument as a dynamic process in which a network of beliefs, linked by support and attack dependencies is incrementally elaborated during the course of an argument. This *argument* graph is an episodic memory (Tulving, 1972) of the argument. Here, we briefly examine ABDUL/ILANA's use of such graphs for argument analysis. Consider this argument about responsibility for conflict in the Middle East:

A: Who started the 1967 war?

I: The Arabs did, by the blockade.

A: But Israel fired first.

I: According to international law, blockades are acts of war.

A: Were we supposed to let Israel import arms through the Straits of Tiran?

I: Israel was not importing arms through the straits...

ABDUL/ILANA models the process of participating in this argument, including natural language analysis, argument analysis and strategy choice, reasoning, and language generation. In different modes it can model either the Arab or the Israeli point of view. The argument's strategic content, which can be expressed graphically (see Figure 2), consists of various propositions or beliefs, justified or attacked by other propositions.

Our notion of belief is a psychological, pragmatic one, distinct from the formal propositional sense (Abelson, 1973). Beliefs consist of a premise, or belief content, an attribution of who believes it, and various kinds of supporting or attacking information. Both support structures and attack structures are themselves built out of beliefs. Support structures consist of a belief, an inference rule[4] whose result is the assertion of the belief, and another belief that is the basis of the inference rule and serves to support the original belief. For example, in Figure 2, the belief that "Israel started the 1967 war" is supported by the belief that "Israel attacked first."

Because we are taking a psychological rather than formal approach to modeling belief, a particular belief may have many kinds of support. Consequently, retraction of one kind of support does not automatically mean that the belief is retracted. In a political domain, positions are often determined on emotional grounds. So, for example, from the Israeli point of view, retracting the existing support for the assertion that the Arab blockade caused the war does not cause the Israeli to retract the belief that the Arabs caused the war.

Attack structures also are relationships between two beliefs, except that they contradict one another according to some inference rule. Attack and support may exist simultaneously for a belief. For example,

---

4. These are left implicit in Figure 2. They are also known as "warrant" (Toulmin, 1958; Toulmin, Reike, & Janik, 1979).

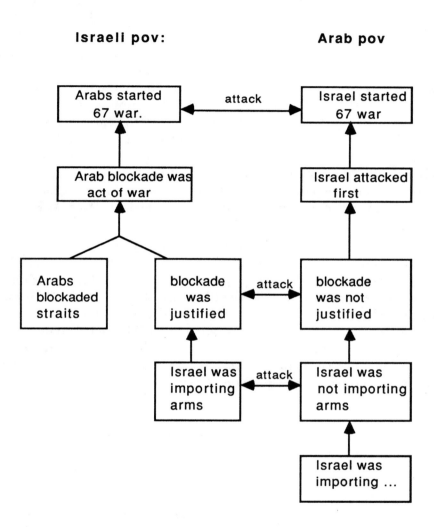

Figure 2. Argument graph illustrating support and attack relationships in an ABDUL/ILANA dialog.

the Israeli point of view, that the Arabs are responsible for the war includes a representation of the Arab belief that attacks this notion. Thus, whether or not a proposition is believed depends on inferences about the kinds of attacks and supports. This approach to belief representation is intended to facilitate modeling the kinds of phenomena that support beliefs, such as selective attention to evidence, and pursuit of hypotheses that are consistent with currently-held theories. Several computer models have been constructed using this approach to representing beliefs: OpEd in the domain of editorial comprehension (Alvarado & Flowers, 1985; Alvarado, Dyer, & Flowers, 1986), JULIA in the domain of editorial analogies (August & Dyer, 1985), SHERLOCK for teaching students inference strategies (Feifer, 1989; Feifer, Dyer, & Flowers, 1986), and AQUA for giving advice on UNIX commands (Quilici, Dyer, & Flowers, 1986, 1989) .

The approach taken by these computer models can be useful in modeling scientific reasoning. Hypotheses in science also can be viewed as beliefs because their validity depends on the balance and relationships of confirming and attacking evidence (Thagard, this volume). For example, the phlogiston theory at one time functioned as a theory to explain observed phenomena, until new experimental evidence led to its retraction. Thus, the processes of science itself can be viewed as arguments, in that they involve competing analyses of data and the assertion, contradiction, retraction, and revision of hypotheses.

5.   Examples of Bias in Learning and Evaluating Evidence and Theories

In this section, we provide several examples of biases in both the layperson and the scientist. The purpose of these examples is to illustrate how these biases can be utilized in models of hypothesis generation and confirmation.

5.1   Rickettsialpox

In 1946, a new disease (today known as *Rickettsialpox*) was first encountered in New York City (Rouche, 1984). The first known case was observed by Benjamin Shankman, who encountered an unusual patient with symptoms that included fever, a small lesion, rash, and swollen glands. The initial diagnosis for this case was chicken pox. The doctor initially dismissed the fact that this patient had already had chicken pox, assuming that perhaps the first case of chicken pox was misdiagnosed. Later examinations revealed that the lesions differed from those of chicken pox, and the symptoms were more severe. The patient was hospitalized, and tests ruled out chicken pox, typhus, smallpox, and Rocky Mountain spotted fever. The patient was given penicillin and recovered in a few days. Shortly thereafter, Dr. Shankman encountered another case, which he treated with sulfonamides, and the patient recovered in a few days. Since the sulfonamide treatment appeared to be no less or more effective than penicillin, Dr. Shankman speculated that neither drug affected the disease. This hypothesis was strengthened when a third patient, treated with aspirin, recovered in about the same length of time.

Soon more cases were encountered, and Dr. Shankman and other physicians noticed that all the patients lived in the same apartment complex. The United States Public Health Service investigated and eventually discovered that a member of the genus *Rickettsia* was responsible for the infections. However, the method of transmission remained unknown.

Meanwhile, the media picked up the story and reported the similarity of the disease to Rocky Mountain spotted fever. Charles Pomerantz, an exterminator, read reports and decided to try to help. He canvassed the neighborhood looking for ticks, which are known to carry Rocky Mountain spotted fever. When he found no ticks outside, and the local kennel reported that there was no tick problem in the neighborhood, he began to consider other possible mechanisms that might carry the disease. He knew that the apartment complex was infested with mice and rats, but they were not known to transmit rash-producing diseases. He considered the possibility that mice might be the host of mites. He asked for permission to inspect the building and found some mites that were subsequently found to be carrying *Rickettsia*.

This case study illustrates several points about the process of discovery. First, after the doctor came up with an initial diagnosis, he dismissed evidence that did not agree with his hypothesis. Only when the evidence against the initial hypothesis was overwhelming did he consider alternative hypotheses. Second, the exterminator's background knowledge focused the generation of hypotheses to a small number that could potentially explain the concentration of the disease in a certain location (e.g., the water supply of an apartment was contaminated). The exterminator's background knowledge put him in a position to advocate the hypothesis that turned out to be correct.

5.2   Piltdown Man

In 1912, a fossil that became known as the Piltdown man was discovered in Sussex. The fossil appeared to be the "missing link" since it had an apelike jaw but human molars. Later, additional fossils were found in Africa and China that differed substantially from Piltdown man. If both these later findings and Piltdown man were accepted, it would imply that there were two distinct evolutionary paths to modern man. Newly developed fluorine tests on the Piltdown man revealed that the fossil was less than 500 years old, suggesting a hoax. Further examination revealed that a molar had been filed and painted brown, and the skull had been stained with iron. Although the dating technique that revealed the hoax was not available at the time Piltdown man was first discovered, it was possible that other signs of fraud could have been detected at the time of discovery.

In this case, the scientists who examined the fossil may have been more lax than usual in their standards of scrutinizing evidence, but the phenomenon is a common one (Lord, Ross, & Lepper, 1979). At the time it was discovered, the fossil of the Piltdown man appeared to satisfy a prediction of evolutionary theory. The evidence was more closely scrutinized only when it presented complications for that theory.

We claim that the general principle of scrutinizing evidence extremely carefully only when it contradicts an expectation is an important part of the discovery process. When an experiment yields results that are expected, the experimental apparatus is often not examined as carefully as when the experiment yields an unexpected result. This closer scrutiny is needed to eliminate error (or fraud) as a possible explanation for the unobserved results. Similarly, when a computational model produces an expected outcome, the average researcher does not (and should not) verify that the compiler converted his program to machine code correctly. Only when a program produces unexpected results is it necessary to take such extreme measures. The benefit of this strategy is that research has more time to spend on new discoveries and additional implications of the theory. Testing these implications is an alternative, and more fruitful, means of validating the theory. One possible difference between the layperson and the scientist is the amount of care that a scientist takes to control an experiment.

5.3   Meteorites

The previous section provided an example of too easily accepting false evidence, but in some cases scientists have too easily dismissed evidence against a current theory. A classic example of this was the failure of the scientific community to accept evidence that meteorites have fallen from the sky. Even when a few scholars finally accepted the evidence on the origins of meteorites, the belief was not universally held.

> *I could more easily believe that two Yankee professors would lie than stones fall from heaven. Thomas Jefferson, 1807 (Cerf & Navasky, 1984.)*

Of course, evidence contrary to existing theories can eventually become overwhelming, and beliefs can eventually change. However, extraordinary claims require extraordinary amounts of evidence to support them.

6.   The Role of Knowledge in Discovery

The process of discovery can utilize both theory-driven learning and explanation-based learning. In this section, we provide examples that show how OCCAM integrates these two processes, and we discuss some limitations of the approaches.

6.1   Theory-Driven Learning in OCCAM: Balloons Rising

OCCAM learns from a number of examples of children playing with balloons. One example is from a colleague who was babysitting her three-year old niece. She decided to entertain the child by inflating some balloons. The child demanded that she tie a string to the balloon. When she complied, the child asked "How come it doesn't go up?" OCCAM's theory-driven component can acquire the hypothesis that balloons with strings on them rise by observing a number of examples in which balloons with strings on them rise while those without strings do not. These examples are consistent with children's experiences since strings are usually tied to helium balloons. Theory-driven learning in OCCAM can suggest the strings as a causal candidate because it covaries with the rising and it is in spatial proximity to the balloon. OCCAM contains a causal pattern to deal with this sort of example, which can be paraphrased:

> If an initial action on an object always occurs before a
>   second action that precedes a state change,
> Then the initial action results in an intermediate state
>   that enables the second action to result in the state change.

In this example, the initial action involves tying a string to a balloon, and the second action is letting go of the balloon, which precedes the rising of the balloon. The causal pattern hypothesizes that tying the string on the balloon results in some state that enables the balloon to rise when released.

6.2   OCCAM on Catalysis: Explanation-Based Learning and Theory-Driven Learning

A catalyst is a substance that modifies the rate of a chemical reaction without being consumed in the process. In some cases, the catalyst changes the rate of a reaction to such an extent that the reaction occurs only in the presence of the catalyst. Inorganic reactions involving catalysis and organic reactions involving enzymes became the subject of study in the early nineteenth century. The process of catalysis was identified by Jons Jacob Berzelius in 1835 (Jorpes, 1966):

> *Platinum, when heated to a certain temperature, had the property of supporting the combustion of alcohol vapors when mixed with atmospheric air and gold and silver lack this property...*
>
> *Thus it is certain that substances, both simple and compound, in solid form as well as in solution, have the property of exerting an effect on compound bodies which is quite different from ordinary affinity, in that they promote the conversion of component parts of the body they influence into other states without necessarily participating in the process with their own component parts...*
>
> *I shall ... call it the catalytic power of the substances, and decomposition by means of this power catalysis.*

Summerlin and Ealy (1985) describe an example of catalysis with common household items. Nothing happens when a sugar cube is touched by a flame. However, if ash from a cigarette is rubbed on the sugar cube first, the sugar cube bursts into flame. The carbon in the ash serves as a catalyst.

Occam has been applied to the problem of catalysis and can identify the rubbing of ash on a sugar cube as an action that enables the sugar cube to burst into flame. The input to Occam is a conceptual dependency representation (Schank, 1973) of a series of actions and state changes. The system tries to account for a state change as the effect of a known causal relationship. Occam's hierarchy of schemata serves as a discrimination net that allows known causal relationships to be recognized. If there is no schema that predicts a state change, the system tries to chain together the effects of different schemata to explain a state change. If an explanation can be found by chaining, a new schema is created by explanation-based learning that allows the state change to be explained in the future by recognition. If an explanation cannot be created with the existing set of schemata, the system tries to create a new schema with theory-driven learning. If all else fails, Occam attempts an empirical method derived from Unimem (Lebowitz, 1986).

To understand the role of explanation-based learning in Occam, consider how explanation-based learning applies to the following example from Summerlin and Ealy (1985). Hydrogen peroxide, iodine, and liquid detergent are added to water, producing a large quantity of soap bubbles. Here, we assume there is a background theory that predicts (1) that in the presence of iodine, hydrogen peroxide will produce water and oxygen and (2) that a gas escaping from a solution of liquid soap will create soap bubbles. These two rules can be chained together to explain how the soap bubbles were produced. The result of this chaining can be generalized and saved, so that in the future, the effect of this reaction can be found by recognition rather than by chaining.

Theory-driven learning differs from explanation-based learning in that the former involves learning at the knowledge-level (Dietterich, 1986). New causal mechanisms (rather than combinations of existing causal mechanisms) are created by this learning process. Consider the example of the sugar cube. There are two training examples used by Occam. The first example describes the situation without the use of the catalyst (see Table 1).

This describes a blond child touching a flame to the sugar cube. There is no state change after this example, so this is nothing for Occam to explain. The example is simply indexed in memory under the `propel`

Table 1. Conceptual dependency representation of touching a sugar cube with a flame.

```
(ACT TYPE (PTRANS)
     ACTOR (HUMAN NAME (LYNN)
                  HAIR (BLOND)
                  EYES (BLUE))
     OBJECT (FLAME)
     TO (P-OBJ TYPE (SUGAR-CUBE)
               COLOR (WHITE)))
```

schema. Next, the example with the catalyst is presented to OCCAM (see Table 2).

When this example is added to memory, OCCAM attempts to explain how the cube burst into flame. In this case, there are no schemata in memory that recognize this situation, nor are there any that can be chained together to produce an explanation. In an attempt to postulate a causal mechanism, similar events are retrieved from memory. In this example, the only similar event that is found is the previous example. Next, the causal patterns are checked to see if any could explain the different outcome of the two similar events. In this case, the only causal pattern that applies is the one illustrated in Section 6.1 to deal with balloons rising. This causal pattern blames a difference in outcome on a different prior action: positing that rubbing ash on the sugar cube results in some intermediate state that enables the cube to burst into flame when touched by a match. This schema (illustrated in Figure 3) is indexed in memory under the **propel** schema for future use. The next time an event is encountered that meets the pattern described by this schema, OCCAM will be able to explain the outcome by recognition rather than by chaining.

7. Limitations of Theory-Driven Learning

Explanation-based and theory-driven learning are not intended as a complete model of the scientific discovery process. The key idea behind theory-driven learning is that knowledge of configurations that appear to be causal can focus the empirical discovery process. The causal patterns consider only a subset of the attributes that describe a situation

Table 2. Conceptual depending representation of burning a sugar cube.

```
(ACT TYPE (PTRANS)
     ACTOR (HUMAN NAME (LYNN)
                 HAIR (BLOND)
                 EYES (BLUE)
     OBJECT (FLAME)
     TO (P-OBJ TYPE (SUGAR-CUBE)
               COLOR (WHITE))
     BEFORE (ACT TYPE (PTRANS)
                  ACTOR (HUMAN NAME (LYNN)
                              HAIR (BLOND)
                              EYES (BLUE)
                  OBJECT (P-OBJ TYPE (ASH)
                                COLOR (BLACK))
                  TO (P-OBJ TYPE (SUGAR-CUBE)
                            COLOR (WHITE))
     AFTER (STATE OBJECT (P-OBJ TYPE (SUGAR-CUBE)
                                COLOR (WHITE)
                 TYPE (BURNING)
                 VALUE (YES)))
```

as potentially relevant. In this manner, these patterns are an explicit form of bias for predictive relationships. In case this bias rules out the correct hypothesis, OCCAM relies on a purely empirical method that considers all its input features to be potentially relevant.

In some cases, the general theory of causality does not provide enough constraints on relationships. For example, in chemical reactions there is an additional constraint that the number of atoms is not changed by the reaction. Thus, if one knows the chemical composition of the substances involved, this serves as an effective constraint on causal relationships (Langley, Zytkow, Simon, & Bradshaw, 1986).

Theory-driven learning currently does not make use of quantitative information. As a consequence, it does not learn that catalysts generally increase the rate of a reaction. This was discovered shortly after catalysts were discovered (Jorpes, 1966):

Platinum sponge is able to ignite a stream of hydrogen as it escapes into the air. This discovery was soon followed by the mutual investigation of Dulong and Thenard that showed several sin-

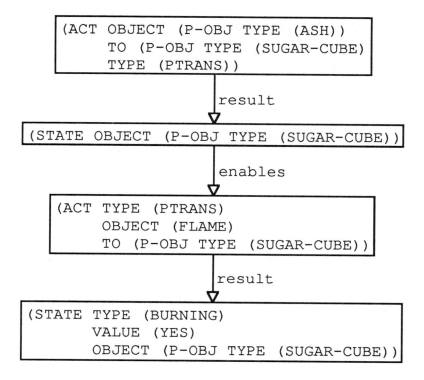

Figure 3. The causal mechanism of a catalytic process.

gle and compound substances have this property, but to differ-
ent degrees... Thus, this power was extended from an exceptional
property to a more general one possessed by substances to different
degrees.

8.   Discovery as Argumentation

Argument analysis can help model the process of scientific reasoning
in two ways. First, it provides a useful framework within which to
interpret the data and steps in the process of scientific reasoning. But
more importantly, heuristics that operate on argument structures can
be applied in the domain of science to help specify productive strategies
for research and theory formation. We illustrate this below with two
examples, the ontogeny of white dwarfs and the chemical structure of
insulin.

8.1 White Dwarfs

One contemporary case, the origin of hydrogen in white dwarf stars, is an example in which new evidence forces revision of a current theory. There are two types of white dwarfs: DB's, with a surface layer composed almost entirely of helium, and DA's, with a surface layer composed almost entirely of hydrogen. A current hypothesis, which we refer to as the *accretion* theory, is that all white dwarfs are created as DB's, which then accrete hydrogen from the interstellar medium. A generalization supported by the accretion theory is a less specific hypothesis that DB's become DA's (by some unspecified method), which we will call the *transformation* theory. This theory predicts that intermediary dwarfs (between DA and DB) should exist. A different prediction follows from the accretion theory, that when binary white dwarfs are found, each of them will be of the same class because, if accretion does occur and both stars are in the same area of interstellar space, then they both should acquire the same type of matter. These predictions, if substantiated, each support in different ways the basic theory that DB's become DA's. Figure 4 depicts these skeletal relationships, with the predicted support assertions outlined and attributions of hypotheses and prediction marked. One can use these relationships, to organize the analysis and incorporation of subsequent data regarding these theories.

Recently, evidence in favor of the transformation theory has been found. Kenyon (1988) reports a white dwarf (G200–39 in Bootes) that has both helium and hydrogen in its surface layer. In addition, the ratio of calcium to hydrogen is approximately the same as that in the interstellar medium. This provides specific evidence to support the prediction of intermediary dwarfs, thus supporting the theory that DB's become DA's. Thus, the prediction of intermediary dwarfs becomes an assertion supported by the specific observational evidence.

Another recent discovery (Oswalt, 1988), that of a binary white dwarf system, calls the accretion theory into question. In this system (L151–81A and L151–81B), one of the stars is of type DB and the other is of type DA. A mechanism that could explain how only one of the binary stars could accrete hydrogen from the interstellar space occupied by them both has not been proposed. As a consequence, the theory that white dwarfs accrete hydrogen is attacked by this evidence.

The argument structure resulting from the evidential role of these two observations is depicted in Figure 5 (with the new observational

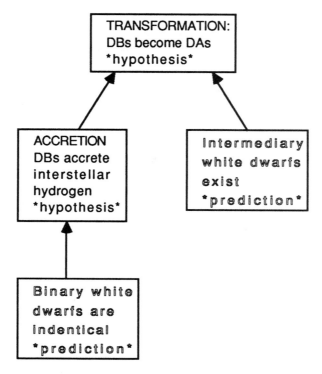

Figure 4. Theories about the origin of white dwarfs and their predictions.

evidence highlighted in bold). The prediction of intermediary dwarfs is confirmed, but finding nonidentical binary dwarfs attacks the prediction of identical binary dwarfs. Because this prediction is the only support for the accretion theory, the entire theory is called into question. Because the accretion theory is called into question, it no longer serves as support for the transformation theory, but, since alternative support for the transformation theory has been found, that theory is still a viable one. Thus, determining the current validity of any particular theory is context sensitive, depending on the full set of possible supports and attacks.

At this point, argument heuristics utilizing this kind of argument analysis can suggest appropriate responses to problematic observations in the scientific domain and focus attention on potentially fruitful research directions. For example, a general argumentation heuristic is that, when contradictory evidence is found for a theory, several responses can be made: attack the validity of the evidence itself, find an alternate expla-

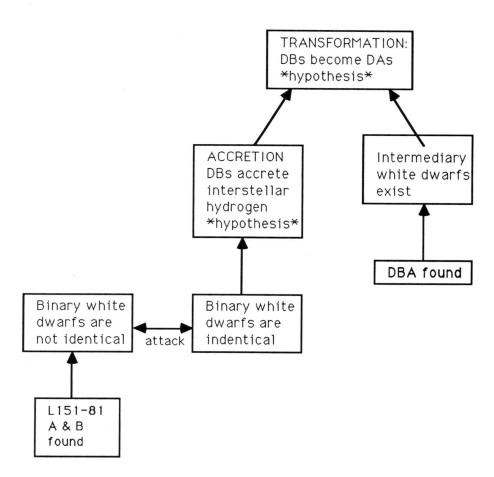

Figure 5. Evidence for white dwarf theories.

nation for the evidence that does not compete with the theory, or accept the contradiction and find an alternate theory. This general argument heuristic provides three reasonable research directions in response to the discovery of L151–81 A and B as a counterexample to the accretion theory. The researcher can look for ways to attack the validity of this finding, such as instrumentation error or data interpretation error. A second direction is to find an alternative explanation for L151–81 A and B that does not contradict the accretion theory. Finally, the researcher might accept that the accretion theory is incorrect but, since there is still valid evidence that the transformation theory holds, develop a de-bugged accretion theory or an entirely different theory that is consistent with the transformation theory and is not contradicted by L151–81 A and B (nor by the existence of other evidence, such as the presence of intermediary dwarfs). A theory-driven learning component, such as that used by OCCAM, may be used to patch the theory or find a new theory.

8.2   Insulin

In the 1950s, research into the disposition of insulin in the body led to the discovery that small peptides could stimulate the production of antigens. This research, recounted in Yalow (1981), illustrates the rhetorical interplay between theory formation and the introduction of new data, showing how techniques of argumentation participate in the process of scientific discovery.

Yalow's group was investigating Mirsky's hypothesis that diabetes, a disease in which the patient suffers the effects of insufficient insulin, was caused by an abnormally rapid degradation of insulin. The first step of this investigation was to document the specific turnover of insulin in the body. Drawing an analogy between albumin and insulin, they borrowed the technique of labeling from prior studies on the turnover of labeled albumin in the body and measured the rate of turnover of la-beled insulin. The rhetorical device of analogy—recognizing similarities between current scientific problems and prior ones and then adapting known techniques to the new situation—is fundamental to scientific in-vestigation.

Contrary to expectation, the rate of insulin turnover in the body was found to be slower in diabetics than in normals. This contradicted the initial hypothesis being investigated, with two results. The hypothesis

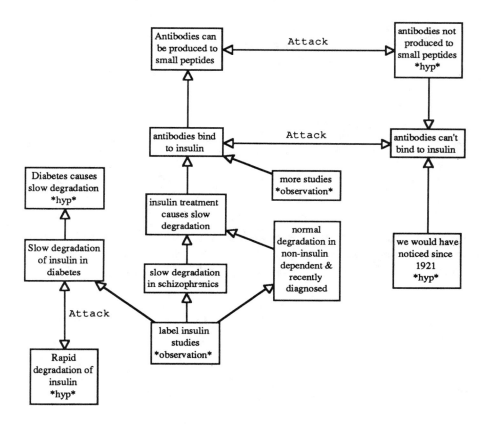

Figure 6. Argument analysis for competing theories of the disposition of insulin.

that diabetes is caused by rapid degradation of insulin had no other support, so it was effectively retracted. In addition, the researchers hypothesized a new assertion supported by the observations of the labeled insulin studies, that something about diabetes itself retards the disappearance of insulin. The argument relationships here and following are depicted in Figure 6.

At this point, selecting what data to focus on was crucial. The retarded insulin hypothesis was also being investigated by another group, who viewed the situation as:

```
If diabetes
    then insulin degradation is slow
    else insulin degradation is normal.
```

Table 3. Insulin degradation rates from Yalow (1981).

	INSULIN DEGRADATION RATE?	
DIABETIC?	NORMAL RATE	SLOW RATE
yes	non-insulin dependent or recently diagnosed	insulin-dependent
no	healthy subject	schizophrenic

However, Yalow's group developed a competing hypothesis based on additional observations. They found two populations that did not fit this pattern. *Normal* turnover was found in diabetics who were only recently diagnosed or who were not insulin-dependent, and *slow* turnover was found in nondiabetic schizophrenics. One subject was found who had been recently diagnosed as diabetic. Initially he had normal insulin degradation, but a few months later he demonstrated delayed insulin degradation.

Selective focusing on the new data (organized in Table 3) forced a shift in the analysis of the data. As it turns out, one kind of treatment given to schizophrenics was "insulin shock therapy." Thus, retarded insulin degradation is associated not with whether or not a subject has diabetes but with whether or not the subject has ever been exposed to exogenous insulin. This can be stated as two rules:

```
If prior insulin exposure
   then insulin degradation is slow
   else insulin degradation is normal.

If insulin degradation is slow
   then prior insulin exposure
   else no prior insulin exposure.
```

This relationship between insulin exposure and slow insulin degradation forced the assumption of a new hypothesis, that prior insulin exposure led to the development of antibodies which bound to circulating insulin, producing a larger molecule that was thus eliminated from circulation more slowly.

There were two problems with taking this approach. First, insulin therapy had begun in 1921, and the antigenicity of insulin had never been observed. Second, standard immunology said that peptides smaller than 10,000 molecular weight could not be antigenic, yet insulin's molecular weight was about 6,000. Each observation provides support for the assertion that small peptides cannot provoke antigen formation, which contradicts the hypothesis based on the population observations.

At this point, the research focused on two contradictory assertions regarding whether or not antibodies can bind to insulin. Argument heuristics utilizing the supports for the two contradictory assertions specify appropriate subsequent research steps. The objection that "we would have noticed" can be countered only indirectly. However, one can contradict the nonantigenicity of small peptides by demonstrating the binding in insulin (or other small peptides). This in fact is what happened—insulin binding only in those treated with insulin was demonstrated in a number of ways. At this point, one side of the contradiction was grounded in actual observation, while the other was grounded only in tenet and thus was invalid. Consequently, by inferential analysis, one could assert in general that small peptides can produce antigens.

Assertions that attack the beliefs of others are often rejected, and establishing the validity of novel beliefs that contradict accepted knowledge can be difficult. Yalow's results were rejected by numerous journals, in spite of the numerous studies demonstrating insulin binding, because they violated basic tenets of immunology. It was published only after rewording compromises, such as "insulin binding globin" versus "antigen," thus establishing the antigenicity of small peptides.

Much of the argumentation process consists of the interaction of competing hypotheses. In this example, there are three major cases: slow versus fast degradation of insulin, diabetic versus insulin treatment as causes of slow insulin degradation, and whether or not antibodies can bind to small peptides (including insulin). Numerous other belief-oriented rhetorical steps were involved: assertion of new beliefs, finding support for beliefs, attacking beliefs of others, inferential analysis of beliefs, rejection of beliefs that violate other beliefs, and retraction of old beliefs shown to be incorrect. Viewing data and theories as beliefs, and applying argument heuristics to them, is a useful technique modeling the process of scientific research.

9. Conclusions

Two immediate points have been made in this chapter. First, models of
scientific discovery should incorporate a number of kinds of knowledge:
empirical laws, understanding of process, specification of intermediate
states, causal constraints, and argument structure and heuristics. Second, models of scientific discovery need to incorporate everyday "layperson" techniques of reasoning and creativity. What is the potential benefit of taking such an approach?

Over a decade ago, Lenat (1982) reported on AM, a model of discovery
that used heuristics of "interestingness" as a focusing technique in the
analysis of data. We claim that additional focusing techniques, such as
those provided by prior knowledge, play a major role in discovery in
the intuitive scientist and the practicing scientist. Yet much subsequent
work in scientific discovery has concentrated primarily on what to do
with given experimental data, rather than on techniques for deciding
which experimental data to focus on. The first step of hypothesis and
theory formation is not explanation of data but noticing or selection
of what particular data is worth looking at. In this chapter, we have
identified some kinds of knowledge and techniques that embody some of
the biases people use to help in this focusing task: causal analysis, use of
prior knowledge and generalizations, and techniques of belief formation
and analysis of argumentation structure. In an integrated, ecologically
valid model of scientific discovery, decisions must explicitly be made
about the meaning, import, and disposition of raw data and developed
theories. The features we have discussed contribute to making such
decisions for productive scientific reasoning.

References

Abelson, R. P. (1973). The structure of belief systems. In R. C. Schank
    & K. M. Colby (Eds.), *Computer models of thought and language.*
    San Francisco: Freeman.

Alvarado, S., Dyer, M., & Flowers, M. (1986). Editorial comprehension
    in OpEd through argument units. *Proceedings of the Fifth National
    Conference on Artificial Intelligence* (pp. 250–256). Philadelphia,
    PA: Morgan Kaufmann.

Alvarado, S., & Flowers, M. (1985). Memory representation and retrieval for editorial comprehension. *Proceedings of the Seventh Annual Conference of the Cognitive Science Society* (pp. 228–235). Irvine, CA: Lawrence Erlbaum.

August, S., & Dyer, M. (1985). Understanding analogies in editorials. *Proceedings of the Ninth International Joint Conference for Artificial Intelligence* (pp. 845–847). Los Angeles, CA: Morgan Kaufmann.

Birnbaum, L., Flowers, M., & McGuire, R. (1980). Towards an AI model of argumentation. *Proceedings of the First Annual National Conference on Artificial Intelligence* (pp. 313–315). Stanford, CA: Morgan Kaufmann.

Bruner, J. S., Goodnow, J. J., & Austin, G.A. (1956). *A study of thinking.* New York: Wiley.

Bullock, M., Gelman, R., & Baillargeon, R. (1982). The development of causal reasoning. In W. Friedman (Ed.), *The developmental psychology of time.* New York: Academic Press.

Cerf, C., & Navasky, V. (1984). *The experts speak: The definitive compendium of authorative misinformation.* New York: Pantheon Books.

Chapman, L. J., & Chapman, J. P. (1969). Illusory correlation as an obstacle to the use of valid psychodiagnostic signs. *Journal of Abnormal Psychology, 74,* 271–280.

Cohen, L. (1981). Can human irrationality be experimentally demonstrated? *Behavioral and Brain Sciences, 4,* 317–331.

DeJong, G. & Mooney, R. (1986). Explanation-based learning: An alternative view. *Machine Learning, 1,* 145–176.

Dietterich, T. G. (1986). *Learning at the knowledge level. Machine Learning, 1,* 287–315.

Eddy, D. (1982). Probabilistic reasoning in clinical medicine: Problems and opportunities. In D. Kahneman, P. Slovic, & A. Tversky (Eds.), *Judgment under uncertainty: Heuristics and biases* (pp. 249–267). New York: Cambridge University Press.

Elstein, A., Shulman, L., & Sprafka, S. (1978). *Medical problem solving: An analysis of clinical reasoning.* Cambridge, MA: Harvard University Press.

Faust, D. (1984). *The limits of scientific reasoning.* Minneapolis: University of Minnesota Press.

Feifer, R. G. (1989). *An intelligent tutoring system for graphic mapping strategies.* Doctoral dissertation, Department of Computer Science, University of California, Los Angeles.

Feifer, R. G., Dyer, M., & Flowers, M. (1986). Teaching inferencing strategies. *Annual Meeting of American Educational Research Associations* (pp. 1–12).

Flowers, M., McGuire, R., & Birnbaum, L. (1982). Adversary arguments and the logic of personal attacks. In W. G. Lehnert & M. G. Ringle (Eds.), *Strategies for Natural Language Understanding.* Hillsdale, NJ: Lawrence Erlbaum.

Gorman, M. E. (1986). How the possibility of error affects falsification on a task that models scientific problem solving. *British Journal of Psychology, 77,* 85–96.

Greenwald, A., Pratkanis, A., Lieppe, M., & Baumgardner, M. (1986). Under what conditions does theory obstruct research progress? *Psychological Review, 93,* 216–229.

Hadamard, J. (1954). *The psychology of invention in the mathematical field.* New York: Dover.

Hovey, D. (1962). *Experience and insight: Experiments on problem-solving and a case study of scientific thinking.* Doctoral dissertation, Department of Psychology, University of Colorado, Boulder.

Jenkins, R. V., (1983). Elements of style: Continuities in Edison's thinking. *Annals of the New York Academy of Sciences, 424,* 149–162.

Jenkins, R. V., & Jeffrey, T. E. (1984). Worth a thousand words: Nonverbal documents in editing. *Documentary Editing, 6,* 1–8.

Jennings, D., Amabile, T. & Ross, L. (1980). Informal covariation assessment: Data-based vs. theory-based judgments. In A. Tversky, D. Kahneman, & P. Slovic (Eds.) *Judgments under uncertainty: Heuristics and biases.* New York: Cambridge University Press.

Jorpes, J. E. (1966). *Jac. Berzelius: His life and work.* Stockholm: Almquist & Wiskell.

Kelley, H. H. (1971). Causal schemata and the attribution process. In E. E. Jones, D. E. Kanouse, H. H. Kelley, R. E. Nisbett, S. Valins, & B. Weiner (Eds.), *Attribution: Perceiving the causes of behavior.* Morristown, NJ: General Learning Press.

Kelley, H. (1973). The process of causal attribution. *American Psychologist, 28*, 107–128.

Klahr, D., & Dunbar, K. (1988). Dual space search during scientific reasoning. *Cognitive Science, 12*, 1–48.

Kuhn, D., Amsel, E., & O'Loughlin, M. (1988). *The development of scientific thinking skills.* San Diego: Academic Press.

Langley, P., Bradshaw G. L., & Simon, H. A. (1983). Rediscovering chemistry with the BACON system. In R. S. Michalski, J. G. Carbonell, and T. M. Mitchell, (Eds.), *Machine learning: An artificial intelligence approach.* San Mateo, CA: Morgan Kaufmann.

Lebowitz, M. (1986). Concept learning in an rich input domain: Generalization-based memory. In R. S. Michalski, J. G. Carbonell, and T. M. Mitchell (Eds.), *Machine learning, an artificial intelligence approach* (Vol. 2). San Mateo, CA: Morgan Kaufman.

Lenat, D. (1982). AM: An artificial intelligence approach to discovery in mathematics as heuristic search. In R. Davis & D. B. Lenat (Eds.), *Knowledge-based systems in artificial intelligence.* New York: McGraw-Hill.

Lord, C., Ross, L., & Lepper, M. (1979). Biased assimilation and attitude polarization: The effects of prior theories on subsequently considered evidence. *Journal of Personality and Social Psychology, 37*, 2098–2109.

Mahoney, M., & Kimper, T. (1976). From ethics to logic: A survey of scientists. In M. Mahoney (Ed.), *Scientist as subject: The psychological imperative* (pp. 187–194). Cambridge, MA: Ballinger.

McGuire, R., Birnbaum, L., & Flowers, M. (1981). Opportunistic processing in arguments. *Proceedings of the Seventh International Joint Conference on Artificial Intelligence* (pp. 58–60). Vancouver, BC: Morgan Kaufmann.

Mitchell, T. M., Keller, R. M., & Kedar-Cabelli, S. T. (1986). *Explanation-based learning: A unifying view. Machine Learning, 1*, 47–80.

Nisbett, R., & Ross, L. (1980). *Human inference: Strategies and shortcomings of social judgment.* Englewood Cliffs, NJ: Prentice-Hall.

O'Brien, D. P., Costa, G., & Overton, C. (1986). Evaluations of causal and conditional hypotheses. *Quarterly Journal of Experimental Psychology, 38A*, 493–512.

Osowski, J. V. (1986). *Metaphor and creativity: A case study of William James.* Doctoral dissertation, Department of Psychology, Rutgers University, New Brunswich, NJ.

Oswalt, T., Hintzen, P., Liebert, J., & Sion, E. (1988). *L151-81A/B: A unique white dwarf binary with DB and DA components. Astrophysical Journal, 333,* L87–L89.

Pazzani, M. (1990). *Creating a memory of causal relationships: An integration of empirical and explanation-based learning methods.* Hillsdale, NJ: Lawrence Erlbaum.

Pazzani, M., Dyer, M., & Flowers, M. (1986). The role of prior causal knowledge in generalization. *Proceedings of the National Conference on Artificial Intelligence* (pp. 545–550). Philadelphia, PA: Morgan Kaufmann.

Pazzani, M., & Schulenberg, D. (1989). The role of prior knowledge in the acquisition of conjunctive and disjunctive concepts. *Proceedings of the Eleventh Meeting of the Cognitive Science Society* (pp. 812–819). Ann Arbor, MI: Lawrence Erlbaum.

Quilici, A., Dyer, M., & Flowers, M. (1986). AQUA: An intelligent UNIX advisor. *Proceedings of the Third European Conference on Artificial Intelligence* (pp. 33–38). Brighton, UK.

Quilici, A., Dyer, M., & Flowers, M. (1989). Recognizing and responding to plan-oriented misconceptions. *Computational Linguistics.*

Rouche, B. (1984). *The medical detectives* (Vol. 2). Washington Square Press.

Schank, R. C. (1973). Identification of conceptualizations underlying natural language. In R.C. Schank & K.M. Colby (Eds.), *Computer models of thought and language.* San Francisco: Freeman.

Schank, R. C. (1982). *Dynamic memory: A theory of reminding and learning in computers and people.* Cambridge: Cambridge University Press.

Shultz, T., & Kestenbaum, N. (1985). Causal reasoning in children. *Annals of Child Development*, 195–249.

Stitch, S., & Nisbett, R. (1980). Justification and the psychology of human reasoning. *Philosophy of Science, 47,* 188–202.

Summerlin, L., & Ealy, J. (1985). *Chemical demonstrations: A source book for teachers.* American Chemical Society.

Toulmin, S. (1958). *The uses of argument.* Cambridge, MA: Cambridge University Press.

Toulmin, S., Reike, R., & Janik, A. (1979). *An introduction to reasoning.* New York: Macmillan.

Tulving, E. (1972). Episodic and semantic memory. In E. Tulving & W. Donaldson (Eds.), *Organization of memory.* New York: Academic Press.

Wertheimer, M. (1959). *Productive thinking* (Enlarged ed.). New York: Harper & Row. Original work published 1945.

Yalow, R. S. (1981). Biomedical investigation. In W. Shropshire (Ed.), *The joys of research.* Smithsonian.

CHAPTER 14

Commonsense Perception and the Psychology of Theory Formation

JEFF SHRAGER

1. Introduction

Sensation and cognition are closely related aspects of human psychology, but their precise relationship has never been clear. Psychological theories addressing this issue range from theories in which sensation is viewed in a peripheral role, coding the environment into symbolic memories (e.g., Anderson, 1983; Laird, Rosenbloom, & Newell, 1986), to theories in which cognition has properties that are fundamentally sensory (e.g., Langacker, 1987). A variety of interim positions rely on spatial imagery (e.g., Kosslyn, 1980; Miller, 1986) or "perceptual buffers." None of these theories has been given a detailed computational account. Where the perceptual aspects are carefully addressed (e.g., Kosslyn, 1980; Ullman, 1984), the cognitive aspects are neglected, and where the cognitive aspects are carefully addressed (e.g., Anderson, 1983; Laird, Rosenbloom, & Newell, 1986), the sensory aspects are neglected. Most theories that do some justice to both (such as Langacker, 1987) provide computational details of neither. One goal of the present research is to build a psychological theory in which sensation and cognition are closely interwoven and which can plausibly subsume both sensory and cognitive aspects of intelligence.

This goal has both empirical and theoretical roots. For several years my colleagues and I have been studying complex instructionless learning, a form of naive scientific reasoning (Shrager & Klahr, 1986). Recent extensions to this work suggest that, in instructionless learning about systems with significant dynamic aspects, sensory content (e.g., vision or analog imagery) is prior to, and more important than, nonsensory content (e.g., quasi-linguistic models[1]). This chapter begins with a summary of our work on instructionless learning and introduces some of the recent results that have led us in this direction.

The theoretical reasons for moving toward perceptually oriented models of cognition and away from models based on quasi-linguistic representations arise from difficulties in constructing computational models of instructionless learning. Briefly put, I have previously proposed that theory change in instructionless learning involves a process called *view application* (Shrager, 1987), whose role is to *reinterpret knowledge of the domain in terms of previously known views (abstractions)*. Unfortunately, implementing view application in a quasi-linguistic representational framework leads to problems with the meaning of terms in the representation, which I refer to as *the paradox of recognition* and *the framework alignment problem*. In the third section of this chapter I elaborate these two problems.

In Section 4 I explain the framework of *grounded representation* and describe a new theory of reinterpretation in this framework, which I call *commonsense perception*. I then show how commonsense perception can be used to simluate the dynamics underlying gas lasers in a psychologically plausible manner.

In the course of this chapter, I hope to convince the reader that: (1) reinterpretation plays a central role in theory formation, at least in the domains that I have studied, that (2) sensory content is a fundamental and inextricable aspect of the theories formed and their formation by reinterpretation, and that (3) commonsense perception (or something very much like it) is a central psychological mechanism, that plays a key role in both theory formation and complex reasoning.

It is worth pausing to explain what I mean by the terms *sensation*, *perception*, and *cognition*. An agent acts in an environment to which it has accesses via *sensation*: vision, hearing, and touch, along with the motor activity that affects the environment and that makes different aspects of that environment available to the agent's senses. What the

---

1. I use the term *quasi-linguistic* to indicate representations that approximate linguistic (e.g., English) words or phrases.

agent adds to its sensory world, leading it sooner or later to action, is cognition. Thus inference, reasoning, memory, and all other mental activity that is not directed externally is here considered *cognitive*. For cognitive activity to make use of the agent's sensations, they must be *interpreted*—put into correspondence with the cognitive aspects of the agent's activity. Perception is the act of making this correspondence; thus *perception connects cognition to sensation*. In these terms, commonsense perception is an interpretive activity that lies between cognition and sensation and that is itself neither purely cognitive nor purely sensory.[2]

2. Psychological Aspects of Theory Formation

To study theory formation, we performed a series of studies loosely referred to as *instructionless learning* (Shrager & Klahr, 1986). In most of these studies, college students from a variety of intellectual backgrounds were given a novel device and were asked to figure out how it operates. By this technique, we extended the theory formation process and so had some opportunity to study its stages.

In this section, I briefly summarize our original instructionless learning work and describe a similar recent study. Because computational models are the focus of this volume, I have omitted the details of our data and analyses, some of which are in Shrager and Klahr (1986). The reader should take from this section that *reinterpretation* plays a central role in theory formation.

2.1 Instructionless Learning About the BigTrak

Shrager and Klahr (1983, 1986) gave college students a programmable toy vehicle called the BigTrak and asked them to figure out how it operates without instructions or advice. In the course of about one-half hour and about 30 to 50 interactions with the device, subjects arrived at a relatively complete understanding of how it worked.

During the session, subjects undertook numerous steps of theory refinement and reformulation. In some of these events, subjects seemed

---

2. It is interesting to note that the term *common sense* derives from Aristotle's term *koine aisthesis*, by which he meant the sense that allows us to perceive those aspects of things that are not properties of any specific sense, such as number, movement, shape, and size (De Anima, 418a10; Ross, 1977, p. 138).

to reformulate their theory of the BigTrak in ways that introduced new terms and representational principles. Consider the segment of protocol in Table 1, taken from Shrager and Klahr (1986). By this point in the protocol, subject FC knew the BigTrak programming *syntax* completely correctly.[3] By programming the toy with

CLR, >, 1, FWD, 2, GO   (FC122-127)

FC caused it to move six degrees clockwise and then move two feet forward. From his previous behavior, we can infer that FC thought that this would make BigTrak move one foot to the right and then one foot forward. Figure 1 shows what we think FC expected and what the device really did.

We hypothesize that a reformulation step took place at the point of mismatch between FC's expectation and the behavior of the BigTrak. In FC128 in Table 1 we see that he *recognized* the actual behavior of the device as approximately the vector sum of the two one-foot vectors that he expected the BigTrak to execute. He said: "I see that's the resultant thing maybe." Around FC127 we hypothesize that the subject did the following:

1. *Recognized* that the behavior of the BigTrak matched his expectations when mediated by the concept of vector addition;

2. *Introduced the concept* of vector addition into his theory of the toy, including "resultant" and "component" terms and their associated representational principles; and

3. *Reformulated his theory* of the BigTrak and its operation in accord with the new terms and representational principles of vector addition introduced in the preceding step.

This reformulation resulted in FC having a theory of the device that was a *combination* of his previous theory and his general knowledge of vector addition.

FC's introduction of vector addition seems very *rapid*, to apply *as a unit* to his understanding of the functioning of the BigTrak, and to *augment* and serve as *reorganizing principles* for his interpretation of the device and its behavior. Many similar reformulations take place in the

---

3. The BigTrak is programmed with up to 16 movement commands, such as "move forward two feet, turn right 15 minutes." Subjects entered commands by pressing buttons on a keypad attached to its back. The program steps are run sequentially when the GO button is pressed.

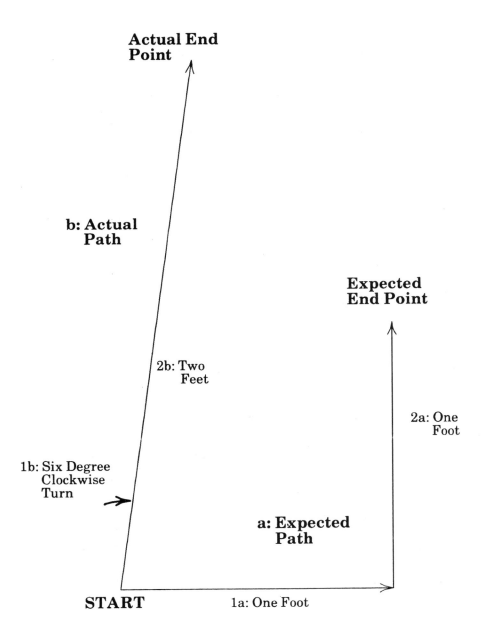

Figure 1. The BigTrak's movements in response to the program: "> 1 FWD 2 GO", (a) as FC expected them and (b) what the BigTrak really did.

Table 1. Part of FC's BigTrak instructionless learning protocol. A "+" after
the line number indicates a syntactically correct button press. The
square brackets are time stamps. The braces show the BigTrak's
movements.

```
115:       Does it...I don't know maybe it remembers things or
116:       something so that...it just did the same thing I told it
117:       to do last time even though I pushed different buttons.
118:+CLR
119:+CLR    Alright.  I guess you can like oh I see (?) program
120:       steps into it or
121:       something like that. So if you push ummm...
[8:30]
122:+CLR
123:+>      right one
124:+1      then forward
125:+FWD    two
126:+2
127:+GO     Went straight and right a little bit.
{right 6 degrees forward 2 feet}
128:       Oh. I see that's the resultant
[9:00]
129:       thing maybe.  I don't know.
```

---

data collected in the BigTrak study. The process of reformulation in
accord with commonsense concepts (here called "views") accounts for
most of the large theory changes and many of the small theory changes
that learners undergo while learning about the device.[4] Subjects understand the BigTrak as a toy, as an electronic device, and at some point
they learn that it has a memory and so that it is a programmable device
of some sort. These views combine into what might be called a "programmable toy" theory of the BigTrak, but the learners must refine the
details of this theory. FC's vector episode provides an example of the
use of commonsense views in service of this refinement.

---

4. Vector addition may not seem very "commonsense," but many of the reformulations that we observed were simpler, such as reformulating the concept of "electronic device" and "toy." By "commonsense" I mean that the subject has gained
expert familiarity with the concept and so operates with it skillfully. Vector addition, or at least its general form, is very familiar to most technically oriented
undergraduates.

2.2 Instructionless Learning About the DigiComp

Although all subjects proposed complex mechanisms for the BigTrak, there was no way for them to verify their hypotheses about its components and internal machinery. Even if the subject could open up the device, most of the interesting electronics reside in a single special-purpose VLSI chip whose workings are invisible. Furthermore, the only dynamic aspects of the BigTrak are the fairly simple movements of the device. Although the above protocol demonstrates that even these simple aspects can lead to interesting interpretive difficulties, only a few such problems arise in the study. There are certainly dynamic aspects of the learner's *theory* of the device in operations such as "pushing instructions onto the memory stack" or "erasing the last instruction in memory," but the BigTrak study provided no clear record of these imaginary dynamics.

These problems led us to examine domains with overt dynamic aspects. In a recent study we observed instructionless learning about a programmable machine called the *DigiComp*, shown in Figure 2. This device is a programmable three-bit state-transition machine whose simple physical structure belies the complexity of its programming model. Very briefly, the DigiComp is programmed by placing short straws on pegs on the front and back of the device. The straws and "logic rods" on the front implement simple logical tests whose input is the current binary state of the machine. Additional rods connect the front to the back of the machine, transmitting the result of the front logic to the clock rods on the back. These change the binary state of the three flip-flops by moving them to the left or right via clock straws. One can see that there are a very large number of ways to arrange the straws into different boolean functions.

In a warm-up task, we asked subjects to place straws on the computer in a predetermined arrangement that causes the device to implement a binary counter (000, 001, 010, 011, 100, 101, 110, 111, 000, and so forth) and to test that this program behaves as specified. This familiarized the subjects with all the manipulable aspects of the device. After the subject succeeded in this task, he or she was given standard protocol instructions and asked to "figure out" how to reprogram the device so that it would count from 0 (000) to 4 (100) and then stop, that is, to implement this sequence: 000, 001, 010, 011, 100, 100, 100,..., so that if the device continues to be cycled it will not change state away from 100 (4).

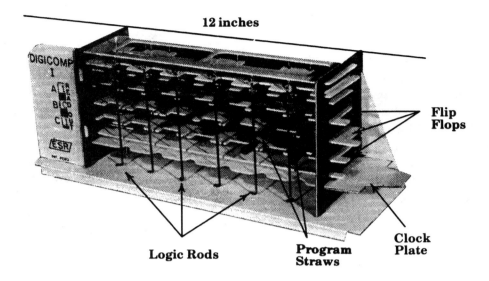

Figure 2. The DigiComp, a product of ESR Corporation, Montclair, New Jersey. The device is programmed by pushing straws onto the pegs on the front and back of each horizontal "flip-flop" plate. It is operated by moving the "clock" plate once in and out for each cycle. The logic rods on the front compute the programmed function and transmit the results to the clock rods on the back, which in turn set and reset the flip-flops.

Four subjects were given this task, including a college junior in marine biology with one semester of programming experience, and a research systems programmer with more than fifteen years of programming experience and knowledge of logic design. As expected, all our subjects first addressed the reprogramming task by trying to understand how the DigiComp implements the normal counting activity, presumably so that they could use that knowledge to modify the program as required. Only one subject (the research programmer) succeeded in the reprogramming task, but all subjects went through the same basic steps in their understanding of the device.

One can view these subjects as engaged in *interpolation* between the structure and function of the DigiComp. The function in this case is the activity of counting in binary, and the "structure" is the physical construction of the DigiComp and its particular programming arrangement. The steps of interpolation include a variety of sorts of knowledge changes, including (but not exclusively):

- *Causal composition:* The subject observes that separate parts of the device work in unison, and he or she mentally combines these into a complex object with composite behavior. The subject then observes how this new composite part acts with others, thereby building up an extended causal chain.

- *Functional refinement:* The subject follows the physical path by which the flip-flops move and thereby refines the function of counting into terms that refer to actions of the device. Often subjects search for physical aspects (or actions) whose pattern is suggestive of activity such as counting upwards.

The aspect of our observations that is most important for the present discussion is that very often composition and refinement steps involve the introduction of new vocabulary, which the subject subsequently uses to refer to those phenomena.[5] In the case of the DigiComp, subjects introduced such terms as *blocking* (for example, the straws block the control rods from moving), *operations* (at various levels), and the *setting* and *resetting* of the flip-flops.

Interpolation between function and structure is also an accurate description of the behavior of subjects learning about the BigTrak. For instance, subjects typically introduce the concept of memory as a structural implementation of the programmable aspect of the device. However, in the case of the BigTrak, our subjects were not free to confirm their structural hypotheses.

It is interesting to consider how the task of the real scientist can be understood in these simple terms. Scientific instruments are often developed with the goal of discovering structural aspects of the system under study, and one role of data recording and analysis is to describe the activity of the system in more detail, leading to the discovery of regularities in its behavior and structure.

---

5. Sometimes actual words were not used, but the subject would make reference to "*that* action" and make hand movements, as if he or she had some specific complex action of the device in mind. Such are the difficulties that arise from the use of dynamic systems; actions are often hard to articulate.

2.3 Summary Observations from the Psychological Work

In the process of learning about the BigTrak or Digicomp, our subjects passed through a number of reformulations of their theory of the function of the device, its construction, or its mechanism. Four phenomena characterize this learning:

1. The reformulation may introduce new causal systems or subsystems into the learner's theory. For instance, once the learner discovers that the BigTrak has a memory, he begins programming the device instead of using it as a simple electronic toy.

2. The learner's interpretation of the observed and inferred physical and behavioral aspects of the system may differ as a result of the reformulation. That is, new representational principles are introduced into the learner's interpretation. We saw this in FC's vector addition interpretation of the BigTrak and in the logical interpretation of the DigiComp's physical structure.

3. The reformulation may introduce new attentional, problem-solving, or inferential skills. Once a subject understands that the BigTrak is programmable, he no longer expects it to move whenever a button is pressed. A person studying the DigiComp begins to focus on the blocked versus free movement of the rods once he understands that they play the role of sensory elements in the computational interpretation of the device.

4. All these phenomena may involve the introduction of new terms and relations into the learner's theory, and they may require reformulation to a greater or lesser extent—understanding the BigTrak as a programmable device, the DigiComp as a programmable logic array, the BigTrak's behavior in terms of vector addition, and so forth.

Reformulations often cover large parts of the learner's theory or introduce entire systems of terms and representational principles. These reformulations apparently take place with little mental effort, as contrasted with, say, the piecemeal introduction of concepts in a process of incremental design or the incremental fusion or fission of concepts, all of which are characterized by a large number of relatively simple changes. If piecemeal reformulation were involved, the learner should be considering each piece of his current theory individually, and so the reformulation should take some time. Also, in accord with Ericsson and Simon's (1984) theory of verbal report, we would expect some of these steps to be reported. On the contrary, reformulation was rapid and took place without verbal report of individual steps in the reformulation activity.

The notion of a view and the processes of view application arise in part from verbal reports in which our subjects used such terms as *resultant* and in part from the logical requirement of theory consistency. If one introduced ten independent changes into a theory, there is no reason to believe that the result would be consistent, much less that it would make better predictions. How can one make massive changes in one's knowledge and interpretation of the world and yet have his knowledge remain consistent?[6] The changes that a learner makes to his knowledge at points of reformulation can generally be expressed in terms of some coherent and self-consistent view or commonsense concept. View application reformulates one's theory in accord with a chosen view. We now turn our attention to that process.

3. View Application and Its Limitations

We have seen that theory formation can be understood, in part, as a process of incremental elaboration, refinement, and reformulation of the learner's knowledge in accord with selected commonsense concepts or "views." Wide-ranging reformulations of knowledge take place as a result of reinterpreting the system under study in terms of these self-consistent views. I refer to the process of reformulation as *view application* (Shrager, 1987). In this section, I summarize and criticize view application. Along with the results of our studies of instructionless learning in dynamic systems (such as the DigiComp), these criticisms led to the present theory of commonsense perception, which I discuss in Sections 4 and 5.

View application involves viewing the device under study in a different way—specifically, as an instance of a commonsense abstraction. It thus implements "incremental conceptual combination" between abstract schemas, called *views*, and the learner's present theory of the device. View application involves introducing new terms and new rep-

---

6. The notion of consistency here requires some clarification. It seems plausible that one's knowledge, taken "as a whole," is not and *need not be* consistent. As the person never inspects his knowledge "as a whole," only small parts need remain consistent. However, the argument here stands because any subpart selected for reasoning at a given moment must be self-consistent. Therefore, either there must be processing at selection time to check consistency or else the whole selected subpart of the subject's knowledge must be self-consistent.

resentational principles into the learner's theory of the device. These novel features are contained in the view being applied. Views are presumed to be a part of the learner's prior knowledge. The view application process takes place in two major steps: *view selection* (retrieving the relevant view) and *view combination* (reformulating the learner's present theory in accord with the selected view).[7]

The most common way that subjects select views is by noticing actions or objects in the situation that "suggest" a view—as in FC's noticing the relationship between vector addition and his observations of the Big-Trak's behavior. In FC's case, the vector addition view introduces the concept of vectors, the operation of vector addition, and some specifics, such as the addend and resultant vectors. View application involves associating parts of FC's previous theory of the device with the new view. Thus he associates the actions he thought that he had programmed with the addends and associates the resulting movement of the BigTrak with the vector resultant. The introduction of new terms and representational principles might require the reformulation of parts of the person's theory in terms of these new representations.[8] In the case of FC's vector addition theory, he had to reformulate his understanding of how the BigTrak computed movements so that the actual movement of the device was the vector resultant and so that each entered command represented a one-foot component vector, all of which were subsequently summed to form the resultant.

All schema-based interpretive theories face two fundamental difficulties. I have called these the *paradox of recognition* and the *framework alignment problem* (Shrager, 1987):

- How can an appropriate view be selected for application to a particular theory if some of the terms that compose the view are not already present in the learner's theory? If learners depend on descriptions in particular terms for view selection, they must wait for the framework given by a view in order to obtain these terms.

- How can a theory that is represented in some set of quasi-linguistic terms be *reformulated* in accord with the new selected view, which may be cast in terms that are not the same as the terms in which the

---

7. The theory of view application is similar to Piaget's (1953) process of assimilation, but whereas Piaget gave a vague and ever-shifting theory of assimilation, view application has a very specific computational model.

8. See Shrager (1987) for details of the reformulation method.

theory is cast? In the worst case, the terms composing the new view may be simply incommensurable with those composing the theory to be reformulated in accord with that view.

For example, since the vector addition view is the locus of the vector/resultant terms and representations, the subject must have noticed these terms and representations in the activity of the BigTrak *before* choosing the vector addition view. However, I claimed just above (and in Shrager, 1987) that view application itself introduced these terms and representations into the learner's theory. This is an example of the paradox of recognition.

When view application begins to reformulate the learner's current theory according to the new view, it is necessary to make representational contact between terms in the view (say, the individual vectors) and terms in the learner's current theory (say, movement commands). This step suffers from the framework alignment problem.

These difficulties stem from thinking of views and theories in terms of schematic internal knowledge in the form of "mental models" composed of quasi-linguistic propositions. These entities are generally represented in computational systems by scripts, frames, schemas, or views. The connection that must hold between the world and these structures in order that they be operational is usually ignored or relegated to the "peripheral" roles of sensation and motor action. This reliance on internal and ungrounded knowledge has resulted in overly rigid theories of mental model formation. These theories lack the ability to reinterpret experience, as *experience is nowhere to be found!*

The remainder of this chapter describes a framework for understanding theory formation and reformulation that provides the important features of view application but does not suffer from the paradox of recognition and the framework alignment problem. We will see that this requires a novel theory of conceptual representation based primarily on sensory rather than on purely propositional content. I call this *grounded representation.* The process of reformulation within this framework is a perceptual one, which I call *commonsense perception.*

4.   Grounded Representation and Commonsense Perception

I have argued that, in order to learn by a process of reformulation in accord with commonsense concepts, knowledge cast in different representational systems must be sensibly inter-utilized (combined, compared, etc). In moving toward such representational inter-utilization, I propose here *grounded representation* as a representational principle and *commonsense perception* as a replacement for view application. In this section, I outline these two aspects of my new theory of learning. In the next section, I discuss the role of commonsense perception in the formation and execution of a simulation model of gas laser physics.

4.1   Grounded Representation

The theory of grounded representation rests upon a fundamental claim:

> *Sensation and sensory experience form the basis of conceptual knowledge.*

Specifically, this theory replaces symbolic representation in other views and theories with a set of *representation-specific coordination procedures* that mediate between traces in one modality (echoic, visual, or motor traces) and traces in another modality. For instance, an expert has *skills* for identifying and often naming things that are relevant to his or her domain, as well as skills for executing appropriate actions with respect to experiences in that domain. The present theory does not distinguish identification and inference from action; these are simply skills of association between traces in one modality and another or the same modality.[9]

It is important to recognize the difference between a *modality* such as quasi-linguistic propositions, vision, imagery, or sound, and a *representation*, a specific class of structures that are cast in some particular

---

9. Science has an explicit analog to grounded knowledge: Some theoretical terms are "operationalized." This is just a procedure for coordinating theoretical entities with the observable world. The scientific theory itself, used as a predicting tool, also carries out coordination between current observations and those in the future (predictions). In physics, mathematics is the carrier of this coordination burden, and the success of a mathematically based science depends in large part on the success of mathematics in this role.

modality. For instance, an algebraic *representation* might be based in a propositional *modality* where each equation is cast as a LISP-like expression; a static spatial *representation* of real-world objects might be cast in a bitmap *modality*. Given this distinction, one can see that there can be several entirely distinct sets of "representation-specific coordination procedures" operating within or between the same modalities. Most production systems (Klahr, Langley, & Neches, 1987) operate entirely within a propositional modality, yet they represent many sorts of knowledge within that modality.

4.2   Inter-Utilization

Grounded representation supports an approach to the framework alignment problem and the paradox of recognition by giving a central cognitive role to experiences themselves via direct sensation or via sensory traces of experiences. Not only is a picture worth ten thousand words, it may be described in ten different ways at a thousand words per description. If each of these thousand-word descriptions is "grounded" on the picture, then we can compare these different descriptions with one another by reference to the picture itself. This is the basic approach to inter-utilization provided by the grounded representation framework:

> *Pieces of knowledge that are cast in different representational frameworks can be compared by understanding how they differently interpret the experiences that compose their grounding.*[10]

Grounding may be direct to sensation, but it is typically indirect through a number of intermediate representations. Section 5 discusses these points in more detail.

The grounded representation approach to the paradox of recognition is similar: As all knowledge is grounded in experience or traces of experience, the desired features for selection can be found in the experience traces themselves, and so recognition does not rely on finding terms in the learner's present theory.

---

10. This approach cannot work if the representational systems are truly and wholly incommensurable, as the definition of *incommensurability* is precisely that no such comparison can be made. However, I believe that there are no truly incommensurable yet valid ways of understanding the same experiences.

Let us return to the point in Table 1 at which FC recognizes the Big-Trak's movements as a vector resultant. Note that the triangle made by the two component arms (1a and 2a in Figure 1) and the resultant (2b) have approximately the geometry that most of us who have been through high school trigonometry associate with vector addition. Thus, the claim is that FC's knowledge of vector addition *includes* some quasi-sensory representation of this image and that it is through this representation that he came to recognize the possible use of the process of vector addition in the BigTrak's activity. The principal form of quasi-sensory representation in the present theory lies in the coordination routines. In the case of vector addition, these include the skills of finding vector heads and tails and of finding the point in the current space that is their sum.

It is less clear how perceptually grounded representation helps with the framework alignment problem at the point of view application. Consider the problem of introducing the notion of "memory" into one's theory of the BigTrak. This commonsense concept is often suggested to subjects in the BigTrak study by the observation that the device is doing "the same thing as before even though I pressed different buttons," to quote FC. The application of the memory view involves reinterpreting the function of pressing command keys on the BigTrak as storing things in the memory and the function of the GO key as reading out the contents of that memory, and executing it.

One explanation for such cases is that the representational modality of our understanding of the BigTrak is quasi-sensory and is grounded by coordination routines that act between an image and a propositional modality. However, the image is quite far away from vision itself, being some cross between deep motor representation and animated imagery. However, as I have no detailed ideas about the nature of this modality, I model this case using bitmap animations of moving schematic diagrams of the BigTrak's actions into memory slots that have spatial organization with respect to one another.

A further benefit of the coordination procedure theory of knowledge is that it provides an elegant framework for understanding attentional action and abstraction. The coordination procedures will lead the agent to attend to certain aspects of the information in the target modality that these procedures see as relevant to the representation that they serve and to ignore other aspects. For instance, in many of the instructionless learning protocols, people simply do not notice small actions of the BigTrak or DigiComp unless they are looking for them. At one point the BigTrak actually turns to the right six degrees at the end of a long forward movement, but FC seems oblivious to this movement, presumably because he was not expecting it, and so truly did not *see* it. At this point, his skill of observation resulting from his understanding of the BigTrak did not include axial turns.[11]

4.3 Commonsense Perception

I must now rebuild the mechanism of view application within the grounded representation framework. I call the new mechanism "commonsense perception" in order to emphasize that it brings commonsense knowledge to bear and that the act of interpretation is a perceptual one operating in unit time and coordinating sensation and cognition.[12]

Like view application, commonsense perception permits reinterpretation of the learner's knowledge in accord with selected abstractions. To accomplish this, as we have seen, the agent's knowledge cannot exist exclusively in quasi-linguistic terms but must be grounded in sensation or imagery via domain-specific and modality-specific coordination procedures. Thus, whereas the commonsense knowledge in view application was carried in terms of frames with slots and symbolic contents, in the present theory the "content" of commonsense knowledge arises from the *activity* of the coordination procedures themselves.

In brief, once a view is selected by the matching activity of the domain-specific coordination procedures that compose the view, the rest of the coordination procedures associated with that view are brought into action. These are included in the set of active procedures that compose one's current theory of the domain, perhaps overriding some of those

11. Presumably, a large axial turn would have been difficult to avoid noticing. Certain phenomena "pop out" of the visual field regardless of whether or not one is looking for them (Treisman & Gormican, 1988).

12. Dretske (1969) provides an excellent philosophical treatment of aspects of this phenomenon, which he calls "epistemic seeing," and it plays a central role in the thinking of Feyerabend (1975, particularly chapters 6 and 17) on scientific theory change. Feyerabend refers to this as "natural interpretation."

with whose action they conflict, but more often simply making one's cognitive machinery, as it were, *operate differently* with respect to the system under study.[13]

5.   A Computational Approach to Commonsense Perception

I have argued that many of the steps in theory change, at least during instructionless learning, can be understood as resulting from the process of commonsense perception. This process brings prior knowledge in the form of inter-modality coordination procedures to bear in labeling and drawing inferences about a domain of observation, in accord with what one can view as a commonsense concept. However, I also argued that these concepts must be grounded in sensation by inter-modal and intra-modal coordination procedures and that the process of commonsense perception primarily involves the introduction of such procedures in accord with the selected commonsense concept.

To refine this theory, I have developed a simplified computational implementation in the domain of gas laser physics. This demonstration takes the form of a computer program that forms a working understanding of laser operation and that engages in continual interpretation as it simulates the behavior of the device. This is not quite the task of theory formation that was carried out by view application. I have not yet been able to accomplish the full-blown reformulations required for instructionless learning. However, the cognitive operations taking part in this simulation of laser activity are very close to those required for theory formation. Specifically, we will see that reinterpretation is required in this task as well. It is accomplished by the introduction and action of domain-specific procedures that operate to coordinate an iconic memory (a visual modality) with a quasi-linguistic memory (a propositional modality).

---

13. I am consciously using action terms such as *operate* and *procedure* to emphasize that commonsense knowledge is carried primarily in the *skill* of inter-modality coordination, and not in fixed representations. In the next section, I retreat from this assumption for ease of implementation.

5.1 Purpose and General Procedure

The purpose of the example implementation is to model the way that a student might think about the activity inside a laser cavity. To do this, processes must be invoked when they are appropriate for the situation in the simulated laser cavity, and traces of the simulated activity must be retained and reexamined for explanation. Both of these tasks require the sort of reformulation capacity that I have argued is the hallmark of commonsense perception.

Figure 3a shows a cutaway view of a laser cavity with mirrors (M) at either end and containing two molecules, one in an excited state (left) and one in its ground state (right). In this simplified operation of a laser, several processes play a role:

- *Absorption:* A ground state molecule absorbs a photon that passes close to it. Thus, the left (excited) molecule must have absorbed a photon, presumably emitted by an unseen source.

- *Spontaneous emission:* Laser activity begins when an excited molecule spontaneously emits a photon toward one of the mirrors. The left molecule could now do this. In fact, spontaneous emission is taking place all the time, but most of the emitted photons do not happen to travel in a direction that is along the main axis of the cavity; these errant photons are absorbed by the walls of the cavity.

- *Reflection:* Photons can be reflected by the mirrors in the obvious way. A photon that happens to be aligned directly along the main axis of the laser cavity will bounce back and forth between the mirrors.

- *Stimulated emission:* This is the fundamental process in laser operation. When a photon passes close to an *excited* molecule—one that has presumably previously absorbed a photon—the excited molecule is *stimulated* to emit its energy in the form of a new photon. This new photon is *coherent* with the incoming one: They are in the same phase and spatial location, and they are traveling in the same direction. The result of this stimulation is two photons for the price of one, so to speak.

One can now see how lasing takes place: Energy from the external source "pumps up" the molecules in the cavity (absorption). Once one of them happens to emit a photon (spontaneous emission) into the mirror, the photon begins to bounce back and forth (reflection). As it

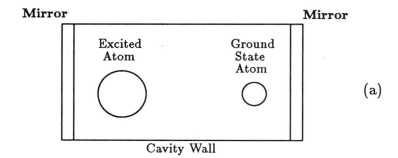

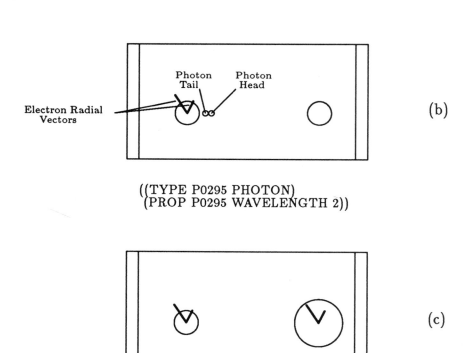

Figure 3. Three stages in the operation of the laser model. (a) The initial state, in which the left atom is excited and the right atom is in its ground state. (b) After spontaneous emission (at $t = 0$) a photon has been emitted from the left atom, leaving the atom in its ground state. Information describing the photon has been entered into propositional memory by the emission process. (c) After absorption of the photon by the right atom (at $t = 19$). (Not drawn to scale of the actual inconic memory in the model.)

passes other excited molecules, it will collect their energy (stimulated emission), and thus the beam will become more energetic and will collect more photons to continue traveling back and forth between the mirrors. A "photon avalanche" thus ensues, as if each pumped molecule were a rock, teetering on the edge. When a rock (photon) from above (going along in the cavity) strikes a teetering one (an excited molecule), the lower one is dislodged (stimulated emission), and they both continue down the mountain (bouncing back and forth in the cavity).

In this way, lasers collect and focus much of the energy that is poured into them from the outside. The laser beam that comes out of the cavity usually represents a small fraction of the energy of the entire beam. Some of the energy is allowed to escape by constructing the laser with one partial mirror. For instance, five percent of the photons might escape from the end of the cavity to form the familiar laser beam. Of course, there are many subtleties to real laser operation, but the foregoing explanation is enough to understand the present model.

5.2   Grounded Knowledge and Working Memories

The laser simulation system consists of a set of working memories and a database of commonsense concepts representing the views to be applied. Both of these include coordination procedures that can operate within and between the working memories.

There are two working memories, an *iconic store* containing a 100×100 two-dimensional array of label lists and a *propositional store* containing quasi-linguistic propositions represented as LISP lists with no recursive substructure.[14] The contents of the iconic store are lists of tokens representing each object that is understood to be "in" that location of the image. I generally refer to such two-dimensional labeled arrays as "images." Each frame of Figure 3 shows a state of the iconic and propositional memories.

Each view may be composed of three parts: a sequence of images; quasi-linguistic propositional content, such as might compose production-system rules; and coordination procedures that operate within and

---

14. There are also a number of other atheoretical temporary memories and the usual assortment of computational appendages required to make an implementation work.

between the image and propositional memories. If the system had a true vision component, coordination procedures could use this component to operate between the world and either of these memories.

The coordination procedures are arbitrary LISP procedures that can read from and write into either memory. They may be conceptually divided into four classes: *finders*, which read from the iconic store and write into the propositional store; *generators*, which read from the propositional store and write into the iconic store; *animators*, which work entirely within the iconic store; and *inference rules*, which operate entirely in the propositional store.[15]

For instance, a "molecule finder" that simply locates circles in the bitmap image would process Figure 3 and enter into the propositional store that there are two molecules, giving each a generated name (e.g., MOLECULE-A and MOLECULE-B) and the type labels MOLECULE. Finders for higher-level relations such as LEFT-OF might process the information delivered up by the molecule finder, along with the contents of the iconic store, and report that MOLECULE-A is LEFT-OF MOLECULE-B.[16] Generators are the inverse of finders. Combining the generator for MOLECULE and LEFT-OF, and guided by the propositional store, one can generate an image of one molecule to the left of another in the iconic store. The more information in the propositional memory, the more precise an image the generation procedure will produce.

Together, the coordination procedures and the iconic and propositional memories form the grounded working knowledge of the laser model; the contents of the iconic and propositional working memories form the grounded momentary state of knowledge that the system has about the laser.

The most important simplification from the theory of grounded knowledge to the present implementation is that some of the knowledge that should theoretically be represented by animators and inference procedures is actually represented by sequences of labeled arrays and in quasi-

---

15. The inference rules are really just a kind of finder, and the animators are a kind of generator. In fact, for the sake of efficiency, these various procedures are mostly implemented by cached input-output values that do the work of the procedures without having to compute the relation every time. Thus, they look more like production rules than procedures.

16. Objects that lie across temporal sequences are located by compressing the sequence into a single frame. This method was also proposed by Langacker (1987).

linguistic preconditions and postconditions. More details and an example of this are provided below.

5.3 Specific Views for the Laser Model

The coordination procedures that compose a particular view are specific to the domain of that view.[17] For the present simulation of the operation of the laser, I have provided the model with three sorts of knowledge: *process views*, containing animations (bitmap sequences), propositional preconditions and postconditions, and coordination procedures to stand between the animations and propositional information; and *object types*, consisting of static images and procedures for locating things of the appropriate type (such as molecules, mirrors, and photons) in the iconic store.

In the present model, the most interesting contents of the knowledge base involve the relationship between the energy carried in molecules of the lasing medium and the energy carried in photons. The energy of a molecule is understood (by the model) as the radius of the electron orbital of the images that are interpreted as molecules of the lasing medium. These orbitals are located by finders that look for circles labeled as composed of electrons, that is, whose label is "electron" in the iconic store. The energy of a photon is a numerical parameter, but it is understood as the wavelength of the photon, where the wavelength is literally a numerical length in image-space units. This length is computed by the spontaneous emission view by carrying out image vector computations. The finder for wavelength literally puts vectors into the image, one for the radius of the orbital before emission and one for the radius of the orbital after emission. It then *visually* aligns their tails and finally *visually* subtracts them to derive a value for wavelength in image-space units. This value is associated with the emitted photon, which is also represented in the image as a spatially extended vector in order to maintain its direction and wavelength. This association is recorded in the propositional store. Once computed, the wavelength can be used in a variety of visual computations, such as determining whether a particular molecule can absorb this photon or, conversely, whether stimulated

---

17. Presumably, one initially composes complex procedures from a primitive set, perhaps by internalizing imitated attentional procedures, guided by one's parents, peers, and teachers.

emission will take place. These computations are carried out by *visually* comparing the photon's wavelength with the vectors representing the permissible electron orbitals for the target atom.

The astute reader will note that the model's present "knowledge" about quantum physics is *not formally correct*; it has a confusion between wavelength and frequency. I have modeled wavelength as directly proportional to the change in the molecule's energy, whereas in reality wavelength is *inversely* proportional, and frequency is directly proportional. However, in ongoing studies of human subjects learning about lasers, this particular confusion arises regularly. Part of the present theory that I do not have space to describe here involves the origin of the views that are used in the present model. Briefly put, the diagrams that a student studies while learning about lasers structure his or her views relevant to the laser, and these diagrams can be easily misinterpreted. (For a short description of these studies, see Shrager, 1989.)

5.4   The Interpretive Cycle

The processing cycle of the system is similar to that of Forbus' (1988) Qualitative Process Engine. First the system locates all the process views that can apply to the present whole state of the laser, as represented in the working memories. Next it instantiates one or more of the possible process views by finding or constructing appropriate objects in memory to associate with the abstract objects in the process view. Finally, it computes the implications of the selected and instantiated process view. It is important to observe that in the present model *each processing step involves reinterpretation of the system's working memories with respect to each possible or selected view.*

A process view is composed of a grounded animation sequence, a propositional sequence, and coordination procedures (finders and generators). The first image and its attachments constitute the preconditions for application of the process. The first of these images is matched against the iconic memory, and the sequence is then "executed" by successively installing the next image. The same takes place in the propositional memory. Thus, instantiation involves associating the terms in the view image with matched terms in the iconic and propositional stores and then "running" the process by replacing the first image with the second, the second with the third, and so on until all the steps in the process view have been run.

In the laser model, matching and execution are done in an alternative "dynamic" image (also a 100×100 cell map). At each step the possible matching individuals are copied onto the dynamic image from the iconic store, and the match takes place between the first image of the process and the entire contents of the dynamic image.[18] This gives all the possible process instances. After each step, changes are merged back into the iconic store. This process can be thought of as "mental animation," in which one imagines something moving across a static background. The background is the iconic store, and the action takes place in the dynamic store.

After each step of a multi-step process is taken, the match and instantiation cycle repeats so that other processes that become viable during the execution of an ongoing process have an opportunity to apply. A conflict-resolution mechanism decides which processes to apply if more than one can take place at a particular point or if some new process tries to interrupt a running one. In most cases, either these conflict-resolution decisions are made manually or they all apply, thus searching the whole space of possible process instantiations. This search is assisted by a truth maintenance system, which records changes in the working memories due to process applications and conflict-resolution decisions.

5.5 Behavior and Functionality

Because much of the action of the program involves sequential animations and the operation of finders and generators between the symbolic and iconic memories, it is difficult to describe its behavior. An enormous amount of information is carried in the iconic and propositional memories as they are justified databases that record all their history as well as their present state. Table 2 shows a trace of a part of the model's activity, simplified for presentation here. It describes Figure 3, which shows what is taking place in the iconic memory during the stages of this trace.

Processes are recognized by image matching and process sequences are run (lines prefixed by "→" in Table 2). Simultaneously, when proposi-

---

18. This seemingly tedious process is done very quickly by special pattern-matching code due to Bernard Mont-Reynaud of Stanford University. Some post-match checking is required, but in general the match runs in time on order of the number of filled cells in the view's pattern image.

tional information is involved, assertions are made in the propositional working memory (lines prefixed by "Asserting" in Table 2). In this case I have only shown one such interaction, in which the "wavelength" of the photon (P0295) ejected from the spontaneous-emission process, which takes place at $t = 0$, 1, and 2, is computed *from the image* by computing a vector that is the difference of the initial excitation diameter of the electron shell of the emitting molecule and the final excitation diameter. The shell drops in diameter as it gives off energy in the form of this photon.

Table 2. A partial trace of the laser model in operation. Lines beginning with an arrow indicate instantiation of an image. Lines beginning with "Asserting" indicate instantiation of propositional content. The first line ("Running...") indicates that the spontaneous emission process was instantiated at this point.

```
Running ((30 . 30) . #seqw:SPONTANEOUS-EMISSION).
--> At 0 :: Image 0 of #seqw:SPONTANEOUS-EMISSION (30, 30).
Asserting Bind INITIAL-EXCITATION-STATE for this process as
         the vector (VECTOR-291) given by Head: HEAD-VECTOR-291
         and Body: BODY-VECTOR-291.
--> At 1 :: Image 1 of #seqw:SPONTANEOUS-EMISSION (30, 30).
Asserting Bind FINAL-EXCITATION-STATE for this process as
         the vector (VECTOR-300) given by Head: HEAD-VECTOR-300
         and Body: BODY-VECTOR-300.
Asserting Assign the property of WAVELENGTH for the photon
P0295 as 2 image units.
--> At 2 :: Image 2 of #seqw:SPONTANEOUS-EMISSION (30, 30).
--> At 3 :: Image 1 of #seqw:PHOTON-MOVEMENT (36, 34).
  . . .
--> At 18 :: Image 1 of #seqw:PHOTON-MOVEMENT (51, 34).
--> At 19 :: Image 0 of #seqw:STIMULATED-ABSORPTION (50, 30).
--> At 20 :: Image 0 of #seqw:SPONTANEOUS-EMISSION (50, 30).
--> At 21 :: Image 1 of #seqw:SPONTANEOUS-EMISSION (50, 30).
--> At 22 :: Image 0 of #seqw:PHOTON-MOVEMENT (56, 34).
  . . .
--> At 34 :: Image 0 of #seqw:REFLECTION (68, 34).
--> At 35 :: Image 1 of #seqw:REFLECTION (68, 34).
--> At 36 :: Image 1 of #seqw:PHOTON-LEFT-MOVEMENT (67, 34).
  . . .
```

I have constructed a simple back-chaining reasoner that can use the laser model to predict which combinations of gases will lase and which will not. For example, suppose that we want to know what gas will produce blue laser light. The term *blue* is grounded by a generator as a constraint on the vectors that represent the wavelengths that the laser model uses when it constructs photons. The reasoner looks back through the trace left by the simulation and computes which state transitions must occur in the atoms of the gas to produce the required wavelengths. This computation is done with the vector/image representation of the present model.

5.6   Summary

This section has briefly described the structure and operation of a new kind of qualitative representation, based on the theory of grounded representation, and a new kind of qualitative simulation, whose action arises primarily by commonsense perception—the active and ongoing interpretation of the contents of the agent's working knowledge of the system under study.

Because of the inter-modality grounding of propositional knowledge and the procedural nature of the interpretive process, this model is not as sensitive to the paradox of recognition and the framework alignment problem as view application. This insensitivity is evident from the fact that the system can determine which processes should be activated without having to program the relevant terms in a propositional database—it initially processes just the image and from this determines the dynamics of the laser.[19] Furthermore, the system can notice serendipitous interactions, such as when a photon encounters another molecule and absorption takes place or when a photon encounters a mirror and is reflected. These interactions, noticed in the iconic store, are very difficult to model in an entirely propositional representational system. The iconic store essentially provides a fine-grained representational modality upon which the system can base its interpretive activity.

---

19. I have cheated somewhat by labeling the items in the image, such as the electron orbital. This would seem to use terms that would reintroduce the paradox of recognition. However, this is merely a computational shorthand for a spatially organized object-type memory. The labels play no role in the match that takes place in the image stores. However, certain human reasoning errors could be attributed to failures of this association, leading to recognition slips. The present model is not able to model such difficulties because of the use of type labels.

The model I have described only approximates an actual implementation of commonsense perception and grounded representation. For instance, it does not form the process views that are used in the simulation. However, I think that it is very important to recognize commonsense perception as taking place *all the time in almost all intelligent activity.* I intend to extend the model to acquire an initial understanding of the laser's activity, just as the reader presumably did in reading Section 5.1.

6.   Discussion

In this chapter I have moved from studies of "intuitive" theory formation to a theory of commonsense perception that relies on a grounded representational system. I have demonstrated how the theory can be instantiated in a "mental model" of how one might understand laser physics where articulate and imaginal knowledge operate hand-in-hand. In this section, I compare the present theory with similar cognitive models of complex learning and reasoning and then return to the domain of science, asking how the present theory relates to explicit theory formation, instrument construction, and experimentation.

6.1   Related Approaches

Theory formation, at least in the cases we have studied, is an ongoing activity of interpolation between a system's function and its structure. Commonsense perception is a cognitive mechanism that is central to this activity, enabling the agent to reformulate its working knowledge of the domain (or the state of the system, in the case of the laser model). Such reformulations are required for structure/function interpolation.

The theory of grounded representation is very close to that proposed by Pavio (1971, 1986) in his "dual coding" approach, but there are significant differences. The present theory emphasizes the procedural nature of the inter-modality relationships, whereas Pavio refers primarily to associations between static structures. Also, I am trying to do away entirely with "symbolic" representation, replacing it with what amount to actual verbal traces (quasi-linguistic propositions), whereas specifically symbolic representations are one of Pavio's two codes. Furthermore, my coordination procedures can act between arbitrary modalities (similar to Pavio's codes), so my theory is not restricted to only two codes. This makes the theory more flexible, but also weaker from the standpoint of prediction.

The present approach is also close to Langacker's (1987) theory of "cognitive grammar," which posits a mental system of static and sequential imagelike schemas that people use to form complex scenes as a means of understanding natural language terms and phrases. Our theories differ primarily in the procedural nature of my representational components; Langacker has no analog for coordination procedures. As a result, Langacker's theory should encounter the very same problems that led me away from the purely propositional schematic structures of view application and to the active grounded representational content of commonsense perception.

Finally, Clement (1989) has proposed a theory of expert scientific problem-solving knowledge that rests on "relations between physical intuitions, perceptual-motor schemas, visual imagery, kinesthetic imagery, and simulations." In Clement's theory, a person joins perceptual-motor schemas and imagery to produce imagistic simulations. Furthermore, Clement attributes physical intuitions to the expectations in perceptual-motor schemas. The principles of grounded representation that I have outlined are very similar to Clement's principles of representation, although he has no computational implementation nor any process analogous to commonsense perception. Presumably such a process will be needed eventually to make his theory operational.

6.2 Scientific Knowledge and Commonsense Perception

A central task of science is to make sense of the environment, to find a pleasing and predictive interpretation of the natural world. However, a scientific theory is an artifact, like a finely tuned musical instrument in which all the notes play in harmony, and it certainly requires effort to tune a theory appropriately. We generally think of this tuning process as a problem of theory formation rather than as being similar to the relatively effortless construction of natural interpretations that we saw in instructionless learning and that occurs in our moment-to-moment daily lives (see Heritage, 1984). However, I believe that an understanding of commonsense perception is critical to understanding the process of theory formation.

I have already mentioned that the process of interpolation between structure and function is central to instructionless learning. Instrument construction is a method of refining our descriptions of either the function or the structure of a system.[20] We can view the task of science as maintaining *alignment* between our articulate knowledge and the world. A theory can be said to be properly aligned when the aspects of it that are observed by our instruments accord with the theory's predictions. If we discover that the theory is not well aligned, then something must change to bring the theory into accord with observation. There are many ways that this can be done, ranging from abandoning the theory to abandoning the instrument. Between these extremes lies actual scientific practice.

Grounded representation is the way that our knowledge is aligned with the world, and commonsense perception coordinates this alignment. In science, we use the terms *operationalization* and *operational definition* to indicate this relation, and instruments mediate the alignment. The meaning of terms and relations such as *energy, mass, moment, up, before,* and *point,* which seem to compose our functioning knowledge and theories, are in part based on their foundational content. On the other hand, researchers studying conceptual structure and acquisition have hypothesized that seemingly foundational knowledge is structured according to implicit theories (e.g., Murphy & Medin, 1985) and that knowledge change is a process akin to theory change (Carey, 1985). From the viewpoint of learning, theory formulation and reformulation go hand in hand with the formulation and reformulation of the foundational terms from which functioning theories are composed. In a sense, the physical concept of energy was developed to underwrite classical mechanics. However, terms are rarely coined afresh without a prior popular meaning. Thus, our prior knowledge plays a part in our use and understanding of the "real" nature of energy, aside from its description in physics textbooks. This is even more apparent for common terms such as *near, before, location,* and *movement.* If the process of commonsense perception serves to bring conceptual knowledge to bear in interpretation, it must also play a role in the interpretation of the experiences we use to give meaning to theoretical terms. Thus,

---

20. Recording devices generally address function, whereas probing instruments like the microscope generally address structure, although this is perhaps incidental. Recently Clark and Pavio (1989) have noted the analogy between the dual coding theory and scientific operationalizations.

an understanding of science requires an understanding of commonsense perception.[21]

Another way that commonsense perception can play an important role in learning about physical systems is the selection and inclusion of specific "phenomenological primitives" (DiSessa, 1983) represented by views. These are experientially learned combinations of analog and articulate information that capture the dynamics of physical or presumed physical processes. Thus, the notion of perceptually grounded commonsense knowledge, containing both articulate and sensory aspects, can enrich our theoretical understanding of qualitative models and may help to unify several parts of the literature on qualitative understanding of physical systems (DiSessa, 1983; Forbus, 1985; Larkin & Simon, 1987; McCloskey, 1983).

People construct interpretations from moment to moment in their daily lives, explicitly, as when they try to figure out a complex device (Bott, 1974; Doyle, 1988; Shrager & Klahr, 1986); and implicitly, as when they meet a new person and behave toward him or her in accord with the situation (see the work of Garfinkel and Schutz, summarized by Heritage, 1984; Nisbett & Ross, 1980) and interpret natural utterances (Collins, Brown, & Larkin, 1980; Schank & Abelson, 1977; Smith, Osherson, Rips, & Keane, 1988) and instruction (Kieras & Bovair, 1984). The classic work of Heider and Simmell (1944) demonstrated that people quite rapidly and easily form intricate interpretations of fairly simple systems.

In this chapter, I have shown that theory formation, as well as other complex cognitive activities, require us to reinterpret our knowledge of the environment. I have argued that conceptual activity is sense-grounded, and I have shown how models stated within this framework escape the paradox of recognition and the framework alignment problem, which are inherent in nonperceptual models of learning and reasoning. Finally, I have outlined a new mechanism called commonsense perception that I argue functions to *continuously* interpret knowledge and experience in commonsense terms.

---

21. This point was made forcefully by Feyerabend (1975, especially chapter 17). One might view the current work as a computational account of Dretske's (1969) and Feyerabend's ideas.

Acknowledgements

John Batali, David Chapman, Mary Hegarty, Pat Langley, Paul Thagard, and Ryan Tweney provided constructive comments on this chapter. Many others, especially Phil Agre, Mike Dixon, Ann Faye, David Klahr, Johan de Kleer, and Dan Weld have helped me to sort out commonsense perception. Special thanks go to Sharon Lunt, who patiently explained lasers to me.

References

Anderson, J. R. (1983). *The architecture of cognition*. Cambridge, MA: Harvard University Press.

Carey, S. (1985). *Conceptual change in childhood*. Cambridge, MA: MIT Press.

Clark, J. M., & Pavio, A. (1989). Observational and theoretical terms in psychology: A cognitive perspective on scientific language. *American Psychologist*, *44*, 500–512.

Clement, J. (1989). Use of physical intuition in expert problem solving. In D. Tirosch & S. Strauss (Eds.), *Explicit and implicit knowledge in education*. Norwood, NJ: Ablex Publishers.

Collins, A., Brown, J. S., & Larkin, K. (1980). Inference in text understanding. In R. J. Rand, B. C. Bruce, & W. F. Brewer (Eds.), *Theoretical issues in reading comprehension*. Hillsdale, NJ: Lawrence Erlbaum.

DiSessa, A. (1983). Phenomenology and the evolution of intuition. In D. Gentner & A. Stevens (Eds.), *Mental models*. Hillsdale, NJ: Lawrence Erlbaum.

Dretske, F. (1969). *Seeing and knowing*. Chicago, IL: University of Chicago Press.

Doyle, R. J. (1988). *Hypothesizing device mechanisms: Opening up the black box*. Doctoral dissertation, Laboratory for Artificial Intelligence, Massachussetts Institutue of Technology, Cambridge, MA.

Ericsson, A., & Simon, H. A. (1984). *Protocol analysis*. Cambridge, MA: MIT Press.

Feyerabend, P. (1975). *Against method*. London: Verso.

Forbus, K. (1985). Qualitative process theory. In D. Bobrow (Ed.), *Qualitative reasoning about physical systems.* Cambridge, MA: MIT Press.

Forbus, K. (1988). QPE: Using assumption-based truth maintenance for qualitative simulation. *Artificial Intelligence in Engineering, 3,* 200–215.

Heider, F., & Simmell (1944). An experimental study of apparent behavior. *Journal of Psychology, 57,* 243–259.

Heritage, J. (1984). *Garfinkel and ethnomethodology.* Cambridge, England: Polity Press.

Kieras, D. E., & Bovair, S. (1984). The role of a mental model in learning to operate a device. *Cognitive Science, 8,* 255–273.

Klahr, D., Langley, P., & Neches, R. (Eds.) (1987). *Production system models of learning and development.* Cambridge, MA: MIT Press.

Kosslyn, S. (1980). *Image and mind.* Cambridge, MA: Harvard University Press.

Langacker, R. W. (1987). *Foundations of cognitive grammar* (Vol. 1). Stanford, CA: Stanford University Press.

Laird, J. E., Rosenbloom, P. S., & Newell, A. N. (1986). Chunking in SOAR: The anatomy of a general learning mechanism. *Machine Learning, 1,* 11–46.

Larkin, J., & Simon, H. A. (1987). Why a diagram is (sometimes) worth ten thousand words. *Cognitive Science, 11,* 65–100.

McCloskey, M. (1983). Naive theories of motion. In D. Gentner & A. Stevens (Eds.), *Mental models.* Hillsdale, NJ: Lawrence Erlbaum.

Miller, A. I. (1986). *Imagery in scientific thought.* Cambridge, MA: MIT Press.

Murphy, G. L., & Medin, D. L. (1985). The role of theories in conceptual coherence. *Psychological Review, 92,* 289–316.

Nisbett, R. E., & Ross, L. (1980). *Human inference: Strategies and shortcomings of social judgment.* Englewood Cliffs, NJ: Prentice-Hall.

Pavio, A. (1971). *Imagery and verbal processes.* New York: Holt, Rinehart & Winston.

Pavio, A. (1986). *Mental representations: A dual coding approach.* Oxford: Oxford University Press.

Piaget, J. (1953). *The origins of intelligence in children.* New York: International Universities Press.

Ross, D. (1977). *Aristotle* (rev. ed.). London: Methuen.

Schank, R., & Abelson, R. P. (1977). *Scripts, plans, goals, and understanding.* Hillsdale, NJ: Lawrence Erlbaum.

Shrager, J. (1987). Theory change via view application in instructionless learning. *Machine Learning, 2,* 247–276.

Shrager, J. (1989). Reinterpretation and the perceptual microstructure of conceptual knowledge: Cognition considered as a perceptual skill. *Proceedings of the Eleventh Annual Conference of the Cognitive Science Society* (pp. 876–883). Ann Arbor, MI: Lawrence Erlbaum.

Shrager, J., & Klahr, D. K. (1983). Learning in an instructionless environment: Observations and analysis. *Proceedings of the National Conference on Computer Human Interaction.* Boston, MA: Association for Computing Machinery.

Shrager J., & Klahr, D. K. (1986). Instructionless learning about a complex device: The paradigm and observations. *International Journal of Man-Machine Studies, 25,* 153–189.

Smith, E. E., Osherson, D. N., Rips, L. J., & Keane, M. (1988). Combining prototypes: A selective modification model. *Cognitive Science, 12,* 485–528.

Treisman, A., & Gormican, S. (1988). Feature analysis in early vision: Evidence from search asymmetries. *Psychological Review, 95,* 15–48.

Ullman, S. (1984). Visual routines. In S. Pinker (Ed.), *Visual cognition.* Cambridge, MA: MIT Press.

CHAPTER 15

Five Questions for Computationalists

RYAN D. TWENEY

1. Introduction

In just a very few years, computational models of scientific thinking have grown from a few vaguely sketched proposals to a large number of carefully specified, powerful systems, each of which is constantly being modified, expanded, and integrated with yet other systems. There would be room for self-satisfaction were it not the case that everyone in the field seems too busy for anything except the briefest of backward looks.

I am not a computationalist, but I am something of a professional "backward-looker," my largest concern for the last ten years having been with the cognitive history of science. Perhaps I can contribute to the goals of the present volume by providing a bit of perspective. In effect, I would like to posit myself as a "consumer" of computational models. I cannot really provide a "consumer's guide," as such, since that implies a review of specific systems, their advantages, disadvantages, and costs. What I can do, however, is provide a general view of certain issues, posed in the form of questions, that the consuming public will want to see addressed. These issues must be dealt with before any specific model will be taken as a reasonable theoretical account of the nature of scientific thinking. In generating these questions, I have deliberately included some that I think are beyond the capabilities of any present system. In at least one case (that concerning dynamic imagery, discussed below), I do not see how current computational models can address the issue. I include it nonetheless, on the view that we had better be aware of what our models cannot do. Perhaps also (and even better), I will be shown wrong. To place the questions in context, I will first briefly review my own concern with scientific thinking.

I became interested in the psychological nature of scientific thinking through two of my colleagues, Michael Doherty and Clifford Mynatt. In the mid 1970s, they became interested in the general issue of whether Kuhn's (1962) notion of a paradigm shift could be studied in the laboratory. I joined in their discussions as they began to develop computer software for a series of "artificial universe" studies of scientific inference (e.g., Mynatt, Doherty, & Tweney, 1978). In these studies, people were asked to determine the laws that governed "particle motion" in a non-Newtonian "universe" consisting of shapes displayed on a computer screen. Subjects were able to locate a "particle gun" anywhere within the universe and could then fire particles in any direction, observing the resulting motions. They were free to conduct such experiments over quite long periods of time. Computerized records of their activities and think-aloud protocols were gathered for analysis. In effect, we were using human subjects to simulate some aspects of the task facing the scientist. Most of the subjects were college students, though a few active research scientists also participated.

The studies generated a wealth of data and led to some interesting results. The most general finding had less to do with Kuhn than with Popper (1959); we found that subjects who employed a strategy of trying to disconfirm their own hypotheses generally performed very badly in terms of the completeness and accuracy of the laws they discovered. This result seemed to us, and to many others, to be inconsistent with the Popperian injunction to "seek disconfirmation" and to avoid "confirmation bias." Naturally, subjects who manifested a strategy of seeking only confirmation fared very badly—the surprise was that subjects who relied on disconfirmation also did badly. Who then fared well? Only those subjects who focused their early efforts on seeking confirmation and only later switched to deliberately seeking disconfirmation. At the time, it appeared that we had stumbled across a much richer, more complex heuristic for scientific thinking than was generally described in the psychological literature on problem solving.

In the intervening years, studies of confirmation bias have multiplied. It is now clear that it is a complex phenomenon, and I will not review it here.[1] What counts is that a logically simple, very precise formulation

---

1. See Tweney and Doherty (1983) for a broader review, and Evans (in press).

of a constraint on scientific thinking, namely, "seeking disconfirmation is more fruitful than seeking confirmation," turns out, in the psychological real world, to be overly simple. We must think instead of a higher-order heuristic, "confirm early; disconfirm late," a heuristic that guides the use of other heuristics. Put in these terms, it is not so strange.

By the late 1970s, my colleagues and I had completed a very lengthy literature search for work that could be considered to represent the "psychology of science." We published an anthology of some of the best of what we found (Tweney, Doherty, & Mynatt, 1981), and we included extensive interpretive and introductory comments. There was relatively little empirical work in our book and even less that was computational. Simon (1966) is represented, and the suggestive pre-BACON computer simulations of Gerwin and Newsted (1966) and of Huesmann and Cheng (1973). But there is plenty in the book that suggests a close convergence between historians of science and psychologists of science. This convergence seems natural when one realizes that it is precisely on the cognitive level that history has the most to offer psychology, and vice versa. Many of the contributors to the present volume have long realized that historians of science have been providing us with detailed case studies of scientific thinking that provide a rich resource for the cognitive scientist.

In the late 1970s, I came across what still strikes me as an ideal source of testing any cognitive account of science, namely, the lengthy detailed laboratory diaries of Michael Faraday (1791–1867), a British physicist best known for his discovery of electromagnetic induction in 1831. For more than 40 years, Faraday kept almost daily records of his laboratory researches and theoretical ideas, part of which has been published (Faraday, 1932–1936). When I discovered the diaries, they seemed to me a natural place to explore the generalizability of our laboratory simulations. To make a long story short, the hunch was a good one. In particular, there was evidence that Faraday manifested some of the same complexities in his use of confirmation and disconfirmation that we had observed in our artificial universe studies. It was clear that no simple characterization of his research as merely confirmatory or merely disconfirmatory would be adequate (Tweney, 1985). Beyond that, there was a good deal in the diaries that was directly relevant to cognitive science on other levels as well. I have explored some of this (Tweney, 1989), and Gooding (in press) has elaborated on other aspects.

Since 1980, I have worked with Faraday's diaries and some of his other unpublished materials. If nothing else, the experience has forced me to consider the prerequisites for making justifiable cognitive claims with historical data and for justifying cognitive science to historians. Furthermore, as an experimentally-oriented cognitive psychologist trying to make sense of issues in the history of science, I have had to think about our precise goals in seeking a cognitive science of science (Tweney, 1989). These considerations underlie the specific questions that I address to computationalists in the remainder of this chapter.

2.   The Five Questions

Now we can turn to the questions. Each is posed as a query to any computational model that claims to capture at least some aspects of scientific thought.

2.1   Can the Model Handle Higher-Order Heuristics?

This question is an extension of the issue raised earlier and stems from my belief that the "early-late" heuristic is a powerful, sophisticated one. Exactly why it is powerful seems fairly clear. In our earliest work with the artificial universe, when we looked at the performance of those subjects who fared poorly, it became clear that those who relied on seeking disconfirmatory evidence were prematurely rejecting promising hypotheses. In effect, promising but unrefined ideas were being rejected because they did not, initially, account for all the available data. Likewise, those who sought only confirmatory data seemed always able to find it, no matter how weak their hypotheses, particularly if they were willing to ignore disconfirmatory evidence ("Well, that doesn't fit, but I am not going to worry about it.") Only those who ignored disconfirming data at the outset and allowed their initial hypotheses to become established as fitting at least some aspects of the available data were in a position to later refine those hypotheses, and to seek disconfirmation of a sort that would allow them to either reject a poor hypothesis or refine a promising one.

Here, as I mentioned earlier, our most effective subjects mirrored Faraday's procedure. Over and over again in the diaries, we see him blatantly ignoring disconfirmatory results *until* he is fairly sure he has a strongly supported hypothesis. Then, and sometimes the switch is quite sudden, he turns his attention to the question of possible artifacts or alternative hypotheses and designs experiments that pit his hypothesis against a range of potential alternatives. Some of his most cherished hypotheses went down in flames as a result; those that did not were published.

One important consideration that emerges from this analysis is that models of scientific discovery need to pay attention to the need for organizing *multiple* experiments. Faraday never discovered anything with a single experiment, and neither, I suspect, did anyone else. Experimentation is a powerful way of gaining knowledge, but the unreality of our methodological textbooks must be acknowledged. In general, we do not gain much from the textbook method of "testing" a hypothesis by deriving a prediction, then running an experiment to confirm or falsify it. Faraday conducted 134 experiments following his discovery of electromagnetic induction in 1831 before he felt ready to publish his results (Faraday, 1832). Thus, insofar as our computational models deal with experimentation, they must be able to organize a series of experiments if they are to be of interest, as some do (e.g., Kulkarni & Simon, 1988; this volume).

2.2   Can the Model Handle Chancy Interactions?

My earliest forays into Faraday's diaries immediately confirmed something that Gruber (1974) had observed in Darwin's notebooks; both Faraday and Darwin worked simultaneously on a variety of problems. One's first inclination (since none of us can really think of two things at once) is to postulate some kind of "time sharing." For example, in 1831 Faraday worked on the problem of accounting for "acoustical figures," the regular patterns that a fine powder will assume when sprinkled on a clamped plate made to vibrate with a violin bow. He completed these studies in the summer and immediately thereafter, in August of 1831, made his famous discovery of electromagnetic induction. It is easy but wrong to imagine him, having completed one task, as "closing a file" and "opening a new file."

As I have shown elsewhere (Tweney, 1985; also Williams, 1965), Faraday's induction experiments were directly influenced by his acoustical research. Far from being independent tasks, aspects of the seemingly unrelated acoustic experiments appeared in the electricity experiments. In fact, all of Faraday's work in 1831 seems permeated by what was then a new element in his research, namely, a strong tendency to seek

out transient phenomena, the effects of which could be made visible
by repeating them over and over again. For the bowed metal plate,
this was achieved by the periodic vibrations, and for induction it was
achieved by running a current through a coil. Gruber argued strongly
for a similar interpretation of seemingly disparate problems in Darwin's
work. For Gruber, such "chancy interactions" form the groundwork of
creative endeavors (see also Wallace & Gruber, 1989).

Clearly time-sharing is the wrong metaphor to use in such cases. If I
run an analysis of variance on a mainframe computer, I do not want any
chance interaction with another user's regression program! In modeling
scientific discovery, however, things may need to be different. Nearly all
discovery systems sharply segregate one area of endeavor from another.
However, this is not an essential characteristic of a computational model.
In fact, it may be that one real advantage of a connectionist system,
such as that used by Thagard (this volume), may be precisely its ability
to handle chance interactions. Falkenhainer's similarity model (this
volume) makes direct use of knowledge that might reside in a different
research area, and Shrager's use of a "view" or a "way of seeing" in
discovery systems (this volume) could perhaps allow for some aspects
of a previous view to be retained in a later version of a theory. These
issues are worth pursuing.

2.3   Can the Model Handle Visual Imagery?

This question is an old one for computationalists (e.g., Kosslyn, Brunn,
Cave, & Wallach, 1985), and it is explicitly acknowledged as a problem
by a number of contributors to this volume, such as Shrager. However,
I want to suggest that the full scope of the problem has not been adequately acknowledged, largely because no attention has been paid to
the *dynamics* of visual cognition.

Understanding the centrality of this point for Faraday's research took
me a long time. That Faraday was a highly visual thinker was clear at
the outset; there are thousands of illustrations in his diaries, and his
richest theoretical concepts are inherently visual, especially those concerned with field notions. But I did not, at first, see quite why James
Clerk Maxwell (1855) had referred to Faraday as a great "intuitive geometrician." The opposite seemed true insofar as Faraday used, literally, *no* mathematical (algebraic) representations in any of his work,
published or unpublished. In fact, Faraday had principled objections

against the use of mathematics (Cantor, 1985). Visual he certainly was, but a geometrician?

I think now that Maxwell's statement derived from the dynamic uses to which Faraday put his imagery. For example, Faraday saw fields as space filling and as time dependent. His images resemble a film strip more than a snapshot, but there is an even deeper sense in which they were dynamic. Faraday's images were used in a constructive way to capture theoretical notions and to derive consequences from them.

Thus, in 1821, when he discovered how to harness the transverse forces produced by a current in a magnetic field (and hence made the first electric motor), he was faced with the problem of how best to summarize and represent the resulting forces. The problem was not trivial, in part because it required a three-dimensional approach (the forces being orthogonal to both the current and the magnetic field) and in part because the actual poles of a bar magnet or a magnetized needle are displaced inward from the ends by a short distance. Gooding (1989) has shown how Faraday resolved these issues in a constructive way by assembling, as it were, the separate observed motions into a larger geometric picture, a quite abstract "mathematical" representation of a simple underlying picture. Such use of imagery is dynamic in a larger sense than that conveyed by the metaphor of a film strip.

Faraday is not unique in this respect. As just one example, consider the following characterization of how one admittedly very mathematical physicist conceptualizes fields (Feynman, 1964, p. 20–10):

> When I talk about the electromagnetic field in space, I see some kind of superposition of all of the diagrams which I've ever seen drawn about them. I don't see little bundles of field lines running about because it worries me that if I ran at a different speed the bundles would disappear. I don't even always see the electric and magnetic fields because sometimes I think I should have made a picture with the vector potential and the scalar potential, for those were perhaps the more physically significant things that were wiggling.

Somehow our theories need to capture such thinking. I speculate that this thinking style may be related to the centrality of differential equations in physics; a differential equation is a shorthand for the kind of imagery that Feynman describes and that Faraday used. It is not the *only* possible shorthand (as the example of Faraday suggests), but it may be the most prevalent today.

This suggestion is related to a common misconstrual of the role of mathematics in science. Too frequently we regard mathematics as a kind of separable intruder to what is really going on. Thus, BACON is frequently criticized for being merely part of a scientific discovery system, as Langley, Simon, Bradshaw, and Zytkow (1987) freely acknowledged. In fact, I think the most interesting uses of mathematics in science stem not from its role as a medium of derivational proof but from its role as a representational device (see also Giere, 1988).

In this view, modern physics is heavily mathematical, in part because it has long since surpassed human capacity for "mere" imagery. Miller (1984) and Nersessian (1984) have argued that modern physics dealt explicitly with this issue and quite self-consciously reconceptualized its notion of the role of visualization in theory. The role of mathematics therefore became further consolidated as a representational medium whose capacity exceeds the usual constraints of "mere" imagery. Since mathematical representations are frequently used by computationalists, perhaps there is, hidden in these considerations, a way to chart a path around the well-known difficulties of dealing with visual representation.

2.4   Can the Model Handle a Large Knowledge Base?

In a sense, this question hardly needs asking since one of the acknowledged tasks of any computational system is to handle a large knowledge base. Even when the actual knowledge base used to implement specific systems is quite small, in general there is serious consideration given to larger amounts of data.

The question cuts deeper than mere size, however, as Faraday's experience makes clear. Faraday recorded many thousands of results during his career; the published portion of his diary (which I estimate to be about half of his entire record) includes some 17,000 numbered entries (each ranging from a few lines to several pages in length). Any database this large is going to require attention to retrieval mechanisms, and it is clear that Faraday gave serious thought to this problem.

Among Faraday's surviving papers, there are a large number of index materials of various sorts. I have dealt with these at length in another context (Tweney, 1988, 1989). Briefly, they represent a variety of special-purpose retrieval aids used in one of two broad contexts, as aids in the generation of new experiments and as aids in the preparation of publications. Faraday's general scheme reached maturity in 1831 when he began to number his diary entries in a strict numerical order, which he held to for 42 years. He constructed, as needed, higher-order indexes to specific entries (these were organized by topic), indexes to the indexes (also organized by topic but at higher levels), and even indexes of category heads. These were frequently cut up into strips, each bearing one or a few numbers. The slips were sorted and rearranged and were sometimes pasted up on large or small sheets of various types. The general lesson for computationalists, I think, is that any truly realistic database ought to be accessible for processing in more than one way. Whether or not scientists other than Faraday are as systematic as he was is not really the issue; he merely made explicit something that everyone needs to worry about. Even my own record keeping (which anchors the other end of the continuum from the systematicity evident in Faraday!) betrays such a need. As a colleague put it, the apparent layered disorder of my desk is in fact "geologically" ordered, which makes me something of an epistemological stratigrapher.

Finally, there is the issue of those entries in a knowledge base that are themselves not really data. Not everything that Faraday recorded was observational or even propositional. For example, at times he would record injunctions to himself (e.g., "Must experiment to see if..."), metacognitive comments (e.g., "Result not satisfactory"), purely episodic material of no immediate relevance to the scientific activity as such ("Went to Mr. Christie's today..."), and so on. It seems likely that such entries are present because they played some functional role in the use Faraday made of his diary, if only as reminders. To my knowledge, no computationalist has ever tried to incorporate such information in a database. Nonetheless, we should not overlook the possible importance of such entries once the database becomes really huge.

2.5 Can the Model Distrust Data?

My colleague Michael Doherty is fond of pointing out that, in contrast to the layperson, scientists are characteristically very *mis*trustful of data. Whereas the layperson is easily swayed by arguments of the form "Studies have shown that ...," the scientist wants to know all sorts of arcane things about the studies—how they were designed, whether the results were statistically significant, and so forth. Whereas the layperson is committed to a "seeing is believing" epistemology, scientists have long since abandoned any such view. How many Copernicans, after all, ever experienced the movement of the Earth through space?

In this respect, Thagard's ECHO (1990) is a step in the right direction since it weights an observation in response to coherence relations derived from other sources. But what is really needed is a far more thorough skepticism. Faraday knew that experiments could go awry in numerous ways and constantly kept this possibility in mind. Perhaps this is one reason why he so readily ignored disconfirming results. Further, as anyone knows who has worked even briefly in a laboratory, the "results" obtained in an experiment rarely look as neat as they sound in their public form.

This point has been eloquently made by Gooding (1989), who has replicated some of Faraday's experiments on the rotation of a current-carrying needle. In fact, when one tries Faraday's actual experiment, the needle does a good deal of chaotic bobbing around. To see the regularity of its orientation, as Faraday did, requires an effort of constructive abstraction. Nearly every existing computational model ignores this problem and treats data as if they were fixed and immutable in the first instance.[2] Although some probabilistic models recognize the problem, in the sense that they allow for probabilistic weights to be placed on data statements, these models hardly capture the scope of the problem facing the scientist in the real world.[3] I see no intrinsic reason why computationalists could not address the issue, and I think they will need to do so before any system becomes truly capable of scientific discovery.

---

2. Shrager (personal communication) has pointed out that, if one is able to get a hint as to the gist of what is going on in the chaos of the needle's activity, then the observer can take that "view" (see Shrager, this volume), can use it to actively filter out the noise, and can later consider what has been ignored by virtue of this initial filtering.

3. The general role of data error in inference has been extensively studied insofar as normally distributed or Gaussian error is concerned, but there are other types that can be more consequential. Thus, a rectangular distribution of noise can be highly disruptive in some circumstances (Doherty, Tweney, Walker, & O'Conner, 1988).

3.   What the Questions Really Mean

My five questions are meant to be provocative. They stem, in effect, from a descriptive context: They are aspects of what I think I have seen in scientific thinking, aspects not generally paid much heed in the current computationalist landscape. In a more general sense, however, it is also necessary to consider the role of computational models in the larger context of our developing theories about scientific thinking.

The fact that a particular computational model can do something is not, in and of itself, much help in assessing its value for theory. There are simply too many ways in which the specific tasks of science can be achieved. Thus, if we consider a system like BACON, but one that finds the gas laws by trying all possible functions until one fits, it would hold little theoretical interest. Systems are interesting because they achieve results in interesting ways. The actual BACON programs are impressive insofar as they achieve so much using so little. Here, as elsewhere in science, simplicity is compelling, in part because it provides a constraint. Newtonian mechanics is intellectually satisfying in part because it reduces to one simple claim, $F = ma$, which is the key to understanding an overwhelming variety of seemingly complex and otherwise disparate phenomena.

However, there is another side to the use of simplicity that I believe is too frequently ignored. To really evaluate a system, one must know what it cannot do as well as what it can. There is, of course, an obligatory section called "Limitations" in nearly every report written by computationalists. That is a start, but it seems to me that such sections most often are unhelpful. The acknowledged limitations are usually things that the system was never designed to do in the first place.

Much more revealing are the cases in which the system fails to give a result when we might reasonably expect it to do so. Such failures are harder to confess publicly, but they are also the ones that the consumer most needs to see. For that matter, I suspect they are also the ones that the designer of computational systems most needs to see because of their potential value in designing even more powerful systems. In the physical sciences, confessing the analogous limitations of a physical theory seems de rigueur these days; one is expected to describe the limits of one's formulations. Why not in computational science as well?

A somewhat related consideration stems from the use of historical case studies to exemplify the power of a computational system. Obviously I think this is the right way to go; my entire concern with Faraday is based on similar reasoning. However, most of the actual examples I have seen in which case studies are used have been based either on brief published scientific results or, what is even worse, on second-hand summary accounts of the published results. In effect, we are frequently given "historical nuggets" rather than history as such. Two problems arise from this approach. First, in Medawar's (1964) words, the published scientific paper is a fraud! Though published papers generally employ a rhetorical style *as if* the discovery process were being described, the actual process is generally far different. Second, even when the accounts are presumably based on the real process (rather than a second-hand account), there is a tendency to deal with them in a very offhand fashion. I am much less impressed by a system that can mimic the highlights of even a large number of case studies than I am by one that captures some interesting level of detail for one real case study.

Let the producers beware!

References

Cantor, G. N. (1985). Reading the book of nature: The relation between Faraday's religion and his science. In D. Gooding & F. James (Eds.), *Faraday rediscovered.* New York: Stockton Press.

Doherty, M. E., Tweney, R. D., Walker, B., & O'Conner, R. (1988). *The role of data and feedback error in inference and cognition* (Technical report). Bowling Green, OH: Bowling Green State University, Department of Psychology.

Evans, J. St.-B. T. (in press). *Bias in human reasoning.* Hillsdale, NJ: Lawrence Erlbaum.

Faraday, M. (1832). Experimental research in electricity. 1. On the induction of electric currents. *Philosophical Transactions, 122*, 125–162.

Faraday, M. (1932–1936). *Faraday's diary* (Martin, Ed.). London: Bell.

Feynman, R. P., Leighton, R. B., & Sands, M. (1964). *The Feynman lectures on physics* (Vol. 2). Reading, MA: Addison-Wesley.

Giere, R. N. (1988). *Explaining science: A cognitive approach.* Chicago: University of Chicago Press.

Gerwin, D., & Newsted, P. (1977). A comparison of some inductive inference models. *Behavioral Science, 22,* 1–11.

Gooding, D. (1989). "Magnetic curves" and the magnetic field: Experimentation and representation in the history of a theory. In D. Gooding, T. Pinch, & S. Schaffer (Eds.), *The uses of experiment.* Cambridge: Cambridge University Press.

Gooding, D. (in press). *The making of meaning.* Dordrecht: Nijhoff/Kluwer.

Gruber, H. E. (1974). *Darwin on man.* New York: Dutton.

Huesmann, L. R., & Cheng, C. M. (1973). A theory for the induction of mathematical functions. *Psychological Review, 80,* 126–138.

Kosslyn, S. M., Brunn, J., Cave, K. R., & Wallach, R. W. (1985). Individual differences in mental imagery ability: A computational analysis. In S. Pinker (Ed.), *Visual cognition.* Cambridge, MA: MIT Press.

Kuhn, T. S. (1962). *The structure of scientific revolutions.* Chicago: University of Chicago Press.

Kulkarni, D., & Simon, H. A. (1988). The processes of discovery: The strategy of experimentation. *Cognitive Science, 12,* 139–176.

Langley, P., Simon, H. A., Bradshaw, G. L., & Zytkow, J. M. (1987). *Scientific discovery: An account of the creative processes.* Cambridge, MA: MIT Press.

Maxwell, J. C. (1855). On Faraday's lines of force. *Transactions of the Cambridge Philosophical Society, 10,* 27–83.

Medawar, P. B. (1964). Is the scientific paper a fraud? In D. Edge (Ed.), *Experiment: A series of scientific case histories first broadcast in the BBC Third Programme.* London: British Broadcasting Corporation.

Miller, A. I. (1984). *Imagery in scientific thought: Creating twentieth-century physics.* Boston: Birkhauser.

Mynatt, C. R., Doherty, M. E., & Tweney, R. D. (1978). Consequences of confirmation and disconfirmation in a simulated research environment. *Quarterly Journal of Experimental Psychology, 30,* 395–406.

Nersessian, N. (1984). *Faraday to Einstein: Constructing meaning in scientific theories.* Dordrecht: Nijhoff.

Popper, K. (1959). *The logic of scientific discovery.* London: Hutchinson.

Simon, H. A. (1966). Scientific discovery and the psychology of problem solving. In R. G. Colodny (Ed.) *Mind and cosmos: Essays in contemporary science and philosophy.* Pittsburgh: University of Pittsburgh Press. Reprinted in H. A. Simon (Ed.) (1977), *Models of discovery.* Boston: Reidel.

Thagard, P. (1990). Explanatory coherence. *Behavioral and Brain Sciences.*

Tweney, R. D. (1985). Faraday's discovery of induction: A cognitive approach. In D. Gooding & F. James (Eds.), *Faraday rediscovered.* New York: Stockton Press.

Tweney, R. D. (1988). *Externally aided memory and Michael Faraday's problem-solving strategies.* Unpublished manuscript, Department of Psychology, Bowling Green State University, Bowling Green, OH.

Tweney, R. D. (1989). A framework for the cognitive psychology of science. In B. Gholson, W. Shadish, R. M. Niemeyer, & A. Houts (Eds.), *Psychology of science: Contributions to metascience.* Cambridge: Cambridge University Press.

Tweney, R. D., & Doherty, M. E. (1983). Rationality and the psychology of inference. *Synthese, 57,* 139–161.

Tweney, R. D., Doherty, M. E., & Mynatt, C. R. (Eds.) (1981). *On scientific thinking.* New York: Columbia University Press.

Wallace, D., & Gruber, H. E. (Eds.). (1989). *Creative people at work.* New York: Oxford University Press.

Williams, L. P. (1965). *Michael Faraday.* London: Chapman and Hall.

Subject Index

A

ABACUS 104, 118, 129

Abduction 32, 157-192, 197-220
    Existential 36
    see also Similarity-driven explanation

ABDUL/ILANA 411-430

AbE (the *Abduction Engine*) 210-220, 250
    see also Abduction

ACT (Adaptive Control of Thought) 189

Activities (and processes) 5-9

Adaptability 157

ADEPT 247, 271

AM 10-11, 14-15, 111, 430

Analogy 13, 157-192
    scenario 160, 180
    verification-based 174, 249
    multiple 188, 191
    see also Similarity-driven explanation

Animation *see* Spatial representation

Anomalies 5, 232-247, 283-290, 386, 389
    Anomaly-driven diagnosis/redesign 319-343

see also Diagnostic reasoning, Theory formation/revision

Argumentation 12, 408-414
    discovery as 422-429
    graph 411-429

Assumption
    implicit 333-334
    scenario 160, 179

ATMS *see* Truth maintenance

Attenuation *see* Theory of attenuation

AUTOCLASS 90-91

B

BACON 10-11, 15-16, 28, 114, 118-120, 130, 155, 313-314, 478, 481

Background Knowledge 4, 99, 167, 406, 417-420, 478-479
    see also Domain theory, Experience

Balloons rising, *see* Theory of

Bayesian methods 14, 73-93

Biases, induction 407-408
    examples of 414-418

BigTrak 357-391, 439-443, 446-449

C

Catalysis, theory of 418-420
Category formation 77-91
    see also Clustering, Concept
        formation
Category utility 122
Causal
    composition 445
    knowledge 408-414
        see also Domain theory,
           Qualitative representation,
           Representation, qualita-
           tive
    processes 348
        see also Qualitative pro-
           cess theory
Chancy interactions
    see Serendipity
Characteristic results 389
Chemistry *see* Qualitative chem-
        istry, Theory of chem-
        istry
Classification
    diagnostic 350
    models 73-93
    unsupervised 79
    see also Clustering, Concept
        acquisition, Similarity-driven
        explanation, View Ap-
        plication Clustering
    conceptual 13, 98, 105-110,
        121-123
    COAST 12, 15, 51, 190, 228-
        251, 310-311
    COBWEB 102-105, 110-111,
        122
Coherence, explanatory 32-33, 48-
    62
Collaboration 17

Commonsense
    concepts 442
    reasoning 405
    see also Qualitative simula-
        tion, Qualitative repre-
        sentation
Commonsense perception 437-467
Communication 17
Communities 18
Complexity, of models 74
Concept acquisition task 406
Conceptual
    change 29-62, 466
    combination 13-14, 32
    systems 29-62
    see also Clustering, Networks,
        Revolution, View Appli-
        cation
Connectionist methods *see* Spread-
    ing activation
Consistency 447
    explanatory 166
Curve fitting 78

D

DALTON 313
Data 3
    theory-driven interpretation
        of 308
    see also Observation, Instru-
        ments, DATMI, Interpre-
        tation
DATMI (Dynamic Across-Time Mea-
    surement Interpretation
    system) 176
Decomposition (of equations), process-
    directed 136-150
Deduction 157-192
    Law formation/revision 7

487

see also Devices; Representation, qualitative; Schema representation
MOLGEN 270
Monster Anomalies 320-343
see also Anomalies
MOP's 410
Multiple hypotheses 387
Multisearch 149-150

N

Negative examples 404
Networks, conceptual 30, 50-54
see also Hierarchy, Spreading activation, Taxonomy
New terms 178, 446
Notebooks 16
Faraday's 473-482

O

Observations 3, 6, 167, 231, 409
see also Data, DATMI, Instruments, Interpretation
Observability strategy 388
see also Magnification strategy
OCCAM 190, 405-430
Ockham's razor (see Parsimony)
Operationalizations 450, 466
Oscillation, theory of 181-183
Overfitting 74
Oxygen theory see Theory, oxygen

P

Parsimony 73-91
Perception see Representation, perceptual

Phenomenological primitives 467
PHINEAS 11, 157-192
see also Similarity-driven explanation
Phlogiston theory see Theory, oxygen
Physical embedding 15-19
PI 35, 220
Piltdown Man see theory of Piltdown Man
Planning 283
Plate tectonics
see Geological revolution, Theory of plate tectonics
Postdictions 5
PRE 308
Predictions 5, 283
Activity 8, 118-119, 277-280
Failure of, 356-369
Model based 88
Value of classification model 75
see also Anomalies, Experimentation
Primitives, phenomenological 467
Problem-spaces see Search
Process diagram 137-150
Process-directed decomposition (of equations) 136-150
Processes
qualitative see Qualitative Process Theory
scientific see Activities
Programmer's Apprentice 309
Propositional networks see Networks, conceptual
memory see Memory, quasi-linguistic

491

Psychological approaches 18, 28, 355-392, 405-407, 439-447

Punnett square 320-342

Q

QPE *see* Qualitative Process Engine

Qualitative chemistry 200-205, 277-283

Qualitative

History 99, 165

Laws 4, 14, 99, 111-113

Representation *see* Representation, qualitative

Simulation 12, 165, 454-463

see also Qualitative Process Engine, Envisionment

understanding 467

Quantitative

Laws 4, 99, 113-118

Model 165

Representation *see* Representation, quantitative

Qualitative Process Theory 11, 100, 137-150, 165-186, 201, 229-247, 278

Engine (QPE) 176, 460

Quantum mechanics, *see* Theory of laser physics

R

Recognition 440

paradox of 438, 448-449

see also Similarity-driven explanation

Reference experiments 240, 304

Reformulation 440-447, 453, 466

Rejection, of theories *see* Theory rejection

Replication experiments 380-381, 405

Representation

External 16

see also Punnett square

Frame *see* schema

Functional 348-352

Grounded 438, 450-453, 457-460

Perceptual 437, 450-467

Qualitative 11-12, 137-150, 198-220, 229-247

Quantitative 11

Schema 327, 359, 410

Sensory-motor 12, 16, 437-467

Spatial 12

Visual/Diagrammatic 327-342, 450-454, 478

Research programs 16, 474-476, 478-480

see also Dual search theory of scientific discovery

Results, characteristic 389

Revision *see* Hypothesis revision, Taxonomy formation/revision, Theory formation/revision

Revolution 28-62

REVOLVER 51, 248

Rickettsialpox *see* Theory of Rickettsialpox

Robotics 19

S

Satisficing 391

see also Search

493

Name Index